9th International Conference on Inverse Problems in Engineering (ICIPE 2017)

AA006879

Journal of Physics: Conference Series
Volume 1047

Waterloo, Canada
23 – 26 May 2017

ISBN: 978-1-5108-6788-8
ISSN: 1742-6588

Printed from e-media with permission by:

Curran Associates, Inc.
57 Morehouse Lane
Red Hook, NY 12571

proceedings.com

Some format issues inherent in the e-media version may also appear in this print version.

Printed by Curran Associates, Inc. (2018)

For permission requests, please contact the Institute of Physics
at the address below.

Institute of Physics
Dirac House, Temple Back
Bristol BS1 6BE UK

Phone: 44 1 17 929 7481
Fax: 44 1 17 920 0979

techtracking@iop.org

Additional copies of this publication are available from:

Curran Associates, Inc.
57 Morehouse Lane
Red Hook, NY 12571 USA
Phone: 845-758-0400
Fax: 845-758-2633
Email: curran@proceedings.com
Web: www.proceedings.com

Table of contents

Volume 1047

9th International Conference on Inverse Problems in Engineering (ICIPE)

23–26 May 2017, Waterloo, ON, Canada

Accepted papers received: 19 June 2018
Published online: 4 July 2018

IOP Conf. Series: Journal of Physics: Conf. Series **1047** (2018) 011001 doi:10.1088/1742-6596/1047/1/011001

Introduction

This proceedings of the Journal of Physics: Conference Series contains a subset of papers presented at the 9th International Symposium on Inverse Problems in Engineering (ICIPE), held at the University of Waterloo, May 23-26 2017. Since the first ICIPE in 1993, this conference has served as the main international venue for collaboration and interaction between mathematical theorists who develop inverse analysis tools, and engineers who use these tools to solve today's problems. Our meeting continued this tradition, with 80 delegates from many sub-disciplines of engineering, science, and applied mathematics.

The conference and its proceedings are dedicated to Professor Graham Gladwell, FRSC. Graham played a key role in establishing this conference series, and was cross-appointed between the Faculties of Mathematics and Engineering during his career at the University of Waterloo, exemplifying the fact that inverse analysis lies at the intersection of these two areas. Sadly, Graham passed away on March 11th, 2017, only two months before our conference. We honoured his memory and spirit throughout the conference.

Our event was opened by His Worship Dave Jaworski, Mayor of Waterloo, and Ms. Catherine Fife, Member of Provincial Parliament for Waterloo. The conference featured a keynote address from Professor Margaret Cheney from Colorado State University on "The radar inverse problem", a workshop on "Inverse problems in the Bayesian framework" led by Dr. Aku Seppänen from the University of Eastern Finland, and a talk from the acclaimed author and journalist Malcolm Gladwell in honour of his father. A number of Graham's friends, colleagues, and former students also spoke about Graham throughout the conference, including Marc Chamberland, Wayne Broadland, George Dixon, Sivabal Sivaloganathan, and Ed Vrscay.

We are grateful for the financial support of the Fields Institute for Research in Mathematical Sciences, The University of Waterloo Office of Research, the Faculty of Mathematics, the Faculty of Engineering, the Departments of Applied Mathematics, Civil and Environmental Engineering, and Mechanical and Mechatronics Engineering, the University of Alabama College of Engineering, as well as corporate sponsorship from MathWorks and MapleSoft. We also convey our sincere thanks to Ms. Susanne Keppler, University of Waterloo Conference Services Manager, for her extraordinary efforts in facilitating this event. Finally, we thank all the attendees for your support and participation.

We especially appreciate the involvement of Joyce Gladwell and her family, as well as Graham's students, friends, and colleagues.

Kyle Daun
Waterloo, Ontario

IOP Conf. Series: Journal of Physics: Conf. Series **1047** (2018) 011001 doi:10.1088/1742-6596/1047/1/011001

Dedication

Graham Gladwell was born in Otford, Kent, England on February 21, 1934, the second child of Basil and Doris Gladwell. He graduated first in mathematics at University College London (UCL) in 1954 at the age of twenty and later received a Ph.D. and a D.Sc. from the same institution.

After completing his Ph.D., Graham was a lecturer at UCL until 1960. This was followed by two years of teaching at the University of the West Indies in Jamaica. In 1962 Graham was back in England at the University of Southampton and joined the new Institute of Sound and Vibration Research. After several years, he became a professor of Civil Engineering at the University of Waterloo, Canada. He was also given a cross-appointment to the Department of Applied Mathematics. He remained at Waterloo until his retirement in 1999.

Graham left his mark as a researcher. Over the course of his career he authored or co-authored five books: *The Matrix Analysis of Vibration* (1965), *Contact Problems in the Classical Theory of Elasticity* (1980), *Inverse Problems in Vibration* (1986), *Inverse Problems in Scattering* (1993), and *Functional Analysis with application to Mechanics and Inverse Problems* (1996). He supervised 30 graduate students in both engineering and mathematics. He published over 130 papers, both single-authored and with many collaborators. Graham's research prowess was recognized in many ways. He was a fellow of the Royal Society of Canada, the American Academy of Mechanics, the Institute of Mathematics and its Applications, and the recipient of a lifetime achievement medal from the Canadian Congress of Applied Mechanics. He retired as Distinguished Professor Emeritus from the University of Waterloo.

As a student, Graham met Joyce Nation, a Jamaican student who came to study in England. They married and raised three sons: Graham Jr.; Geoffrey; and Malcolm. Graham was committed to his Christian faith, including taking a leadership role in the University of Waterloo's Pascal Lectures on Christianity. He was also wholly devoted to his extensive gardens.

— Marc Chamberland

Prof. Marc Chamberland
Department of Mathematics and Statistics
Grinnell College
(student of GML Gladwell)

IOP Conf. Series: Journal of Physics: Conf. Series **1047** (2018) 011001 doi:10.1088/1742-6596/1047/1/011001

A Tribute to Professor G.M.L. Gladwell

As a mathematician would say, Professor Graham Gladwell was my academic father. Somehow, this is much more accurate than saying he was simply my PhD supervisor. Upon his passing, I have been engulfed by a flood of memories with strong feelings of respect, affection, and gratitude.

Perhaps first and foremost is the rare opportunity I had to watch a true master at work. Often I would sit beside Professor Gladwell, at his desk, watching him develop a proof associated with his work. This was much more than a demonstration of elegant mathematics for the sake of a young academic. It was a life experience I'll always cherish. Each pen stroke had meaning and clarity. The final result was always beautiful and inspiring. Flowing out of these sessions were fundamental lessons that were to guide me in my career. Perhaps of greatest impact was the Gladwellian rule: "If you can obtain a simple result in a complicated way, you should be able to obtain that simple result in a simple way." I can still hear his voice ringing with this message phrased so poetically. Now that I look back on my 26 years as a faculty member at the University of Saskatchewan, I realize how many of my own contributions have sprung from this lesson, whether it be a paper on unraveling the Coriolis effect, a new approach to teaching the mathematical foundations of an undergraduate class, or a session with graduate students towards understanding the mathematics underlying their stream of research. This has been pure joy to me, and Professor Gladwell set the table for the mathematical banquet I was to enjoy.

During my apprenticeship, Professor Gladwell provided all sorts of gems. For example, he taught me that hyperbolic trigonometric functions are not to be feared, but savored. I was present on one occasion when an undergraduate student, referring to cosh(x) and sinh(x), said "I hate those things." Professor Gladwell replied with a smile and said, "Really, I salivate when I see them."

It was from Professor Gladwell that I especially learned how mathematical thinking could permeate every facet of a person's life. During one of many visits to the Gladwells' lovely home in the country, the after-dinner activity included a board game called Chaos. In this game, each player must keep track of an evolving mind-stretching pattern of checkers. Prof. Gladwell suggested that each of us play the pieces for two players. For some reason, I considered this to be more than just a game. I felt like my mental capacity was being tested. During the game, Professor Gladwell was calm and collected while smoothly and rapidly making each move. I was sweaty, intense, and apprehensive, pushing my brain to full capacity. I recall how this ended – we completed our final moves in succession, Professor Gladwell got up, looked refreshed, and said, "a tie, what fun". I felt completely exhausted, but excited at the completion of yet another bout with the master. I must say, though, that I probably bent a Gladwellian rule during the process. Professor Gladwell would often say, "never do mathematics when you're tired". From my observations, he was able to follow this rule without exception. I, on the other hand, did feel a bit stretched, whether it be in his course on contact problems, or the oral math quiz he gave me on the way home to Waterloo after an inverse problems conference at Michigan State University.

I could go on with more anecdotes involving math challenges and lessons, but I would like to focus the remainder of this tribute on what I consider to be one of Professor Gladwell's greatest attributes. He was perhaps the most open-minded academic I have ever encountered. His high standard of excellence for himself and others did not rely on adherence to a particular style or a pre-conceived set of superficial rules; his criteria were pure and fundamental – clear, concise descriptions and flawlessly logical arguments. Beyond these cognitive pillars, he was open to various approaches and paths. I recall that, one day, I was working with another Gladwell PhD student, Brad Willms, as we tried to prove Fermat's last theorem in a simple, direct way. This, of course, had absolutely nothing to do with our doctoral research. Unexpectedly, Professor Gladwell popped into our graduate office and I was afraid that he would scold us for taking time away from our work. To the contrary, he was delighted at our attempt (even though it didn't reach fruition). When I felt I was ready to defend my PhD. work, Professor Gladwell's first response was to say, "no, but convince me". Right there, on the spot, I gave my arguments after which he simply said, "you're ready". So often I would hear Graham firmly but fairly challenge other academics to be clear and informative. Upon hearing a meaningful response, he was receptive and gracious without fail.

Perhaps the greatest testament to Professor Gladwell's effect on my life is that the week before his passing, when I had no idea that he was in the hospital, I was talking about him in my graduate class on Finite Elasticity, and teaching by his precepts. During the lecture, a lengthy mathematical development led to a result of unity for a particular quantity. With enthusiasm, I exclaimed, "As Professor Gladwell would say, if you can obtain a simple result in a complicated way, you should be able to obtain the result in a simple way." And this is exactly what we did!

In closing, let me say that I can still hear the melodic humming of Prof. Gladwell down the hallway at the University of Waterloo. It's as if his "work" was an ongoing piece of beautiful classical music which filled his soul with joy and satisfaction. Thankfully, this symphony will continue and flourish - through his many wonderful papers, his graduate and undergraduate students, and everyone who has had the pleasure of his company.

With respect and gratitude,

—Allan Dolovich

Prof. Allan Dolovich
Department of Mechanical Engineering
University of Saskatchewan
(student of GML Gladwell)

IOP Conf. Series: Journal of Physics: Conf. Series **1047** (2018) 011001 doi:10.1088/1742-6596/1047/1/011001

Conference Organizers

Local Organizing Committee (University of Waterloo)

Kyle Daun, Mechanical and Mechatronics Engineering	Chair
Ed Vrscay, Applied Mathematics	Co-Chair
Sriram Narasimhan, Civil and Environmental Engineering	Co-Chair
Paul Hadwin, Mechanical and Mechatronics Engineering	Member
Andrea Scott, Systems Design Engineering	Member
David Siegel, Applied Mathematics	Member
Alexander Wong, Systems Design Engineering	Member
Sam Grauer, Mechanical and Mechatronics Engineering	Volunteer
Tim Sipkens, Mechanical and Mechatronics Engineering	Volunteer
Susanne Keppler, University of Waterloo Conference Services	

ICIPE Steering Committee

Keith Woodbury, University of Alabama
Jim Beck, Michigan State University

IOP Conf. Series: Journal of Physics: Conf. Series **1047** (2018) 011001 doi:10.1088/1742-6596/1047/1/011001

Sponsors

We gratefully acknowledge the contributions of the following sponsors, without whom the ICIPE would not have been possible:

The Fields Institute for
Research in Mathematical Science
FIELDS

UNIVERSITY OF WATERLOO
FACULTY OF ENGINEERING

UNIVERSITY OF WATERLOO
FACULTY OF MATHEMATICS

UNIVERSITY OF WATERLOO
FACULTY OF ENGINEERING
Department of Civil &
Environmental Engineering

UNIVERSITY OF WATERLOO
FACULTY OF ENGINEERING
Department of Mechanical &
Mechatronics Engineering

THE UNIVERSITY OF ALABAMA
College of Engineering
Mechanical Engineering

College of Engineering

MathWorks

Maplesoft
Mathematics • Modeling • Simulation

IOP Conf. Series: Journal of Physics: Conf. Series **1047** (2018) 011001 doi:10.1088/1742-6596/1047/1/011001

Conference Participants

Wojciech Adamczyk Silesian University Technology Gliwice, Poland wojciech.adamczyk@polsl.pl	Wasseem Al Hadad Université de Lorraine Nancy, France waseem.al-hadad@univ-lorraine.fr	Viplove Arora Purdue University West Lafayette, IN, USA arora34@purdue.edu
Jim Beck Michigan State University East Lansing, MI, USA beck@msu.edu	Alistair Boyle University of Ottawa Ottawa, ON, Canada aboyle2@uottawa.ca	Fabio Bozzoli University of Parma Parma, Italy fabio.bozzoli@unipr.it
Ameneh Boroomand University of Waterloo Waterloo, ON, Canada a.boroomand@gmail.com	Victoria Brott University of Guelph Guelph, ON, Canada vbrott@mail.uoguelph.ca	Rickey Caldwell Merrimack College North Andover, MA, USA caldwellr@merrimack.edu
Kaitlyn Casulli Michigan State University East Lansing, MI, USA casullik@msu.edu	Margaret Cheney Colorado State University Fort Collins, CO, USA cheney.margaret@gmail.com	Marc Chamberland Grinnell College Grinnell, IA, USA chamberl@math.grinnell.edu
Kevin Cole University of Nebraska Lincoln NE, USA kcole1@unl.edu	Andrei Constantinescu CNRS - École Polytechnique Palaiseau France constant@lms.polytechnique.fr	Filippo De Monte University of L'Aquila L'Aquila, Italy filippo.demonte@univaq.it
Kyle Daun University of Waterloo Waterloo, ON, Canada kjdaun@uwaterloo.ca	Matthew DiDomizio University of Waterloo Waterloo, ON, Canada matt.didomizio@uwaterloo.ca	Kirk Dolan Michigan State University East Lansing, MI, USA dolank@msu.edu
Ashley Emery University of Washington Seattle, WA, USA emery@u.washington.edu	Kaan Erkorkmaz University of Waterloo Waterloo, ON, Canada kaane@uwaterloo.ca	Maxwell Fitzsimmons University of Guelph Guelph, ON, Canada mfitzsim@mail.uoguelph.ca
Javier Gazzarri MathWorks Novi, MI, USA javier.gazzarri@mathworks.com	Farideh Ghasempour University of Waterloo Waterloo, ON, Canada fghasemp@uwaterloo.ca	Jakub Grabski Poznan University of Technology Poznan, Poland jakub.grabski@put.poznan.pl
Sam Grauer University of Waterloo Waterloo, ON, Canada samuel.grauer@uwaterloo.ca	Asko Hänninen University of Eastern Finland Kuopio, Finland asko.hanninen@uef.fi	Paul Hadwin University of Waterloo Waterloo, ON, Canada pjhadwin@uwaterloo.ca
Graham Hendra University of British Columbia Vancouver, BC, Canada grrhendra@gmail.com	Marcin Janicki Lodz University of Technology Łódź, Poland janicki@dmcs.pl	Trevor Jones Bishop's University Sherbrooke, QC, Canada tjones@ubishops.ca
Hessam Kalbasi University of Waterloo Waterloo, ON, Canada hkalbasi@uwaterloo.ca	Shiro Kubo Setsunan University Neyagawa, Japan kubo@mech.eng.osaka-u.ac.jp	Herb Kunze University of Guelph Guelph, ON, Canada hkunze@uoguelph.ca
Philippe Le Masson Université Bretagne Sud Lorient, France philippe.le-masson@univ-ubs.fr	Ossi Lehtikangas University of Eastern Finland Kuopio, Finland ossi.lehtikangas@uef.fi	Kimberly Levere University of Guelph Guelph, ON, Canada klevere@uoguelph.ca
Thomas Loussouarn Université de Lorraine Nancy, France thomas.loussouarn@univ-lorraine.fr	Avery Ma University of Waterloo Waterloo, ON, Canada avery.mabojie@gmail.com	Artur Maciag Kielce University of Technology Kielce, Poland maciag@tu.kielce.pl
Elisan Magalhães Universidade Federal de Itajubá Itajubá, Brazil elisan@unifei.edu.br	Denis Maillet Université de Lorraine Nancy, France denis.maillet@univ-lorraine.fr	Blake Martin University of Waterloo Waterloo, ON blake.martin@uwaterloo.ca

IOP Conf. Series: Journal of Physics: Conf. Series **1047** (2018) 011001 doi:10.1088/1742-6596/1047/1/011001

Carlos Matt Electric Energy Research Center Rio de Janeiro, Brazil cfmatt@cepel.br	Robert McMasters Virginia Military Institute Lexington, VA, USA mcmasters@1978.usna.com	Magdalena Mierzwiczak Poznan University of Technology Poznan, Poland magdalena.mierzwiczak@wp.pl
Rodrigo Miguel University of Waterloo Waterloo, ON, Canada rbmiguel@uwaterloo.ca	Carlos Minutti National University of Mexico Mexico City, Mexico carlos.minutti@iimas.unam.mx	Andrzej Napieralski Lodz University of Technology Łódź, Poland napier@dmcs.pl
Sriram Narasimhan University of Waterloo Waterloo, ON, Canada sriram.narasimhan@uwaterloo.ca	Rajdip Nayek University of Waterloo Waterloo, ON, Canada rnayek@uwaterloo.ca	Aleksei Nenarokomov Moscow Aviation Institute Moscow, Russia aleksey.nenarokomov@mai.ru
Markus Neumayer Graz University of Technology Graz, Austria neumayer@tugraz.at	Ahmet Okyay University of Waterloo Waterloo, ON, Canada aokyay@uwaterloo.ca	Weizhen Pan Harbin Institute of Technology Harbin, China hitpwz@126.com
Anna Pawińska Kielce University of Technology Kielce, Poland a.pawinska@tu.kielce.pl	Thomas Pierre Université Bretagne Sud Lorient, France thomas.pierre@univ-ubs.fr	Nicholas Polydorides University of Edinburgh Edinburgh, UK n.polydorides@ed.ac.uk
Benjamin Remy Université de Lorraine Nancy, France benjamin.remy@univ-lorraine.fr	Arkadiusz Ryfa Silesian Univ. of Technology Gliwice, Poland arkadiusz.ryfa@polsl.pl	Agnieszka Samson Lodz University of Technology Łódź, Poland asamson@dmcs.pl
Andrea Scott University of Waterloo Waterloo, ON, Canada ka3scott@uwaterloo.ca	Aku Seppänen University of Eastern Finland Kuopio, Finland aku.Seppanen@uef.fi	Athinthra K Sethurajan McMaster University Hamilton, ON, Canada krisha9@mcmaster.ca
Jakub Ševčík Czech Technical University Prague, Czech Republic kubasevca@gmail.com	Timothy Sipkens University of Waterloo Waterloo, ON, Canada timothy.sipkens@uwaterloo.ca	Katarína Šišková Ghent University Ghent, Belgium katarina.siskova@ugent.be
Vít Škvára UTIA Prague, Czech Republic skvara@utia.cas.cz	Vaclav Šmídl University of West Bohemia Pilsen, Czech Republic vsmidl@rice.zcu.cz	Feifei Sun University of Waterloo Waterloo, ON, Canada feifei.sun@uwaterloo.ca
Sina Talebi Moghaddam University of Waterloo Waterloo, ON, Canada stalebimoghaddam@uwaterloo.ca	Kostas Tifkitsis Cranfield University Cranfield, UK k.tifkitsis@cranfield.ac.uk	Thanh Phong Tran Université Angers Angers, France thanhphong.tran@univ-angers.fr
Mustafa Hakan Turhan University of Waterloo Waterloo, ON, Canada mhturhan@uwaterloo.ca	Marko Vauhkonen University of Eastern Finland Kuopio, Finland marko.vauhkonen@uef.fi	Mario Ventresca Purdue University West Lafayette, IN, USA mventresca@purdue.edu
Antti Voss University of Eastern Finland Kuopio, Finland antti.voss@uef.fi	Ed Vrscay University of Waterloo Waterloo, ON, Canada ervrscay@uwaterloo.ca	Yanfei Wang Institute of Geology and Geophysics Beijing, China yfwang@mail.iggcas.ac.cn
Brad Willms Bishops University Sherbrooke, QC, Canada bwillms@ubishops.ca	Keith Woodbury University of Alabama Tuscaloosa, AL, USA keith.woodbury@ua.edu	Alexander Wong University of Waterloo Waterloo, ON, Canada alexander.wong@uwaterloo.ca
Hiroyuki Yamauchi Fukuoka Institute of Technology Fukuoda, Japan yamauchi@fit.ac.jp	Hongmei Zhu York University Toronto, ON, Canada hmzhu@yorku.ca	Lijun Zhuo Harbin Univ. of Technology Harbin, China zhuolijun@hit.edu.cn

IOP Conf. Series: Journal of Physics: Conf. Series **1047** (2018) 011002 doi:10.1088/1742-6596/1047/1/011002

Peer review statement

All papers published in this volume of *Journal of Physics: Conference Series* have been peer reviewed through processes administered by the proceedings Editors. Reviews were conducted by expert referees to the professional and scientific standards expected of a proceedings journal published by IOP Publishing.

IOP Conf. Series: Journal of Physics: Conf. Series **1047** (2018) 012001 doi:10.1088/1742-6596/1047/1/012001

Transfer function identification through total least squares

W. Al Hadad [1], D. Maillet [2]

[1, 2] LEMTA Lab, University of Lorraine & CNRS, ENSEM – 2, avenue de la forêt de Haye- BP 90161 – 54505 Vandoeuvre-lès-Nancy, France the author addresses here

waseem.al-hadad@univ-lorraine.fr

Abstract. When transient diffusion and/or advection occur in a physical system modelled by linear equations whose coefficients are time-invariant, the relationships between a source (a temperature or a thermal power) and its consequences (local temperatures) are convolution products. This study deals with identification of a transfer function when an additive noise not only corrupts the output (response) but also the input (excitation) in a calibration experiment. Regularization is implemented either through Truncated Singular Value Decomposition or by Total Least Squares.

1. Introduction

Inverse input estimation problems (source estimation) or model identification inverse problems (model calibration), based on measurements of the output in a corresponding experiment, are present in many engineering fields (mechanics, bio-physics, heat transfer, ...). If the transfer model they rely on is linear with time-invariant coefficients (LTI model), their common mathematical form is:

$$\boldsymbol{A}\,\boldsymbol{x} = \boldsymbol{y} \tag{1}$$

where $\boldsymbol{A}$ is the matrix of the linear operator of size m x n, with $m \geq n$, $\boldsymbol{x}$ the input column vector of size n x 1 and $\boldsymbol{y}$ the output column vector of size m x 1. We are interested here in the particular LTI model where system (1) is the discrete parameterized form of a convolution product in time. Before dealing with this problem, we will recall here the different ways of solving equation (1), here the discrete form of a Volterra integral equation of the first kind, where errors (noise) are present in both output $\boldsymbol{y}$ and matrix $\boldsymbol{A}$. The field of application of this deconvolution type of problem is very large, from inverse heat conduction (surface heat flux estimation from interior temperature measurements) [3] to inversion of a seismometer signal for estimating a corresponding acceleration [1].

2. The different solutions of the inverse input problem, for a linear model

2.1 Regularized inversion in case of error in the output only

In the absence of regularization, the ordinary least squares (OLS) solution, consists in minimizing the following criterion based on the L2 (Euclidian norm) $\|.\|$ of the output residual vector $\boldsymbol{r}_y(\boldsymbol{x})$:

$$J_{OLS}(\boldsymbol{x}) = \left\|\boldsymbol{r}_y(\boldsymbol{x})\right\|^2 = \boldsymbol{r}_y^T(\boldsymbol{x})\,\boldsymbol{r}_y(\boldsymbol{x}) \qquad \text{with} \qquad \boldsymbol{r}_y(\boldsymbol{x}) = \boldsymbol{y}^{exp} - \boldsymbol{A}\,\boldsymbol{x} \tag{2a}$$

where superscript T designates the matrix transpose and where the true version $\boldsymbol{A}^{true}$ of matrix $\boldsymbol{A}$ is available while the true value $\boldsymbol{y}^{true}$ of the output is corrupted by an additive noise $\boldsymbol{\varepsilon}_y$ of zero stochastic expectation (noted $\mathrm{E}\,(.)$ here), that is:

$$\boldsymbol{A} = \boldsymbol{A}^{true} \qquad \text{and} \qquad \boldsymbol{y}^{exp} = \boldsymbol{y}^{true} + \boldsymbol{\varepsilon}_y \quad \text{with} \quad \mathrm{E}\,(\boldsymbol{\varepsilon}_y) = 0 \tag{2b}$$

In this case the minimization of (2a) with respect to $\boldsymbol{x}$ provides an explicit and unbiased estimation $\hat{\boldsymbol{x}}_{OLS}$ of the true value $\boldsymbol{x}^{true}$ of this vector:

$$\hat{\boldsymbol{x}}_{OLS} = \left(\boldsymbol{A}^T\boldsymbol{A}\right)^{-1}\boldsymbol{A}^T\,\boldsymbol{y}^{exp} \;\Rightarrow\; \mathrm{E}\,(\hat{\boldsymbol{x}}_{OLS}) = \boldsymbol{x}^{true} \tag{2c}$$

The following estimator can be given a strictly equivalent form using the Singular Value Decomposition (SVD) of matrix $\boldsymbol{A}$, given here under its 'full' form version:

$$\hat{\boldsymbol{x}}_{OLS} = \boldsymbol{V}'\,\boldsymbol{S}'^{-1}\,\boldsymbol{U}'^T\,\boldsymbol{y}^{exp} \quad \text{where } \boldsymbol{A} = \boldsymbol{U}'\,\boldsymbol{S}'\,\boldsymbol{V}'^T$$

$$\text{with } \boldsymbol{U}'^T\boldsymbol{U}' = \boldsymbol{U}'\boldsymbol{U}'^T = \boldsymbol{I}_m\,;\; \boldsymbol{V}'^T\boldsymbol{V}' = \boldsymbol{V}'\boldsymbol{V}'^T = \boldsymbol{I}_n \qquad \text{and } \boldsymbol{S}' = \begin{bmatrix} \mathrm{diag}\,(s'_1 \;\; s'_2 \;\; \cdots \;\; s'_n) \\ \boldsymbol{0}_{(m-n)\times n} \end{bmatrix} \tag{2d}$$

where the singular values s_k are positive (we suppose that matrix $\boldsymbol{A}$ has a full rank here) and have subscripts in decreasing order $(s'_1 \geq s'_2 \geq \cdots \geq s'_n > 0)$. $\boldsymbol{S}'$ is a m x n matrix and diag (.) designates a square diagonal matrix whose diagonal is defined by its arguments. $\boldsymbol{U}'$ and $\boldsymbol{V}'$ are square orthonormal matrices of respective dimensions m x m and n x n. $\boldsymbol{I}_k$ designates the identity matrix of dimensions k x k.

With the additional assumption that noise ε is independent and identically distributed (i.i.d.), with a standard deviation σ, this estimator is the unbiased linear estimation of minimum variance, with a variance-covriance matrix equal to:

$$\mathrm{cov}\,(\hat{\boldsymbol{x}}_{OLS}) = \sigma^2\left(\boldsymbol{A}^T\boldsymbol{A}\right)^{-1} = \sigma^2\,\boldsymbol{V}'\,\boldsymbol{S}'^{-2}\,\boldsymbol{V}'^T \tag{2e}$$

However, this solution corresponds to the inversion of a discrete form of an integral equation (a function estimation problem) that may be ill-posed if the condition number of $\boldsymbol{A}$, equal to s'_1 / s'_n, is too high, for a given value of the signal over noise ratio of the output.

So, a regularization of the inversion is compulsory. We only recall here the Truncated SVD (TSVD) regularization [1], even if many other regulation techniques are available (Tikhonov [2], Function specification method with future times for causal problems [3], ...).

This TSVD regularization method consists in writing the SVD of $\boldsymbol{A}$ using partitions of its matrices:

$$\boldsymbol{A} = \boldsymbol{U}'\boldsymbol{S}'\ \boldsymbol{V}'^T = \left[\boldsymbol{U}'_\alpha\ \boldsymbol{U}'_\beta\ \boldsymbol{U}'_\gamma\right] \begin{bmatrix} \boldsymbol{S}'_\alpha & \boldsymbol{0}_{\alpha \times \beta} \\ \boldsymbol{0}_{\beta \times \alpha} & \boldsymbol{S}'_\beta \\ \boldsymbol{0}_{\gamma \times \alpha} & \boldsymbol{0}_{\gamma \times \beta} \end{bmatrix} \left[\boldsymbol{V}'_\alpha \boldsymbol{V}'_\beta\right]^T = \boldsymbol{U}'_\alpha \boldsymbol{S}'_\alpha \boldsymbol{V}'^T_\alpha + \boldsymbol{U}'_\beta \boldsymbol{S}'_\beta \boldsymbol{V}'_\beta \quad (3a)$$

with $\alpha + \beta = n$ and $\gamma + n = m$

where $\boldsymbol{S}'_\alpha = \text{diag}\ (s'_1\ \ s'_2\ \ \cdots\ \ s'_\alpha)$ and $\boldsymbol{S}'_\beta = \text{diag}\ (s'_{\alpha+1}\ \ s'_{\alpha+2}\ \ \cdots\ \ s'_n)$ (3b)

and
$$\dim(\boldsymbol{U}'_\alpha) = m \times \alpha\ ;\ \dim(\boldsymbol{U}'_\beta) = m \times \beta\ ;\ \dim(\boldsymbol{U}'_\beta) = m \times \gamma$$
$$\dim(\boldsymbol{V}'_\alpha) = n \times \alpha\ ;\ \dim(\boldsymbol{V}'_\beta) = n \times \beta \quad (3c)$$

and keeping only the first α values (with $1 \le \alpha < n$) in the inversion (2d), that is:

$$\hat{\boldsymbol{x}}_{T-OLS}(\alpha) = \boldsymbol{V}'_\alpha\ \boldsymbol{S}'^{-1}_\alpha\ \boldsymbol{U}'^T_\alpha\ \boldsymbol{y}^{exp} \Rightarrow \text{E}\left(\hat{\boldsymbol{x}}_{TSVD}(\alpha)\right) - \boldsymbol{x}^{true} = -\boldsymbol{V}'_\beta\ \boldsymbol{V}'^T_\beta\ \boldsymbol{x}^{true} \neq 0 \quad (3d)$$

This estimator is biased, but since matrix $\boldsymbol{A}$ has been replaced by its first part $\boldsymbol{U}'_\alpha \boldsymbol{S}'_\alpha \boldsymbol{V}'^T_\alpha$ in (3a) , its condition number has improved and is now $s'_1 / s'_\alpha < s'_1 / s'_n$, which yields a more stable inversion.

The hyperparameter is chosen according to the discrepancy principle [4] that is, for an i.i.d. noise of standard deviation :

$$J_{OLS}\left(\hat{\boldsymbol{x}}_{TSVD}(\alpha)\right) \ge m\,\sigma^2 \quad and \quad J_{OLS}\left(\hat{\boldsymbol{x}}_{TSVD}(\alpha+1)\right) < m\,\sigma^2 \quad (3e)$$

2.2 Regularized inversion in case of error in the output and in the matrix of the model

We first consider, in this section that we have more data than unknowns, which means that model matrix $\boldsymbol{A}$ is $m \times n$, with $m > n$. We also assume that both output vector $\boldsymbol{y}$ and matrix $\boldsymbol{A}$ are corrupted by additive noises of zero expectations, noted $\boldsymbol{\varepsilon}_y$ and $\boldsymbol{\varepsilon}_A$ respectively:

$$\boldsymbol{A}^{exp} = \boldsymbol{A}^{true} + \boldsymbol{\varepsilon}_A \quad \text{and} \quad \boldsymbol{y}^{exp} = \boldsymbol{y}^{true} + \boldsymbol{\varepsilon}_y \quad \text{with} \quad \text{E}(\boldsymbol{\varepsilon}_A) = 0 \quad \text{and} \quad \text{E}(\boldsymbol{\varepsilon}_y) = 0 \quad (4a)$$

In the inverse analysis terminology, this type of assumption corresponds to an EIV (Error In Variable) model, where the structural parameters (matrix $\boldsymbol{A}$ here) are not errorless: not only the dependent variable (the output $\boldsymbol{y}$ here), but also the independent variables are corrupted by errors.

This inverse problem is solved by orthogonal regression, or Deming regression, or Total Least Squares (TLS) method. We assume that the physical units of all the the coefficients of $\boldsymbol{A}$ are the same and that they are identical to the units of $\boldsymbol{y}$, while the coefficients of $\boldsymbol{x}$ are dimensionless. So the sum to be minimized, the TLS criterion [5] is defined as:

$$J_{TLS}(\boldsymbol{A}, \boldsymbol{x}) = \left\|\boldsymbol{r}_G(\boldsymbol{A}, \boldsymbol{x})\right\|_F^2 = \left\|\boldsymbol{r}_A(\boldsymbol{A})\right\|_F^2 + \left\|\boldsymbol{r}_y(\boldsymbol{A}, \boldsymbol{x})\right\|^2 \quad (4b)$$

with $\boldsymbol{G}(\boldsymbol{A}, \boldsymbol{x}) = \left[\boldsymbol{A}, \boldsymbol{A}\boldsymbol{x}\right]\ ;\ \boldsymbol{G}^{exp} = \left[\boldsymbol{A}^{exp}, \boldsymbol{y}^{exp}\right]\ ;\ \boldsymbol{r}_G(\boldsymbol{A}, \boldsymbol{x}) = \boldsymbol{G}^{exp} - \boldsymbol{G}(\boldsymbol{A}, \boldsymbol{x})$ (4c)

IOP Conf. Series: Journal of Physics: Conf. Series **1047** (2018) 012001 doi:10.1088/1742-6596/1047/1/012001

or $\boldsymbol{r}_G(\boldsymbol{A},\boldsymbol{x}) = \left[\boldsymbol{r}_A(\boldsymbol{A}), \boldsymbol{r}_y(\boldsymbol{A},\boldsymbol{x})\right]$ with $\boldsymbol{r}_A(\boldsymbol{A}) = \boldsymbol{A}^{exp} - \boldsymbol{A}$; $\boldsymbol{r}_y(\boldsymbol{A},\boldsymbol{x}) = \boldsymbol{y}^{exp} - \boldsymbol{A}\boldsymbol{x}$ (4d)

Subscript F designates the Frobenius norm of a matrix, that is the sum of the squares of all its coefficients, $\boldsymbol{G}$ is the "augmented" matrix, of size m x $(n+1)$, and $\boldsymbol{r}_G$, $\boldsymbol{r}_A$ and $\boldsymbol{r}_y$ the residuals of the augmented matrix, of the system matrix, and of the output vector respectively. We can note here that the rank of matrix $\boldsymbol{G}$ is equal or smaller than n

The minimization of J_{TLS} with respect to $\boldsymbol{A}$ and $\boldsymbol{x}$ can be derived from the Eckart-Young theorem [6]. This one considers a matrix $\boldsymbol{D}^{exp}$ of size m x n, with $m \le n$ with its SVD decomposition given below:

$$\begin{aligned} &\boldsymbol{D}^{exp} = \boldsymbol{U}\ \boldsymbol{S}\ \boldsymbol{V}^T \text{ with rank } (\boldsymbol{D}^{exp}) \ge r \text{ and } \boldsymbol{U} = [\boldsymbol{U}_r, \boldsymbol{U}_c];\ \boldsymbol{S} = [\boldsymbol{S}_r, \boldsymbol{S}_c]\ ;\ \boldsymbol{V} = [\boldsymbol{V}_r, \boldsymbol{V}_c] \\ &\text{where } \dim(\boldsymbol{D}^{exp}) = \dim(\boldsymbol{S}) = m \text{ x } n\ ;\ \dim(\boldsymbol{U}) = m \text{ x } m\ ;\ \dim(\boldsymbol{V}) = n \text{ x } n \\ &\qquad \boldsymbol{U}^T\boldsymbol{U} = \boldsymbol{U}\boldsymbol{U}^T = \boldsymbol{I}_m\ ;\ \boldsymbol{V}^T\boldsymbol{V} = \boldsymbol{V}\ \boldsymbol{V}^T = \boldsymbol{I}_n \\ &\text{and } \dim(\boldsymbol{S}_r) = m \text{ x } r\ ;\ \dim(\boldsymbol{U}_r) = m \text{ x } r\ ;\ \dim(\boldsymbol{V}_r) = n \text{ x } r \end{aligned} \tag{5a}$$

The best rank r approximation $\hat{\boldsymbol{D}}$ of $\boldsymbol{D}^{exp}$ corresponds to the first part of the partitioning of the above SVD [7]:

$$\hat{\boldsymbol{D}} = \boldsymbol{U}_r\ \boldsymbol{S}_r\ \boldsymbol{V}_r^T \text{ and } \left\|\boldsymbol{D}^{exp} - \hat{\boldsymbol{D}}\right\|_F = \min_{\text{rank}(\boldsymbol{D}) \le r} \left\|\boldsymbol{D}^{exp} - \boldsymbol{D}\right\|_F = \sqrt{s_{r+1}^2 + s_{r+2}^2 + \cdots + s_m^2} \tag{5b}$$

where the s_k's (for $k = 1$ to m) are the singular values of $\boldsymbol{D}^{exp}$ listed in descending order. The minimizer $\hat{\boldsymbol{D}}$ is unique if and only if $s_{r+1} \ne s_r$.

Application of this theorem to matrix $\boldsymbol{G}^{exp}$, that is replacing $\boldsymbol{D}^{exp}$, $\boldsymbol{D}$ and n by $\boldsymbol{G}^{exp}$, $\boldsymbol{G}$ and $n+1$ respectively, yields:

$$\begin{aligned} &\boldsymbol{G} = \boldsymbol{U}\ \boldsymbol{S}\ \boldsymbol{V}^T = \begin{bmatrix}\boldsymbol{U}_A & \boldsymbol{u}_y\end{bmatrix}\begin{bmatrix}\boldsymbol{S}_A & \boldsymbol{0} \\ \boldsymbol{0} & s_y\end{bmatrix}\begin{bmatrix}\boldsymbol{V}_A & \boldsymbol{v}_y\end{bmatrix}^T = \boldsymbol{U}_A\ \boldsymbol{S}_A\ \boldsymbol{V}_A^T + s_{n+1}\ \boldsymbol{u}_y\ \boldsymbol{v}_y \\ &\text{with } \dim(\boldsymbol{U}) = m \text{ x } (n+1)\ ;\ \dim(\boldsymbol{S}) = (n+1) \text{ x } (n+1)\ ;\ \dim(\boldsymbol{V}) = (n+1) \text{ x } (n+1) \\ &\qquad \dim(\boldsymbol{U}_A) = m \text{ x } n\ ;\ \dim(\boldsymbol{S}_A) = n \text{ x } n\ ;\ \dim(\boldsymbol{V}_A) = n \text{ x} n \\ &\qquad \boldsymbol{S}_A = \text{diag}(s_1\ \ s_2\ \ \cdots\ \ s_n)\ ;\ s_y = s_{n+1} \text{ where } s_1 \ge s_2 \ge \cdots \ge s_n \ge s_{n+1} > 0 \end{aligned} \tag{5c}$$

Here the 'thin' SVD decomposition has been used, where matrix $\boldsymbol{S}$ is square and where matrix $\boldsymbol{U}$ is rectangular with the same dimensions as $\boldsymbol{G}^{exp}$, which means that only the $n+1$ original left singular vectors are present in $\boldsymbol{U}$.

The approximation of rank n of $\boldsymbol{G}^{exp}$ consists in discarding the second part of its decomposition (5c), that is taking $r = n$ in (5b):

$$\hat{\boldsymbol{G}}_{TLS} = \boldsymbol{U}_A\ \boldsymbol{S}_A\ \boldsymbol{V}_A^T \tag{5d}$$

One can show [8, 9] that the corresponding TLS estimation of $\boldsymbol{x}$ is given by the coefficients of the last right singular vector $\boldsymbol{v}_y$, that is a n x 1 vector $\boldsymbol{v}_{Ay}$ and a scalar v_{yy}:

$$\hat{\boldsymbol{x}}_{TLS} = -\frac{1}{v_{yy}}\boldsymbol{v}_{Ay} = \left(\boldsymbol{A}^T\boldsymbol{A} - s_{n+1}^2\,\boldsymbol{I}_n\right)^{-1}\boldsymbol{A}^T\,\boldsymbol{y}^{exp} \quad \text{with} \quad \boldsymbol{V} = \begin{bmatrix}\boldsymbol{V}_A & \boldsymbol{v}_y\end{bmatrix} = \begin{bmatrix}\boldsymbol{V}_{AA} & \boldsymbol{v}_{Ay}\\ \boldsymbol{v}_{yA} & v_{yy}\end{bmatrix} \tag{5e}$$

Let us note that in the case of a square $\boldsymbol{A}^{exp}$ matrix, $m = n$ and the rank of the augmented matrix $\boldsymbol{G}^{exp}$ is equal to m and, as a consequence $s_{n+1} = s_{m+1} = 0$ and the OLS estimator (2c) and the TLS estimator (5e) are the same.

Since the variance of the components of the TLS estimator can be large, it is possible to get its regularized version by truncation, a regularization method called T-TLS here, by discarding more small singular values than the smallest one in (5c). The thin decomposition of $\boldsymbol{G}^{exp}$ is partitioned the following way:

$$\boldsymbol{G}^{exp} = \boldsymbol{U}\,\boldsymbol{S}\,\boldsymbol{V}^T = \begin{bmatrix}\boldsymbol{U}_\alpha & \boldsymbol{U}_\beta\end{bmatrix}\begin{bmatrix}\boldsymbol{S}_\alpha & \boldsymbol{0}\\ \boldsymbol{0} & \boldsymbol{S}_\beta\end{bmatrix}\begin{bmatrix}\boldsymbol{V}_\alpha & \boldsymbol{V}_\beta\end{bmatrix}^T = \boldsymbol{U}_\alpha\,\boldsymbol{S}_\alpha\,\boldsymbol{V}_\alpha^T + \boldsymbol{U}_\beta\,\boldsymbol{S}_\beta\,\boldsymbol{V}_\beta^T \quad \text{with } \alpha+\beta = n+1 \text{ and } \beta > 1 \tag{6a}$$

and the T-TLS estimator is given by:

$$\hat{\boldsymbol{x}}_{T\text{-}TLS}(\alpha) = -\frac{\boldsymbol{V}_{12}\,\boldsymbol{V}_{22}^T}{\left\|\boldsymbol{V}_{22}\right\|^2} \text{ with } \boldsymbol{V} = \begin{bmatrix}\boldsymbol{V}_1 & \boldsymbol{V}_2\end{bmatrix} = \begin{bmatrix}\boldsymbol{V}_{11} & \boldsymbol{V}_{12}\\ \boldsymbol{V}_{21} & \boldsymbol{V}_{22}\end{bmatrix} \text{ where dim}(\boldsymbol{V}_1) = (n+1)\,\text{x}\,\alpha\text{ ; dim}(\boldsymbol{V}_2) = (n+1)\,\text{x}\,\beta \tag{6b}$$

with dim $(\boldsymbol{V}_{11}) = n$ x α ; dim $(\boldsymbol{V}_{12}) = n$ x β ; dim $(\boldsymbol{V}_{21}) = 1$ x α ; dim $(\boldsymbol{V}_{22}) = 1$ x β

We will derive in the next section an exact form of model (1) for 1D heat diffusion in a slab, before considering the corresponding identification problem using both T-OLS and T-TLS.

3. Convolution model in a 1D heat diffusion case

3.1 Solution in Laplace domain

A homogeneous slab of thickness ℓ is considered here. Its initial temperature is supposed to be uniform at a level $T_i = T_\infty$, where T_∞ is the temperature of the ambient air that is supposed to be a constant. A surface heat source $q(t)$, in W.m^{-2} is imposed over its front face ($x = 0$) at time $t = 0^+$, see fig. 1. The other (rear) face ($x = \ell$) exchange heat with the ambient air through a uniform and constant heat transfer coefficient h. The slab is made out of a material of heat conductivity λ, of volumetric heat ρc and of heat diffusivity $a = \lambda/(\rho c)$.

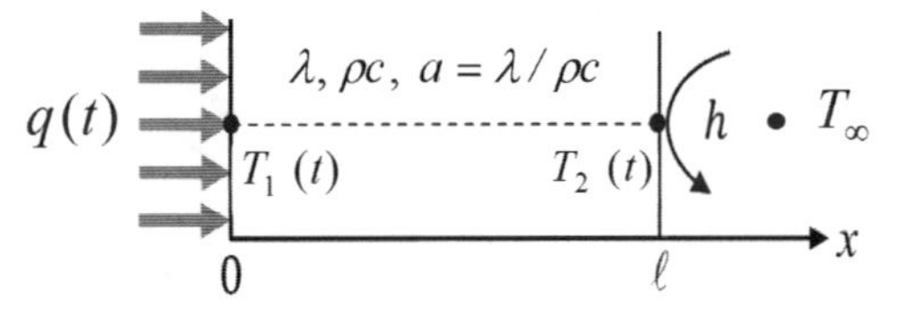

The heat equation, after the change of variable $\theta\,(x, t) = T\,(x, t) - T_i$ is:

$$\frac{\partial^2\theta}{\partial x^2} = \frac{1}{a}\frac{\partial\theta}{\partial t} \tag{7}$$

The initial and boundary conditions are:

IOP Conf. Series: Journal of Physics: Conf. Series **1047** (2018) 012001 doi:10.1088/1742-6596/1047/1/012001

Figure 1 – Slab with corresponding 1D boundary conditions

$$\theta = 0 \quad \text{at } t = 0 \quad \text{for} \quad 0 \le x \le \ell \tag{8}$$

$$\varphi = -\lambda \frac{\partial \theta}{\partial x} = q(t) \quad \text{at} \quad x = 0 \text{ for } t > 0\,; \quad \varphi = -\lambda \frac{\partial \theta}{\partial x} = h\,\theta \quad \text{at} \quad x = \ell \text{ for } t > 0 \tag{9 - 10}$$

where $\varphi\,(x, t) \equiv -\lambda \frac{\partial \theta}{\partial x}$ is the local x-component of the heat flux in the slab and λ its conductivity. This system is solved through integral time (Laplace) transformation of both temperature and flux [10]:

$$\overline{\psi}\,(x, p) \equiv \int_0^t \psi(x, t) \exp\,(-\,p t)\, \mathrm{d}t \qquad \text{for} \quad \psi = \theta \text{ or } \varphi \tag{11}$$

where p is the Laplace parameter. After this transformation, equations (2) and (5) become, with the introduction of the local flux φ:

$$\frac{\partial}{\partial x}\begin{bmatrix} \overline{\theta} \\ \overline{\varphi} \end{bmatrix} = -\begin{bmatrix} 0 & 1/\lambda \\ \lambda\, p/a & 0 \end{bmatrix}\begin{bmatrix} \overline{\theta} \\ \overline{\varphi} \end{bmatrix} \tag{12}$$

Integration of this ordinary differential equation system between $x = 0$ and $x = \ell$ yields, after taking into account conditions (8) and (9) :

$$\begin{bmatrix} \overline{\theta}_1 \\ \overline{q} \end{bmatrix} = \begin{bmatrix} A & B \\ C & D \end{bmatrix}\begin{bmatrix} 1 & 0 \\ h & 1 \end{bmatrix}\begin{bmatrix} \overline{\theta}_2 \\ 0 \end{bmatrix} \quad \text{with} \quad \begin{cases} A = D = \cosh\,(\eta\,\ell) \;\; ; \;\; \eta^2 = p/a \\ B = \lambda\,\eta\,\sinh\,(\eta\,\ell) \;\; ; \;\; C = \sinh\,(\eta\,\ell)/(\lambda\eta) \end{cases} \tag{13a, b}$$

The rear face temperature response θ_2 at $x = \ell$, is linked to the front face temperature response θ_1 at $x = 0$ by a simple product involving a transfer function W, called here a "transmittance", in the Laplace domain:

$$\overline{\theta}_2\,(p) = \overline{W}\;\overline{\theta}_1\,(p) \qquad \text{with} \qquad \overline{W}\,(p) = \frac{1}{A\,(p) + h\,B\,(p)} \tag{14a, b}$$

Return to the time domain makes a convolution product, noted '*' here, appear:

$$\theta_2\,(t) = W\,(t) * \theta_1\,(t) \tag{15a}$$

or

$$\theta_2\,(t) = \int_0^t W(t - t')\,\theta_1\,(t')\,\mathrm{d}t' = \int_0^t W(t')\,\theta_1\,(t - t')\,\mathrm{d}t' \tag{15b}$$

One can consider the input to be either the surface power $q\,(t)$ or the front face temperature $\theta_1\,(t)$ since this one is related to q by a transfer function, an impedance Z here, defined here :

$$\overline{\theta}_1\,(p) = \overline{Z}\,(p)\,\overline{q}\,(p) \quad \text{with} \quad \overline{Z}\,(p) = \frac{A\,(p) + h\,B\,(p)}{C\,(p) + h\,A\,(p)} \quad \Rightarrow \quad \theta_1\,(t) = Z\,(t) * q(t) \tag{16a, b, c}$$

One cannot write the opposite $q(t) = Y(t) * \theta_1(t)$, with $\overline{Y}(p) = 1/\overline{Z}(p)$, since $\overline{Y}(p)$ is not the Laplace transform of any time function: a transfer function has to be a monotonically decreasing function of the Laplace parameter p, which is the case for $\overline{Z}(p)$ but is not true for its inverse $1/\overline{Z}(p)$.

3.2 Solution in finite dimensions

The commutative convolution integrals (15b) can be written after a numerical quadrature involving m discrete times $t_i = i\,\Delta t$ (for $i = 1$ to m), where $\Delta t = t_f / m$ is the time step depending of the upper bound t_f of the time domain considered. This time step has to be very small with respect to the characteristic diffusion time ℓ^2 / a of the slab and with respect of the characteristic time of the temperature input function $\theta_1(t)$. Under these conditions, one can write at any time

$$\theta_2(t_i) = \Delta t \sum_{k=1}^{i} W_{i-k+1}\,\theta_{1,k} \quad \text{or} \quad \theta_2(t_i) = \Delta t \sum_{k=1}^{i} W_i\,\theta_{1,i-k+1} \tag{17a, b}$$

Let us note that quadrature (17a or b) is not an application of the trapezoidal rule to equation (15b). In order to be unbiased $\theta_2(t_i)$ is an instantaneous sampled value of function $\theta_2(t)$, while both coefficients W_k and $\theta_{1,k}$ are parameterized components of the corresponding functions $W(t)$ and $\theta_1(t)$: they are their projections over a base of unit step functions. These are averaged values of each function over the $[t_{k-1}, t_k[$ interval that precedes time t_k. This is due to the fact that the problem is causal and response $\theta_2(t)$ shall follow excitation $\theta_1(t)$ (and consequently $W(t)$, because of their symmetric form in convolution product (15a)). So, both functions $\theta_1(t)$ and $W(t)$ are piecewise constant functions:

$$\theta_1(t) = \sum_{j=1}^{m} \theta_{1,j}\, f_j(t) \quad \text{and} \quad W(t) = \sum_{j=1}^{m} W_j\, f_j(t) \quad \text{with} \quad f_j(t) = \mathrm{H}(t-t_{j-1}) - \mathrm{H}(t-t_j) \text{ for } j=1 \text{ to } m \tag{18a, b}$$

where $\mathrm{H}(t)$ is the Heaviside function and where:

$$\theta_{1,j} = \frac{1}{\Delta t}\int_{t_{j-1}}^{t_j} \theta_1(t)\,\mathrm{d}t \quad \text{and} \quad W_j = \frac{1}{\Delta t}\int_{t_{j-1}}^{t_j} W(t)\,\mathrm{d}t \tag{19a, b}$$

If all the observation times t_k over the $]t_1, t_m = t_f]$ are considered, equations (17) can be put under a matrix/column-vector form:

$$\boldsymbol{\theta}_2 = \mathbf{M}(\mathbf{W})\,\boldsymbol{\theta}_1 = \mathbf{M}(\boldsymbol{\theta}_1)\,\mathbf{W} \tag{20a, b}$$

Here $\mathbf{M}(.)$ is a square matrix function, of dimensions m x m whose argument is a column vector. It is a lower triangular Toeplitz matrix defined by:

$$\mathbf{M}(z) \equiv \Delta t \begin{bmatrix} z_1 & & & & \\ z_2 & z_1 & & 0 & \\ z_3 & z_2 & z_1 & & \\ \vdots & \vdots & \vdots & \ddots & \\ z_m & z_{m-1} & z_{m-2} & \cdots & z_1 \end{bmatrix} \quad \text{where} \quad z = \begin{bmatrix} z_1 \\ z_2 \\ z_3 \\ \vdots \\ z_m \end{bmatrix} \quad \text{and} \quad \begin{cases} z_j = \dfrac{1}{\Delta t} \displaystyle\int_{t_{j-1}}^{t_j} z(t)\, dt \\ \text{for } z = \theta_1 \text{ or } W \end{cases} \tag{21a, b}$$

The direct problem consists in calculating the output θ_2 from a given known input θ_1, if the structural parameters gathered in transmittance vector W of the white box model (20a) are known. In order to estimate the components of W, that is to identify model (15a) of convolutive structure, two paths are available:

- either use a numerical inversion algorithm of the exact Laplace transform (14b) of the its transfer function, to calculate $W_j(t)$ at any time t: it is a model reduction procedure,
- or use a measurement of both input and output in a calibration experiment: since input and transmittance commute in the model in its form (20ab), it is an inverse input problem (very similar to the classical Inverse Heat Flux Conduction problem). So, it is also a model identification experiment, that can be polluted by the presence of noise on both output and input.

4. Inverse methods for convolutive model identification

We are interested in estimating the transmittance vector W of model (15a) using measured values of both input θ_1^{exp} and output θ_2^{exp} of a potential calibration experiment. These experimental input and output differ from their exact value because of the presence of 2 random noises called ε_1 and ε_2 respectively. We assume that both noises have 0 mean and are independent and identically distributed (i.i.d), that is:

$$\theta_1^{exp} = \theta_1^{true} + \varepsilon_1 \qquad \text{with} \quad \mathrm{E}(\varepsilon_1) = 0 \quad \text{and} \quad \mathrm{cov}(\varepsilon_1) = \sigma_1^2\, \boldsymbol{I}_m \tag{22a, b, c}$$

$$\theta_2^{exp} = \theta_2^{true} + \varepsilon_2 \qquad \text{with} \quad \mathrm{E}(\varepsilon_2) = 0 \quad \text{and} \quad \mathrm{cov}(\varepsilon_2) = \sigma_2^2\, \boldsymbol{I}_m \tag{23a, b, c}$$

where $\mathrm{E}(.)$ designates the expectation of a random vector, $\mathrm{cov}(.)$ its variance-covariance matrix. We also assume that ε_1 and ε_2 are mutually independent (their covariance matrix, $\mathrm{cov}(\varepsilon_1,\varepsilon_2)$, is equal to zero). In this paper we only consider a synthetized calibration experiment where the exact values of θ_1, W and θ_2 are known.

4.1 Validation of the direct model and of its identification by Ordinary or Total Least Squares (OLS/TLS)

We consider a noiseless input θ_1 that is defined by:

$$\theta_1(t) = \begin{cases} 0 \quad \text{if} \quad t \le 0 \\ (1 - \exp(-t/\tau)\, \theta_1^{ss} \quad \text{if} \quad t > 0 \end{cases} \tag{24}$$

where τ is a constant and θ_1^{ss} is the asymptotic (steady state value) of $\theta_1(t)$. The temperature response $\theta_2(t)$ on the rear face ($x = \ell$) has been calculated at the discrete times defined in section 3.2. The values of the different quantities necessary for the solution of the discrete direct problem are given in Table 1.

Table 1 – Parameters of the simulation (direct problem)

ℓ (mm)	τ (s)	θ_1^{ss} (K)	t_f (s)	Δt (s)	h (W.m^{-2}.K^{-1})	λ (W.m^{-1}.K^{-1})	ρc (kJ.m^{-3}.K^{-1})
50	30	30	700	0.5	10	43	3666

In order to validate our input/output model, the commercial finite elements code COMSOL has been used to solve equations (7) to (10), and to simulate the output $\theta_2(t)$ using the input $\theta_1(t)$ given by (24). The corresponding plots are shown in Figure 2. Transmittance W, see (15a) and (15b) has been calculated using these synthetic $\theta_1(t)$ and $\theta_2(t)$ generated by COMSOL using either OLS or TLS and are shown in Figure 3. A numerical inversion of the analytical expression (14b) of the Laplace transform $\overline{W}$ of W has been implemented, using the de Hoog's algorithm [11] and the corresponding transmittance is shown in the same figure. This shows that without any noise, both OLS and TLS method yield the same estimation of the transmittance.

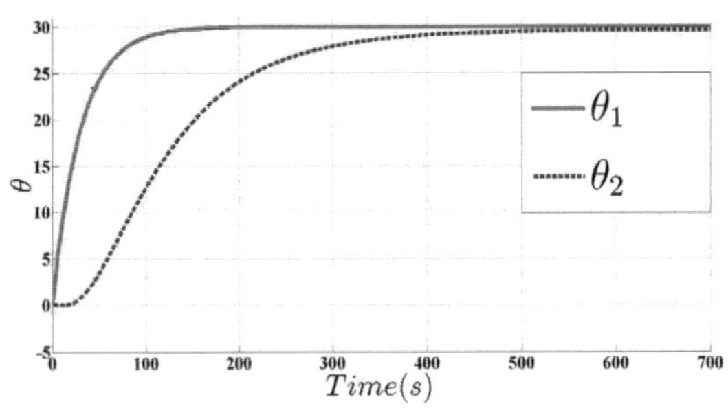

Figure 2 – Input and numerical solution (COMSOL) of the output

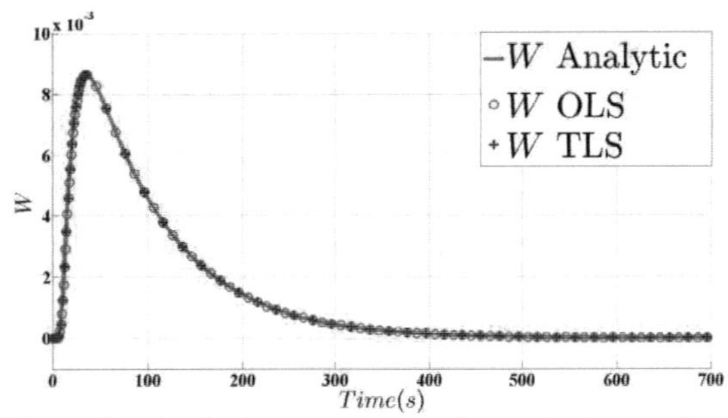

Figure 3 – Analytic transmittance (numerical inversion of Laplace transform), and estimated transmittances identified by OLS and TLS without regularization, starting from inversion of noiseless input θ_1 and noiseless output θ_2 calculated by COMSOL as a direct model

4.2 Identification of the transmittance starting from noisy input and output

Starting from the synthetic 'true' temperature distributions shown in Figure 2, a i.i.d. noise has been added on each input and output signal, see equations (22) and (23) and the transmittance has been estimated using truncated OLS (T-OLS) and truncated TLS (T-TLS) with an optimized truncation hyperparameter α corresponding to the minimum Euclidian norm of the error $\left\|\hat{W}_\alpha - W^{true}\right\|$. The results are shown for 3 cases: a noise of standard deviation 1 K on θ_2, see Figure 4, the same noise on θ_1 only, see Figure 5, and the same noise on both θ_1 and θ_2, see Figure 6.

We can notice that the two identification techniques (T-OLS and T-TLS) yield the same $\hat{W}_{\alpha_{optimum}}$ estimate, whatever the noise. A second remark is that the noise on the input has a smaller impact on the estimate that the noise on the output: this is due to its filtering by the system matrix, see section 4.3 further on.

In order to verify these results, inversions have been made for different values of the standard deviations σ_1 and σ_2. We define here the relative estimation error as:

$$e_W^* = \left\| \hat{W}(\alpha_{opt}) - W^{true} \right\| / \left\| W^{true} \right\| \tag{25a}$$

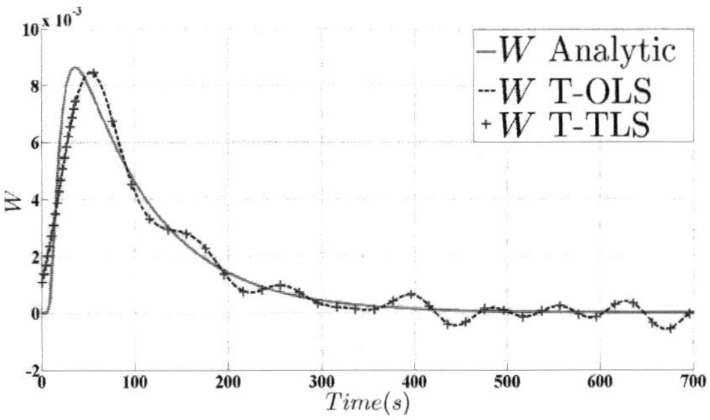

Figure 4 – True transmittance (analytic) and transmittances estimated by T-OLS and by T-TLS for $\sigma_1 = 0$ and $\sigma_2 = 1$ K

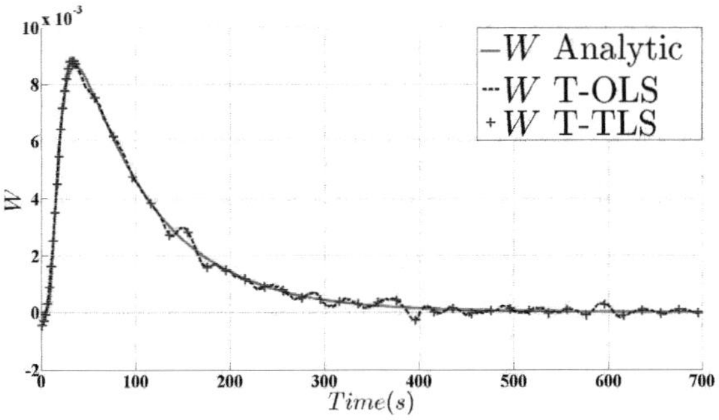

Figure 5 – True transmittance (analytic) and transmittances estimated by T-OLS and by T-TLS for $\sigma_1 = 1$ K and $\sigma_2 = 0$

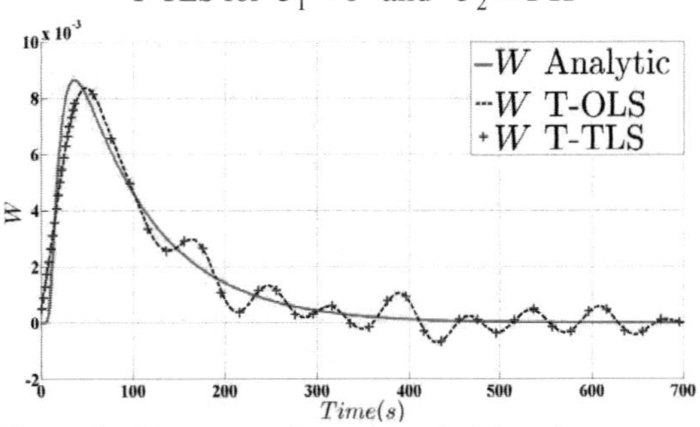

Figure 6 – True transmittance (analytic) and transmittances estimated by T-OLS and by T-TLS for $\sigma_1 = 1$ K and $\sigma_2 = 1$ K

The corresponding results are presented in Table 2, in terms of the optimum value α_{opt} of α, of the Signal over noise ratio SNR of each temperature signal, and of the relative estimation error e_W^*, with the following definitions:

$$SNR_k = \frac{\left\| \theta_k \right\|}{\left\| \varepsilon_k \right\|} \approx \frac{\left\| \theta_k \right\|}{\sigma_k \sqrt{m}} \quad \text{for } k = 1 \text{ or } 2 \tag{25b}$$

Table 2 – Relative error e_W^* and optimum hyperparameter α_{opt} by T-OLS and T-TLS inversion for different standard deviations σ_1 and σ_2 of noise on input and/or output

Configuration	Case	σ_1 (K)	σ_2 (K)	SNR_1	SNR_2	T-TLS		T-OLS	
						e_W^* (%)	α_{opt}	e_W^* (%)	α_{opt}
same noise for both signals	1	0.01	0.01	2907	2563	3.7	57	3.7	57
	2	0.10	0.10	290.2	256.3	6.1	28	6.1	28
	3	1.00	1.00	29.3	25.6	19.3	20	19.0	20
	4	10	10	3.1	2.8	38	4	37.9	4
different noise for each signal	5	0.01	1.00	2907	25.6	18.6	19	18.6	19
	6	1.00	0.01	29.3	2563	5.9	59	5.9	59
	7	0.01	10	2907	2.8	37.7	4	37.3	9
	8	10	0.01	3.1	2563	17.1	467	17.1	467
noise on one signal only	9	0	1.00	∞	25.6	18.6	19	18.6	19
	10	0	10	∞	2.8	37.7	4	37.3	9
	11	1.00	0	29.3	∞	5.5	606	5.5	606
	12	10	0	3.1	∞	17.1	492	17.1	492

This table confirms that the two regularization techniques give nearly the same results, except for cases 3, 4 and 7, which can be explained by the numerical noise of the computer we used. This also shows that output noise degrades the quality of the inversion in a higher way than the input noise.

Since the results of T-OLS and T-TLS inversions are very close, we had a look on the distribution of the singular values of the two exact matrices $A = \mathbf{M}(\theta_1)$ and $G = [\mathbf{M}(\theta_1), \theta_2 = \mathbf{M}(\theta_1)\,W]$ and of their noisy versions $A^{exp} = \mathbf{M}(\theta_1^{exp})$ and $G^{exp} = [\mathbf{M}(\theta_1^{exp}), \theta_2^{exp}]$. These are plotted in Figure 7.

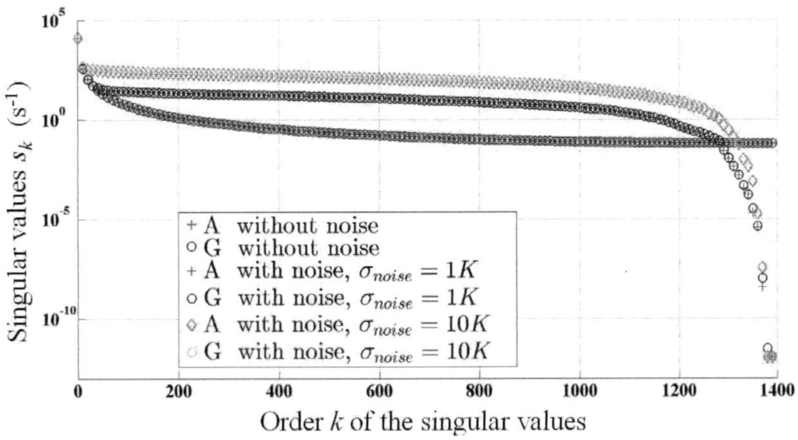

Figure 7 – Distribution of the eigenvalues of the $A = \mathbf{M}(\theta_1)$ and of the augmented matrix $G = [A, \theta_2]$ without and with noise (same standard deviation for both input and output). For clarity reasons, only 1 out of 20 singular values is plotted (k =1, 21, 41, ..., 1400).

These plots show that the singular values of A and G are nearly the same. Presence of noise makes the intermediate singular values increase in the range [$100 < k < 1300$] while they stay closer for the larger s_k's, and it has the opposite effect in the last interval [$1300 < k < 1400$] with an increase of the condition numbers of A and G.

4.3 Identification modelling by a unique output noise in the case of noisy input and output

We assume here that i.i.d. noises corrupt both input and output of model (15b), see equations (22) and (23) and OLS are used to estimate transmittance vector W. Since the dimensions of input and output are the same, the normal equation of the OLS problem can be written the following way:

$$\mathbf{M}(\theta_1 + \varepsilon_1)(W + e_w) = \theta_2 + \varepsilon_2 \tag{26}$$

where $e_w = W - W^{true}$ is the estimation error. Substracting (20b) from (26) yields:

$$\mathbf{M}(\theta_1)\,e_w = \varepsilon_2 - \mathbf{M}(W)\,\varepsilon_1 = K\,\varepsilon \text{ where } K = [I_m, -\mathbf{M}(W)]\,;\ \varepsilon = \begin{bmatrix}\varepsilon_2\\ \varepsilon_1\end{bmatrix};\ \text{cov}(\varepsilon) = \begin{bmatrix}\sigma_2^2\,I_m & 0\\ 0 & \sigma_1^2\,I_m\end{bmatrix} \tag{27}$$

This has been obtained by neglecting second order terms, with the use of the commutativity of the convolution product and with the additional assumption of mutually independent noise.
So, presence of noise ε_1 on the input can be taken into account by replacing the output noise ε_2 by a unique equivalent noise $\varepsilon'_2 = K\,\varepsilon$ that is not i.i.d. anymore. One easily shows:

$$\text{cov}(\varepsilon'_2) = K\,\text{cov}(\varepsilon)\,K^T = \sigma_2^2\left(I_m + \frac{\sigma_1^2}{\sigma_2^2}\mathbf{M}(W)\,\mathbf{M}^T(W)\right) = \sigma_2^2\,\Omega(W) \quad \text{with } \varepsilon'_2 = K\,\varepsilon \tag{28}$$

IOP Conf. Series: Journal of Physics: Conf. Series **1047** (2018) 012001 doi:10.1088/1742-6596/1047/1/012001

This would mean that, before any regularization, a Gauss-Markov estimator, which minimizes the variance of each estimated parameter, within the framework of linear estimators, would be well suited. This estimator corresponds to the minimum of the following criterion

$$J_{GM}(\boldsymbol{W}) = \left(\theta_2^{exp} - \mathbf{M}(\theta_1^{exp})\,\boldsymbol{W}\right)^{\boldsymbol{T}}\ \boldsymbol{\Omega}^{-1}\left(\theta_2^{exp} - \mathbf{M}(\theta_1^{exp})\,\boldsymbol{W}\right) \tag{29a}$$

whose solution is : $$\hat{\boldsymbol{W}}_{GM} = \left[\mathbf{M}^T(\theta_1^{exp})\,\boldsymbol{\Omega}^{-1}\,\mathbf{M}(\theta_1^{exp})\right]^{-1}\mathbf{M}^T(\theta_1^{exp})\,\boldsymbol{\Omega}^{-1}\,\theta_1^{exp} \tag{29b}$$

Unfortunately, when the linear model is square ($m = n$), this estimator, see (29b) in its unregularized version, degenerates into the OLS estimator. However, this deserves to be tested in the future for rectangular models with $m > n$.

5. Conclusions

We have shown, using an example of transient 1D heat diffusion in a wall, that identification of a temperature/temperature transfer function, when noise corrupts both input and output, could be regularized through truncated total least squares (T-TLS) based on a SVD decomposition of the augmented matrix. This TLS regularization of a deconvolution problem does not add anything new when compared to regularization of the ordinary least square problem, achieved here through truncated SVD. Even if the stochastic properties of an equivalent noise on the output only are known, it does not allow the construction of a better estimator than the two previous ones.

This unsuccessful performance of T-TLS here may stem from the fact that the reconstructed matrix, called $\hat{A}_\alpha$ here does not possess the Toeplitz structure of a convolution matrix: this information, a constraint, is not present in the version of T-TLS we have used. Another perspective of this work is to change our target: instead of minimizing the relative (global) estimation error, called e_W^* here, it can rather be more interesting to improve the estimation of transfer function $W(t)$ at the earlier times, that is up to its maximum: once this function estimated in the identification (calibration) experiment considered here, the quality of its early values conditions the quality of the reconstruction of the input, in a different inverse input experiment. This can be used in the design of a virtual sensor that allows to reconstruct here the front face temperature of a wall starting from measurement of the rear face temperature and of prior identification of the associated transfer function. As far as applications are concerned, not only inverse problems in diffusion, but also in advection-diffusion can be the subject of this type of identification through deconvolution, see [12] for example.

References

[1] R. C. Aster, B. Borchers, and C. H. Thurber, *Parameter estimation and inverse problems*, vol. 90. Academic Press, 2011.
[2] A.N. Tikhonov, V.Y. Arsenin, *Solution of Ill-posed Problems*, Winston & Sons, Washington, 1977.
[3] J.V. Beck, B. Balckwell, C.R. St.Clair Jr, *Inverse Heat Conduction – Ill-Posed Problems*, Wiley, New York, 1985.
[4] V.A. Morozov, *Methods for Solving Incorrectly Posed Problems*, Springer- Verlag, Berlin, 1984.
[5] G. H. Golub and C. F. Van Loan, "An analysis of the total least squares problem," *SIAM Journal on Numerical Analysis*, vol. 17, no. 6, pp. 883–893, 1980.
[6] C. Eckart and G. Young, The approximation of one matrix by another of lower rank, *Psycometrika*, 1936, 1 (3), 211-8.
[7] R. D. Fierro, G. H. Golub, P. C. Hansen, and D. P. O'Leary, Regularization by truncated total least squares, *SIAM Journal on Scientific Computing*, vol. 18, no. 4, pp. 1223–1241, 1997.
[8] I. Markovsky, S. Van Huffel, *Overview of total least squares methods.* Signal Processing, vol. 87, pp. 2283–2302, 2007.
[9] H.P. Gavin and G.P. Zéhil, *Total Least Squares*, Course, Duke University, Sept. 10, 2013.
[10] D. Maillet, S. André, J.-C. Batsale, A. Degiovanni, and C. Moyne, *Thermal quadrupoles : solving the heat equation through integral transforms.* Chichester : New York : Wiley., 2000.
[11] F. R. De Hoog, J. Knight, and A. Stokes, "An improved method for numerical inversion of Laplace transforms, *SIAM Journal on Scientific and Statistical Computing*, vol. 3, no. 3, pp. 357–366, 1982.
[12] W. Al Hadad, Y. Jannot, and D. Maillet, Characterization of a heat exchanger by virtual temperature sensors based on identified transfer functions, *Journal of Physics : Conference Series*, vol. 745 (http://iopscience.iop.org/article/10.1088/1742-6596/745/3/032089), p. 032089, IOP Publishing, October 21, 2016.

IOP Conf. Series: Journal of Physics: Conf. Series **1047** (2018) 012002 doi:10.1088/1742-6596/1047/1/012002

Optimum design of complementary transient experiments for estimating thermal properties

Filippo de Monte[1], James V Beck[2] and Donald E Amos[3]

[1] Department of Industrial and Information Engineering and Economics, University of L'Aquila, Italy

[2] Department of Mechanical Engineering, Michigan State University, USA

[3] Sandia National Laboratories, USA

filippo.demonte@univaq.it

Abstract. In the design of transient thermal experiments for estimating the thermal conductivity, *k*, and volumetric heat capacity, *C*, the scaled sensitivity coefficients are utilized. These coefficients should be large and uncorrelated. In the current paper it is proven that they are the largest when the heated surface temperature of the sample is held at the maximum value. Hence, they can be assumed as 'ideal reference sensitivities,' which actual *k*- and *C*-sensitivity coefficients can refer to in order to evaluate their relative magnitude. In addition, by using two complementary experiments, the sensitivity coefficients can be made more uncorrelated than if a single experiment is used. The result of two combined experiments is that the above two parameters are found with the greatest accuracy. The parameters are also of equal accuracy, which is generally not the case. This problem is unique not only because of intrinsic solutions between the small and large time solutions but in the exchanging of the roles in the scaled sensitivity coefficients for *k* and *C*.

1. Introduction

Thermal properties need to be measured using transient methods for several reasons. One of which is to estimate different properties simultaneously [1]. Another is that properties of a medium (such as a biological tissue [2]) can change during heating.

One criterion for optimal experiments involves maximization of the confidence region. See [1, pp. 444 - 456]. One implication of this criterion is that the scaled sensitivity coefficients should be large and uncorrelated. Many geometries can be utilized, including one-dimensional and two-dimensional rectangular bodies and also 1D and 2D radial and axial ones. Various boundary conditions can also be imposed including temperature and heat flux, which are called the first and second kinds. Fourth-kind [3] accounting for thermal inertia of the heater as well as time-dependent boundary conditions are also possible. Another major consideration is the location of the sensors and the duration of the experiment. In this work, in order to obtain the largest and the most uncorrelated sensitivity coefficients when simultaneously estimating thermal conductivity *k* and volumetric heat capacity *C*, two ideal transient experiments are considered. The former is the X21B(X11B10T0)0T0 problem, where "B(X11B10T0)" indicates that the time-dependent heat flux applied to the boundary $x = 0$ is the one coming from the surface heat flux of the X11B10T0 problem (which has prescribed temperature at $x = 0$, zero temperature at $x = L$ and is initially at zero temperature). (See Refs. [1, 2] for the heat conduction numbering system devised by Cole et al.). The latter is a complementary problem denoted

by X22B(X12B10T0)0T0, where the heat flux corresponding to the X12B10T0 problem (which has prescribed temperature at $x = 0$ and zero heat flux at $x = L$) is applied to the sample of interest.
It is proven that the heated surface scaled sensitivity coefficients for these two problems are the largest and, hence, they can be assumed as 'ideal reference sensitivities,' which actual k- and C-sensitivity coefficients can refer to. In addition, by using two combined complementary experiments as was done in Ref. [4], the sensitivity coefficients can be made more uncorrelated than if a single experiment is used. In other words, they allow us to minimize the area or hyper-volume of the confidence region [1, Chap. 8]. Concerning this, the confidence region is recalled to be at a minimum when the determinant of the related matrix $\mathbf{X}^\mathrm{T}\mathbf{X}$ for two parameters is a maximum, where $\mathbf{X}$ is the sensitivity matrix.
When computing the above sensitivity functions, an exact solution for the X21B(X11B10T0)0T0 and X22B(X12B10T0)0T0 problems for the whole time region is derived using Green's functions [5]. The small time solutions are derived too (still using GFs) and the relationship indicated above is demonstrated. Alternatively, the small time solutions can be derived using Laplace transforms [6]. An advantage of the early time solutions [7] is that they are mathematically simpler and the time range of validity is larger than one would expect. Both the early time and whole time solutions are helpful for several reasons. One is the computational advantage for early times of the one and the efficiency of other solution for moderate large time. Another reason is that having two different solutions for the "same" problem" provides an indication of intrinsic verification [8,9]. Actually, this problem is unique not only because of intrinsic verification between the small and large time solutions but in the exchanging of the roles in the scaled sensitivity coefficients for k and C.

2. X21B(X11B10T0)0T0 transient experiment

Consider the 1D transient experiment where the thin heater is located between two samples of the same thickness whose back surfaces are kept at zero temperature. A dimensional mathematical statement for this thermally-symmetric experiment denoted by X21B(X11B10T0)0T0 is

$$\frac{\partial^2 T}{\partial x^2} = \frac{1}{\alpha}\frac{\partial T}{\partial t} \qquad (0 < x < L;\; t > 0) \tag{1a}$$

$$-k\frac{\partial T}{\partial x}(0,t) = \varphi_{\mathrm{X11B10T0}}(0,t) \qquad (t > 0) \tag{1b}$$

$$T(L,t) = 0 \qquad (t > 0) \tag{1c}$$

$$T(x,0) = 0 \qquad (0 < x < L) \tag{1d}$$

where $\varphi_{\mathrm{X11B10T0}}(0,t)$ is the surface heat flux supplied by the thin heater during measurements. It is assumed that this heat flux is the same as the surface heat flux of the X11B10T0 case and, hence, can be fitted by using the following correlation in a series form

$$\varphi_{X11B10T0}(0,t) = \frac{\varphi_0}{\sqrt{\pi\phi t / L^2}}\sum_{n=0}^{\infty}\left\{\exp\left(-\frac{n^2L^2}{\phi t}\right) + \exp\left[-\frac{(n+1)^2 L^2}{\phi t}\right]\right\} \tag{2a}$$

where φ_0 (units of $\mathrm{W\cdot m^{-2}}$) and ϕ (units of $\mathrm{m^2\cdot s^{-1}}$) are empirical coefficients known. The above heat flux tends to infinity for small times and goes to a constant for very large times. Alternatively, another series correlation is given by

$$\varphi_{X11B10T0}(0,t) = \varphi_0\left(1 + 2\sum_{m=1}^{\infty} e^{-\beta_m^2\frac{\phi t}{L^2}}\right) \tag{2b}$$

where $\beta_m = m\pi$ are the "X11" dimensionless eigenvalues, as discussed in next subsection.

2.1 Heat flux supplied by the heater. The X11B10T0 case

The temperature to the X11B10T0 case, valid for all $t > 0$, is given by Carslaw and Jaeger [10] as

- short-time form [10, p. 310, equation (6)]

$$T_{X11B10T0}(x,t) = T_0 \sum_{n=0}^{\infty} \left\{ \operatorname{erfc}\left(\frac{2nL + x}{\sqrt{4\alpha t}} \right) - \operatorname{erfc}\left(\frac{2(n+1)L - x}{\sqrt{4\alpha t}} \right) \right\} \tag{3a}$$

- large-time form [10, p. 313, equation (10)]

$$T_{\mathrm{X11B10T0}}(x,t) = T_0 \left[\left(1 - \frac{x}{L} \right) - 2 \sum_{m=1}^{\infty} \frac{\sin(\beta_m x / L)}{\beta_m} e^{-\beta_m^2 \frac{\alpha t}{L^2}} \right] \tag{3b}$$

The solution for the heat flux at $x = 0$ is, respectively,

$$q_{X11B10T0}(0,t) = \frac{kT_0}{\sqrt{\pi \alpha t}} \sum_{n=0}^{\infty} \left[\exp\left(-\frac{n^2 L^2}{\alpha t} \right) + \exp\left(-\frac{(n+1)^2 L^2}{\alpha t} \right) \right] \tag{4a}$$

$$q_{X11B10T0}(0,t) = \frac{kT_0}{L} \left(1 + 2 \sum_{m=1}^{\infty} e^{-\beta_m^2 \frac{\alpha t}{L^2}} \right) \tag{4b}$$

where the group kT_0 / L (units of $\mathrm{W \cdot m^{-2}}$) corresponds to the coefficient φ_0 of Eqs. (2a) and (2b) as well as α (units of $\mathrm{m^2 \cdot s^{-1}}$) corresponds to ϕ of the same correlations.

The temperature solution to the X21B(X11B10T0)0T0 experiment at $x = 0$ (where the sensor is located) is discussed in next subsections for the whole time region and for early times.

2.2 Exact solution for the whole time region

Using the heat flux defined by Eq. (2b) as being known during measurements, the surface temperature of the specimen can be computed using large-cotime Green's functions [5] as

$$\begin{aligned} T_{\mathrm{X21B(X11B10T0)0T0}}(0,t) &= \frac{1}{C} \int_{\tau=0}^{t} \varphi_{X11B10T0}(0,\tau) G_{X21}(0,0,t-\tau) d\tau \\ &= \frac{\varphi_0}{C} \int_{\tau=0}^{t} \left(1 + 2 \sum_{m=1}^{\infty} e^{-\beta_m^2 \frac{\phi \tau}{L^2}} \right) \underbrace{\frac{2}{L} \sum_{n=1}^{\infty} e^{-\eta_n^2 \frac{\alpha (t-\tau)}{L^2}}}_{G_{X21}(0,0,t-\tau)} d\tau, \quad \eta_n = (n - 1/2)\pi \end{aligned} \tag{5}$$

By some algebra given in Appendix A, Eq. (5) becomes

$$\begin{aligned} T_{\mathrm{X21B(X11B10T0)0T0}}\left(0, t, \frac{\phi}{kC} \right) = \frac{\varphi_0 L}{k} \Bigg\{ & 1 - 2 \sum_{n=1}^{\infty} e^{-\eta_n^2 \frac{kt}{CL^2}} \frac{\cot\left[\eta_n \sqrt{k/(C\phi)} \right]}{\eta_n \sqrt{C\phi / k}} \\ & + 2 \sum_{m=1}^{\infty} e^{-\beta_m^2 \frac{kt}{CL^2} \left(\frac{C\phi}{k} \right)} \frac{\tan\left(\beta_m \sqrt{C\phi / k} \right)}{\beta_m \sqrt{C\phi / k}} \Bigg\} \end{aligned} \tag{6}$$

where, for an accuracy of about 10^{-A} (with $A = 2,3,...,15$), the required number terms for either summation is [11]

$$N = \text{ceil}\left(\frac{1}{\pi}\sqrt{\frac{A\ln(10)}{\tilde{t}}} + \frac{1}{2}\right), \qquad M = \text{ceil}\left(\frac{1}{\pi\sqrt{C\phi/k}}\sqrt{\frac{A\ln(10)}{\tilde{t}}}\right) \tag{7}$$

where $\tilde{t} = \alpha t / L^2 = kt/(CL^2)$. For the extremely accurate value of A = 15 (errors less than 10^{-15}), the required numbers of terms for the n-summation for times $\tilde{t}$ = 0.01, 0.1 and 0.5 are respectively, 20, 7 and 4. Even the number of 20 terms is not large. Therefore, the solution defined by Eq. (6) converges nicely with a relatively small numbers of terms. Furthermore, a more efficient solution is available for small dimensionless times, as discussed in the next subsections.

Before proceeding to calculate the sensitivity coefficients, it may be noted that equation (6) produces $\varphi_0 L / k = (T_{rise})_\infty$ as time goes to infinity, which represents the sample temperature rise for $t \to \infty$. Actually, the same result is obtained for all times equal to or greater than zero for $\phi = k / C$ since $\cot(\eta_n) = 0$ and $\tan(\beta_m) = 0$ in Eq. (6).

The sensitivity coefficients for k and C can be obtained from equation (6) by direct differentiation. However, it is more complicated than utilizing finite differences, which is possible because the computed temperatures can be calculated so precisely. A central finite difference scheme for computing the dimensionless scaled k- and C-sensitivities $X_{k,\text{X21}}$ and $X_{C,\text{X21}}$ is

$$X_{k,\text{X21}} = \frac{k}{(T_{rise})_\infty}\frac{\partial T_{\text{X21}}}{\partial k}(0,\tilde{t}) \approx \frac{T_{\text{X21B(X11B10T0)0T0}}[0,\tilde{t},k(1+\varepsilon),C] - T_{\text{X21B(X11B10T0)0T0}}[0,\tilde{t},k(1-\varepsilon),C]}{2\varepsilon(T_{rise})_\infty} \tag{8a}$$

$$X_{C,\text{X21}} = \frac{C}{(T_{rise})_\infty}\frac{\partial T_{\text{X21}}}{\partial C}(0,\tilde{t}) \approx \frac{T_{\text{X21B(X11B10T0)0T0}}[0,\tilde{t},k,C(1+\varepsilon)] - T_{\text{X21B(X11B10T0)0T0}}[0,\tilde{t},k,C(1-\varepsilon)]}{2\varepsilon(T_{rise})_\infty} \tag{8b}$$

where the subscript "X21" can be used as a short notation. The errors using this equation are about ε^2 so that if $\varepsilon = 0.00001 = 10^{-5}$, the errors would be about 10^{-10}. Numerical experiments with ε values from 0.01 to 0.00000001 have shown that both small values, such as 0.01 or 0.001, can cause significant errors; but extremely small values such as 0.00000001 also can. To get an accuracy of about 8 decimal places when compared with the sensitivity coefficients for k and C obtained from equation (6) by direct differentiation, an ε value of 0.00001 is recommended.

2.3 One-term approximate solution for very short times

For very early times, less than the so-called deviation time defined as [7, 8]

$$t_d^{(1)} = \frac{0.1}{A\alpha}(2L - x)^2 \qquad (A = 2,3,...,15), \tag{9}$$

the temperature and heat flux of the X11B10T0 finite problem treated in Subsection 2.1 are the same (with errors less than 10^{-A}) as the ones of the X10B1T0 semi-infinite case. The solution is [5, 10]

$$q_{\text{X11B10T0}}(x,t) \approx q_{\text{X10B1T0}}(x,t) = \frac{T_0 k}{\sqrt{\pi\alpha t}} \mathrm{e}^{-\frac{x^2}{4\alpha t}} \qquad (t \le t_d^{(1)}) \tag{10}$$

At the boundary surface x = 0, the heat flux is

IOP Conf. Series: Journal of Physics: Conf. Series **1047** (2018) 012002 doi:10.1088/1742-6596/1047/1/012002

$$q_{\mathrm{X11B10T0}}(0,t) \approx q_{\mathrm{X10B1T0}}(0,t) = \frac{T_0 k}{\sqrt{\pi \alpha t}} \qquad (t \le 0.4L^2/(A\alpha)) \tag{11}$$

Therefore, the heat flux applied to the sample during measurements defined by Eqs. (2a) or (2b) may be taken at very early times as

$$\varphi_{\mathrm{X11B10T0}}(0,t) \approx \varphi_{\mathrm{X10B1T0}}(0,t) = \frac{\varphi_0}{\sqrt{\pi \phi t / L^2}} \qquad (t \le 0.4L^2/(A\phi)) \tag{12}$$

The surface temperature for the X21B(X11B10T0)0T0 problem given by Eq. (6) for the whole time region can accurately be replaced at early times (less than the deviation time) with the surface temperature for the X21B(X10B1T0)0T0 case. This can be calculated using Green's functions [5] where a further approximation with errors less than 10^{-A} can be performed for co-times $(t-\tau) \le 0.1 \cdot (2L-x)^2/(A\alpha)$. In fact,

$$G_{\mathrm{X21}}(x,0,t-\tau) \approx G_{\mathrm{X20}}(x,0,t-\tau) = \frac{1}{\sqrt{\pi\alpha(t-\tau)}} \exp\left[-\frac{x^2}{4\alpha(t-\tau)}\right] \tag{13}$$

where $\alpha = k/C$ is the unknown thermal diffusivity of the sample. Therefore,

$$\begin{aligned} T_{X21B(X11B10T0)0T0}\left(0,t,\frac{\phi}{kC}\right) &\approx T_{X20B(X10B1T0)T0}\left(0,t,\frac{\phi}{kC}\right) = \frac{1}{C}\int_{\tau=0}^{t} \varphi_{X10B1T0}(0,\tau) G_{X20}(0,0,t-\tau)\,d\tau \\ &= \frac{\varphi_0 L}{\pi\sqrt{(kC)\phi}} \underbrace{\int_{\tau=0}^{t} \frac{d\tau}{\sqrt{\tau(t-\tau)}}}_{=\pi} = \frac{\varphi_0 L}{\sqrt{(kC)\phi}} = \frac{(T_{rise})_\infty}{\sqrt{C\phi/k}} = (T_{rise})_0 \end{aligned} \tag{14}$$

that is valid with errors less than $2 \cdot 10^{-A}$ for $t \le \min\{0.4L^2/(A\alpha), 0.4L^2/(A\phi)\}$. Also, $(T_{rise})_0$ is the temperature rise at $t=0$; while $(T_{rise})_\infty = \varphi_0 L/k$ is the temperature rise for $t \to \infty$. Equation (14) states that the surface temperature is constant for a while when the sample heating starts. In particular, for $\phi = k/C$, this constant is equal to $(T_{rise})_\infty$.

The k- and C-scaled sensitivity coefficients at the heated surface $x = 0$ (where the temperature sensor is located) can be obtained by direct differentiation as

$$k\frac{\partial T_{X21}}{\partial k}\left(0,t,\frac{\phi}{kC}\right) = C\frac{\partial T_{X21}}{\partial C}\left(0,t,\frac{\phi}{kC}\right) \approx -\frac{1}{2}\frac{\varphi_0 L}{\sqrt{(kC)\phi}} = -\frac{1}{2}\frac{(T_{rise})_\infty}{\sqrt{C\phi/k}} = -\frac{1}{2}(T_{rise})_0 \tag{15}$$

Notice that these two sensitivity coefficients are equal and functions of the product kC. In particular, they are both equal to the negative of the temperature rise divided by 2. Therefore, their sum is equal to the negative of the temperature rise at the heated surface according to the sensitivity equation [3]

$$k\frac{\partial T_{X21}}{\partial k}(0,t) + C\frac{\partial T_{X21}}{\partial C}(0,t) = -T_{X21B(X11B10T0)0T0}(0,t) \tag{16}$$

In a dimensionless form, with errors less than $2 \cdot 10^{-A}$, Eq. (15) becomes

$$X_{k,\mathrm{X21}} = \frac{k}{(T_{rise})_\infty}\frac{\partial T_{\mathrm{X21}}}{\partial k}\left(0,\tilde{t},\frac{\phi}{kC}\right) = X_{C,\mathrm{X21}} = \frac{C}{(T_{rise})_\infty}\frac{\partial T_{\mathrm{X21}}}{\partial C}\left(0,\tilde{t},\frac{\phi}{kC}\right) \approx -\frac{1}{2\sqrt{C\phi/k}} \tag{17}$$

where $\tilde{t} = \alpha t / L^2 = kt / (CL^2)$. Also, for $\phi = k / C$, it results in $X_{k,X21} = X_{C,X21} = -1/2$.

2.4 Three-terms approximate solution for short times
For times less than a certain time given by

$$t_d^{(2)} = \frac{0.1}{A\alpha}(2L+x)^2 \qquad (A = 2,3,...,15), \tag{18a}$$

which is still called deviation time but is longer than the previous one defined by equation (9), the exact solutions defined by Eqs. (3a) and (4a) can be computed with only two terms (n = 0 and 1) with errors less than 10^{-A}. However, as for the location x = 0 (which is here of interest) it results in $t_d^{(2)} = t_d^{(1)}$, another and longer deviation time has to be considered. It is defined as

$$t_d^{(3)} = \frac{0.1}{A\alpha}(4L-x)^2 \qquad (A = 2,3,...,15), \tag{18b}$$

and the exact solutions stated before can be computed with only three terms (n = 0, 1, 2) with errors less than 10^{-A} for $t \le t_d^{(3)}$. The superscripts "(1)," "(2)" and "(3)" appearing in the symbol of the deviation time indicate the number of terms required, which are one, two and three, respectively. In particular, the first three terms for the surface heat flux defined by equation (4a) are

$$q_{X11B10T0}(0,t) \approx \frac{kT_0}{\sqrt{\pi\alpha t}}\left(1 + e^{-\frac{L^2}{\alpha t}} + e^{-\frac{L^2}{\alpha t}}\right) \qquad (t \le 1.6L^2 / (A\alpha)) \tag{19}$$

Therefore, the heat flux applied to the sample during measurements defined by Eqs. (2a) or (2b) may be taken at early times as

$$\varphi_{X11B10T0}(0,t) \approx \frac{\varphi_0}{\sqrt{\pi\phi t / L^2}}\left(1 + 2e^{-\frac{L^2}{\phi t}}\right) \qquad (t \le 1.6L^2 / (A\phi)) \tag{20}$$

Then, the surface temperature of the sample can still be derived using the Green's function solution equation. Also, the Green's function for the X21 case at $x = x' = 0$ can be replaced with the three-term approximate GF with errors less than 10^{-A} for co-times $(t-\tau) \le 1.6L^2 / (A\alpha)$. It is [5]

$$G_{X21}(0,0,t-\tau) \approx \frac{1}{\sqrt{\pi\alpha(t-\tau)}}\left[1 - 2e^{-\frac{L^2}{\alpha(t-\tau)}}\right] \tag{21}$$

The solution for the surface temperature is mathematically given by

$$\begin{aligned} T_{\text{X21B(X11B10T0)0T0}}(0,t) &= \frac{1}{C}\int_{\tau=0}^{t} \varphi_{X11B10T0}(0,\tau) G_{X21}(0,0,t-\tau)\, d\tau \\ &\approx \frac{\varphi_0 L}{\pi\sqrt{(kC)\phi}} \int_{\tau=0}^{t} \frac{1}{\sqrt{\tau(t-\tau)}}\left[1 - 2e^{-\frac{CL^2}{k(t-\tau)}} + 2e^{-\frac{L^2}{\phi\tau}}\right] d\tau \end{aligned} \tag{22}$$

where the cross term was dropped as negligible. Some needed integrals are given in Mathematica. For convenience, one of these integrals is

$$\int_{\tau=0}^{t} \frac{e^{-\frac{L^2}{\sigma\tau}}}{\sqrt{\tau(t-\tau)}} d\tau = \pi \operatorname{erfc}\left(\sqrt{\frac{L^2}{\sigma t}}\right) \qquad (\sigma = k/C \text{ or } \phi) \tag{23}$$

Using these integrals, equation (22) produces

$$T_{\text{X21B(X11B10T0)0T0}}\left(0,t,\frac{\phi}{kC}\right) \approx \frac{\varphi_0 L}{\sqrt{(kC)\phi}}\left[1 - 2\operatorname{erfc}\left(\sqrt{\frac{CL^2}{kt}}\right) + 2\operatorname{erfc}\left(\sqrt{\frac{L^2}{\phi t}}\right)\right] \tag{24a}$$

where

$$\frac{\varphi_0 L}{\sqrt{(kC)\phi}} = (T_{rise})_0 = \frac{(T_{rise})_\infty}{\sqrt{C\phi/k}} = \frac{\varphi_0 L/k}{\sqrt{C\phi/k}} \tag{24b}$$

The above expression is valid with errors less than $2\cdot 10^{-A}$ for $t \le \min\{1.6L^2/(A\alpha), 1.6L^2/(A\phi)\}$. Also, it states that the surface temperature is time-independent for $\phi = k/C$ (as expected) and equal to $\varphi_0 L/k = (T_{rise})_\infty$, which represents the sample temperature rise for $t \to \infty$.
The thermal conductivity scaled sensitivity coefficient obtained by differentiating equation (24a) is

$$\begin{aligned} k\frac{\partial T_{\text{X21}}}{\partial k}\left(0,t,\frac{\phi}{kC}\right) &\approx -\frac{1}{2}T_{\text{X21B(X11B10T0)0T0}}(0,t) - \frac{2\varphi_0 L/k}{\sqrt{\pi\phi t/L^2}} e^{-\frac{CL^2}{kt}} \\ &= -\frac{1}{2}\frac{(T_{rise})_\infty}{\sqrt{C\phi/k}}\left[1 + \frac{4}{\sqrt{\pi kt/(CL^2)}} e^{-\frac{CL^2}{kt}} - 2\operatorname{erfc}\left(\sqrt{\frac{CL^2}{kt}}\right) + 2\operatorname{erfc}\left(\sqrt{\frac{L^2}{\phi t}}\right)\right] \end{aligned} \tag{25}$$

Similarly, the volumetric heat capacity scaled sensitivity coefficient is

$$\begin{aligned} C\frac{\partial T_{\text{X21}}}{\partial C}\left(0,t,\frac{\phi}{kC}\right) &\approx -\frac{1}{2}T_{\text{X21B(X11B10T0)0T0}}(0,t) + \frac{2\varphi_0 L/k}{\sqrt{\pi\phi t/L^2}} e^{-\frac{CL^2}{kt}} \\ &= -\frac{1}{2}\frac{(T_{rise})_\infty}{\sqrt{C\phi/k}}\left[1 - \frac{4}{\sqrt{\pi kt/(CL^2)}} e^{-\frac{CL^2}{kt}} - 2\operatorname{erfc}\left(\sqrt{\frac{CL^2}{kt}}\right) + 2\operatorname{erfc}\left(\sqrt{\frac{L^2}{\phi t}}\right)\right] \end{aligned} \tag{26}$$

The sum of these two scaled sensitivity coefficients is equal to the negative of the temperature rise at the heated surface according to the sensitivity equation (16). In a dimensionless form, they are

$$X_{k,\text{X21}} = \frac{k}{(T_{rise})_\infty}\frac{\partial T_{\text{X21}}}{\partial k}\left(0,\tilde{t},\frac{\phi}{kC}\right) \approx -\frac{1}{2\sqrt{C\phi/k}}\left[1 + \frac{4e^{-1/\tilde{t}}}{\sqrt{\pi\tilde{t}}} - 2\operatorname{erfc}\left(\frac{1}{\sqrt{\tilde{t}}}\right) + 2\operatorname{erfc}\left(\frac{1}{\sqrt{\tilde{t}(C\phi/k)}}\right)\right] \tag{27a}$$

$$X_{C,\text{X21}} = \frac{C}{(T_{rise})_\infty}\frac{\partial T_{\text{X21}}}{\partial C}\left(0,\tilde{t},\frac{\phi}{kC}\right) \approx -\frac{1}{2\sqrt{C\phi/k}}\left[1 - \frac{4e^{-1/\tilde{t}}}{\sqrt{\pi\tilde{t}}} - 2\operatorname{erfc}\left(\frac{1}{\sqrt{\tilde{t}}}\right) + 2\operatorname{erfc}\left(\frac{1}{\sqrt{\tilde{t}(C\phi/k)}}\right)\right] \tag{27b}$$

where $\tilde{t} = \alpha t/L^2 = kt/(CL^2)$. At very early times, it is convenient to use equations (17) based on one-term approximation. As regards Eqs. (8a) and (8b) related to Eq. (6) with the number of terms defined by Eq. (7), they are computationally efficient for $\tilde{t} \ge \min\{1.6/A, 1.6/[A(C\phi/k)]\}$ though they are

IOP Conf. Series: Journal of Physics: Conf. Series **1047** (2018) 012002 doi:10.1088/1742-6596/1047/1/012002

valid for the whole time region. Substituting $\tilde{t}=1.6/A$ in the first of the two Eqs. (7) gives $N\approx\text{ceil}(0.4A+0.5)$, that is, 7 terms for an accuracy of A = 15 (errors less than 10^{-15}). For $\phi=k/C$,

$$X_{k,\text{X21}}\approx-\frac{1}{2}\left(1+\frac{4e^{-1/\tilde{t}}}{\sqrt{\pi\tilde{t}}}\right)\qquad X_{C,\text{X21}}\approx-\frac{1}{2}\left(1-\frac{4e^{-1/\tilde{t}}}{\sqrt{\pi\tilde{t}}}\right)\tag{27c}$$

2.5 Numerical results

Some numerical values (to 8 digits, i.e. A = 8) for the scaled sensitivity coefficients are given in table 1 for $\phi=k/C$. They were obtained using equation (8), with equation (6), Eq. (17) for $\phi=k/C$ and Eq. (27c), each in its own time range of validity. In detail, equations (8) and (6) were used for $\tilde{t}\geq 0.3$, equation (17) for $\tilde{t}=0.01$ and equation (27c) for $\tilde{t}=0.1$ and 0.2. In particular, when dealing with equations (8) and (6), the maximum number of terms defined by equation (6) for $\tilde{t}=0.3$ was only of $N = M = 3$. A plot is shown in figure 1 for visual purposes and for $\phi=k/C$. This figure indicates that, in the current theoretical experiment with $\phi=k/C$, the *k*-sensitivity is the largest and can be assumed as an 'ideal reference sensitivity' which *k*-sensitivity coefficients of real experiments can refer to in order to evaluate their relative magnitude.

Table 1. Dimensionless *k*- and *C*-scaled sensitivity coefficients for the heated surface of the X21B(X11B10T0)0T0 and X22B(X12B10T0)0T0 (discussed ahead) problems when $\phi=k/C$.

$\tilde{t}=\frac{\alpha t}{L^2}$	$X_{k,\text{X21}}=X_{C,\text{X22}}$	$X_{C,\text{X21}}=X_{k,\text{X22}}$	X_k+X_C
0.01	-0.50000000[a]	-0.50000000[a]	-1.00000000[a]
0.10	-0.50016200[b]	-0.49983800[b]	-1.00000000[b]
0.20	-0.51700073[b]	-0.48299927[b]	-1.00000000[b]
0.30	-0.57349303[c]	-0.42650697[c]	-1.00000000[c]
0.40	-0.64644983[c]	-0.35355017[c]	-1.00000000[c]
0.50	-0.71596389[c]	-0.28403611[c]	-1.00000000[c]
0.60	-0.77514144[c]	-0.22485856[c]	-1.00000000[c]
0.70	-0.82321393[c]	-0.17678607[c]	-1.00000000[c]
0.80	-0.86146119[c]	-0.13853881[c]	-1.00000000[c]
0.90	-0.89160143[c]	-0.10839857[c]	-1.00000000[c]
1.00	-0.91524675[c]	-0.08475325[c]	-1.00000000[c]
1.25	-0.95424011[c]	-0.04575989[c]	-1.00000000[c]
1.50	-0.97530407[c]	-0.02469593[c]	-1.00000000[c]
1.75	-0.98667288[c]	-0.01332712[c]	-1.00000000[c]
2.00	-0.99280812[c]	-0.00719188[c]	-1.00000000[c]
2.25	-0.99611896[c]	-0.00388104[c]	-1.00000000[c]
2.50	-0.99790563[c]	-0.00209437[c]	-1.00000000[c]
2.75	-0.99886979[c]	-0.00113021[c]	-1.00000000[c]
3.00	-0.99939009[c]	-0.00060991[c]	-1.00000000[c]
8.00	-1.00000000[c]	-0.00000000[c]	-1.00000000[c]

[a] Numerical values obtained using Eq. (17) for both X21 and X22 cases.

[b] Numerical values obtained using Eq. (27c) for X21 and Eq. (32) with $\phi=k/C$ for X22

[c] Numerical values obtained using Eqs. (6) and (8) for X21 and Eqs. (30) and (31) for X22.

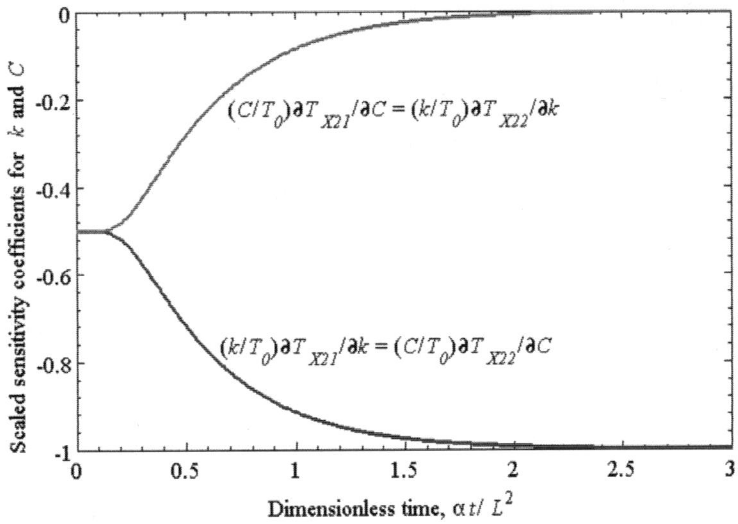

Figure 1. Dimensionless scaled sensitivity coefficients for k and C for the heated surface location of the X21B(X11B10T0)0T0 and X22B(X12B10T0)0T0 (discussed ahead) problems for $\phi = k / C$.

3. X22B(X12B10T0)0T0 transient experiment

Consider another 1D transient experiment similar to the one of Section 2 where, however, the back surfaces of the two samples are thermally insulated. A dimensional mathematical statement for this thermally-symmetric experiment denoted by X22B(X12B10T0)0T0 is

$$\frac{\partial^2 T}{\partial x^2} = \frac{1}{\alpha}\frac{\partial T}{\partial t} \qquad (0<x<L;\ t>0) \tag{28a}$$

$$-k\frac{\partial T}{\partial x}(0,t) = \varphi_{\mathrm{X12B10T0}}(0,t) \qquad (t>0) \tag{28b}$$

$$-k\frac{\partial T}{\partial x}(L,t) = 0 \qquad (t>0) \tag{28c}$$

$$T(x,0) = 0 \qquad (0<x<L) \tag{28d}$$

where $\varphi_{\mathrm{X12B10T0}}(0,t)$ is the surface heat flux of the X12B10T0 case. It may be taken as

$$\varphi_{X12B10T0}(0,t) = \frac{\varphi_0}{\sqrt{\pi\phi t / L^2}}\sum_{n=0}^{\infty}(-1)^n\left\{\exp\left(-\frac{n^2L^2}{\phi t}\right) - \exp\left[\frac{(n+1)^2L^2}{\phi t}\right]\right\} \tag{29a}$$

or, alternatively, as

$$\varphi_{\mathrm{X12B10T0}}(0,t) = 2\varphi_0\sum_{n=1}^{\infty}e^{-\eta_n^2\frac{\phi t}{L^2}} \tag{29b}$$

where $\eta_n = (n-1/2)\pi$ are the "X12" eigenvalues. Following the methodology proposed in Subsections 2.2-2.4, the surface temperature solution and, hence, the sensitivity coefficients may be derived for the whole time region and for short times. However, for the sake of brevity, the solution derivation will not be given here.

3.1 Exact solution for the whole time region

The surface temperature solution is

$$T_{\text{X22B(X12B10T0)0T0}}\left(0,t,\frac{\phi}{kC}\right)=\frac{\varphi_0 L}{C\phi}\left\{1+2\sum_{m=1}^{\infty}e^{-\beta_m^2\frac{kt}{CL^2}}\frac{\tan\left[\beta_m\sqrt{k/(C\phi)}\right]}{\beta_m\sqrt{k/(C\phi)}}\right.$$
$$\left.-2\sum_{n=1}^{\infty}e^{-\eta_n^2\frac{kt}{CL^2}\left(\frac{C\phi}{k}\right)}\frac{\cot\left(\eta_n\sqrt{C\phi/k}\right)}{\eta_n\sqrt{k/(C\phi)}}\right\} \tag{30}$$

where $\beta_m = m\pi$ are the "X22" dimensionless eigenvalues and $\varphi_0 L/(C\phi)=(T_{rise})_\infty$ is the sample temperature rise for $t\rightarrow\infty$. For the case of $\phi=\alpha=k/C$, the term inside the braces is equal to 1 since $\tan(\beta_m)=0$ and $\cot(\eta_n)=0$. It is significant to notice that the two expressions defined by equations (6) and (30) are mirror images in the sense that $k/(C\phi)$ by $(C\phi)/k$ can be interchanged to get identical expressions if also ϕ in the exponentials is replaced by k/C. This mirror image is rather amazing. It has application to the case of optimal experiments for estimating *k* and *C*.

The maximum number of terms for either summation appearing in equation (30) is still given by equation (7). Sensitivity coefficient for the *k* and *C* properties can be found using differences as

$$X_{k,\text{X22}}=\frac{k}{(T_{rise})_\infty}\frac{\partial T_{\text{X22}}}{\partial k}(0,\tilde{t})\approx\frac{T_{\text{X22B(X12B10T0)0T0}}[0,\tilde{t},k(1+\varepsilon),C]-T_{\text{X22B(X12B10T0)0T0}}[0,\tilde{t},k(1-\varepsilon),C]}{2\varepsilon(T_{rise})_\infty} \tag{31a}$$

$$X_{C,\text{X22}}=\frac{C}{(T_{rise})_\infty}\frac{\partial T_{\text{X22}}}{\partial C}(0,\tilde{t})\approx\frac{T_{\text{X22B(X12B10T0)0T0}}[0,\tilde{t},k,C(1+\varepsilon)]-T_{\text{X22B(X12B10T0)0T0}}[0,\tilde{t},k,C(1-\varepsilon)]}{2\varepsilon(T_{rise})_\infty} \tag{31b}$$

where the subscript "X22" can be used as a short notation. They are efficient and accurate since the temperature is known very accurately.

3.2 One-term approximate solution for very short times

For early times, less than the so-called deviation time $t_d^{(1)}$ defined by equation (9), the surface temperature for the X22B(X12B10T0)0T0 problem can accurately be replaced with the surface temperature for the X22B(X10B1T0)0T0 case. The latter can be calculated using Green's functions where a further approximation (with errors less than 10^{-A}) can be performed. In fact, $G_{\text{X22}}(x,0,t-\tau)\approx G_{\text{X20}}(x,0,t-\tau)$ defined by equation (13) for co-times $(t-\tau)\le 0.1\cdot(2L-x)^2/(A\alpha)$. Therefore, $T_{X22B(X12B10T0)0T0}(0,t)\approx T_{X20B(X10B1T0)T0}(0,t)$ defined by equation (14a).

It follows that, for short times, the sensitivity coefficients are still given by equation (15) and, in dimensionless form, by equation (17), where the subscript "X21" has to be replaced with "X22."

3.3 Three-terms approximate solution for short times

The scaled sensitivity coefficients in dimensionless form for $\tilde{t}\le\min\{1.6/A,1.6/[A(C\phi/k)]\}$ are

$$X_{k,\mathrm{X22}} = \frac{k}{(T_{rise})_\infty}\frac{\partial T_{\mathrm{X22}}}{\partial k}\left(0,\tilde{t},\frac{\phi}{kC}\right) \approx -\frac{1}{2\sqrt{C\phi/k}}\left[1-\frac{4e^{-1/\tilde{t}}}{\sqrt{\pi\tilde{t}}}-2\mathrm{erfc}\left(\frac{1}{\sqrt{\tilde{t}}}\right)+2\mathrm{erfc}\left(\frac{1}{\sqrt{\tilde{t}(C\phi/k)}}\right)\right] \quad (32a)$$

$$X_{C,\mathrm{X22}} = \frac{C}{(T_{rise})_\infty}\frac{\partial T_{\mathrm{X22}}}{\partial C}\left(0,\tilde{t},\frac{\phi}{kC}\right) \approx -\frac{1}{2\sqrt{C\phi/k}}\left[1+\frac{4e^{-1/\tilde{t}}}{\sqrt{\pi\tilde{t}}}-2\mathrm{erfc}\left(\frac{1}{\sqrt{\tilde{t}}}\right)+2\mathrm{erfc}\left(\frac{1}{\sqrt{\tilde{t}(C\phi/k)}}\right)\right] \quad (32b)$$

which are the same as equations (27a) and (27b) except for the + and – signs inside the brackets. For $\phi = k/C$, we obtain Eq. (27c) except for the + and – signs inside the brackets. Therefore, the *k*-scaled sensitivity coefficient of the X21B(X11B10T0)0T0 problem is equal to the *C*-scaled sensitivity function for the X22B(X12B10T0)0T0 case. This same reversal occurs for the *C* sensitivities.

3.4 Numerical results

Numerical values are given in table 1. They were obtained using equations (17), (32) for $\phi = k/C$ and equation (30) with Eq. (31), each in its own time range of validity. A plot is shown in figure 1 for visual proposes. This figure indicates that the scaled sensitivity coefficients are large in magnitude as well as in the previous experiment. In particular, in the current experiment the *C*-sensitivity is the largest and can be assumed as an 'ideal reference sensitivity' which real *C*-sensitivity coefficients can refer to in order to evaluate their relative magnitude.

Conclusions

It was shown that the scaled sensitivity coefficients for *k* and *C* can be made largest keeping the heated surface temperature at the maximum value during the transient experiment. This allows us to obtain ideal reference sensitivities, which actual sensitivities can refer to in order to evaluate their relative magnitude. In addition, the sensitivity coefficients can be made more uncorrelated by using two complementary experiments rather than a single one experiment. Also, when computing the sensitivity functions for the heated surface of the X21B(X11B10T0)0T0 and X22B(X12B10T0)0T0 experiments, it was noted that the roles in the scaled sensitivity coefficients for *k* and *C* can be exchanged. In addition, the small and large time solutions for their calculation exhibit intrinsic verification.

Appendix A. Solution derivation for the whole time region

The first part of the integral in Eq. (5) is

$$\int_{\tau=0}^{t}\sum_{n=1}^{\infty} e^{-\eta_n^2\frac{\alpha(t-\tau)}{L^2}}d\tau = \frac{L^2}{\alpha}\left(\underbrace{\sum_{n=1}^{\infty}\frac{1}{\eta_n^2}}_{=1/2}-\sum_{n=1}^{\infty}\frac{e^{-\eta_n^2\frac{\alpha t}{L^2}}}{\eta_n^2}\right) = \frac{L^2}{\alpha}\left(\frac{1}{2}-\sum_{n=1}^{\infty}\frac{e^{-\eta_n^2\frac{\alpha t}{L^2}}}{\eta_n^2}\right) \quad (A.1)$$

Solving the second integral in the same equation yields

$$\int_{\tau=0}^{t}\sum_{m=1}^{\infty} e^{-\beta_m^2\frac{\phi\tau}{L^2}}\sum_{n=1}^{\infty} e^{-\eta_n^2\frac{\alpha(t-\tau)}{L^2}}d\tau = \sum_{m=1}^{\infty}\sum_{n=1}^{\infty} e^{-\eta_n^2\frac{\alpha t}{L^2}}\int_{\tau=0}^{t} e^{-(\beta_m^2\phi-\eta_n^2\alpha)\frac{\tau}{L^2}}d\tau \doteq L^2\sum_{m=1}^{\infty}\sum_{n=1}^{\infty}\frac{e^{-\eta_n^2\frac{\alpha t}{L^2}}-e^{-\beta_m^2\frac{\phi t}{L^2}}}{\beta_m^2\phi-\eta_n^2\alpha} \quad (A.2)$$

Combining the above two integrals in equation (5) gives the surface temperature as

$$T_{\mathrm{X21B(X11B10T0)0T0}}(0,t) = 2\varphi_0 L\left[\frac{1}{k}\left(\frac{1}{2}-\sum_{n=1}^{\infty}\frac{e^{-\eta_n^2\frac{\alpha t}{L^2}}}{\eta_n^2}\right)+2\sum_{m=1}^{\infty}\sum_{n=1}^{\infty}\frac{e^{-\eta_n^2\frac{\alpha t}{L^2}}-e^{-\beta_m^2\frac{\phi t}{L^2}}}{\beta_m^2 C\phi-\eta_n^2 k}\right] \quad (A.3)$$

where the double summation can be expanded to

$$\sum_{m=1}^{\infty}\sum_{n=1}^{\infty}\frac{e^{-\eta_n^2\frac{\alpha t}{L^2}}-e^{-\beta_m^2\frac{\phi t}{L^2}}}{\beta_m^2 C\phi-\eta_n^2 k}=\frac{1}{C\phi}\sum_{n=1}^{\infty}e^{-\eta_n^2\frac{\alpha t}{L^2}}\left(\sum_{m=1}^{\infty}\frac{1}{\beta_m^2-\eta_n^2\alpha/\phi}\right)-\frac{1}{k}\sum_{m=1}^{\infty}e^{-\beta_m^2\frac{\phi t}{L^2}}\left(\sum_{n=1}^{\infty}\frac{1}{\beta_m^2\phi/\alpha-\eta_n^2}\right) \quad (A.4)$$

The two single summations without the exponentials converge slowly. However, the following relations (found using Mathematica) can be employed to remove this convergence problem,

$$\sum_{m=1}^{\infty}\frac{1}{\beta_m^2-\eta_n^2\alpha/\phi}=\sum_{m=1}^{\infty}\frac{1}{(m\pi)^2-\eta_n^2\alpha/\phi}=\frac{1}{2\eta_n^2\alpha/\phi}-\frac{\cot\left(\eta_n\sqrt{\alpha/\phi}\right)}{2\eta_n\sqrt{\alpha/\phi}} \quad (A.5a)$$

$$\sum_{n=1}^{\infty}\frac{1}{\beta_m^2\phi/\alpha-\eta_n^2}=\sum_{n=1}^{\infty}\frac{1}{\beta_m^2\phi/\alpha-[(n-1/2)\pi]^2}=-\frac{\tan\left(\beta_m\sqrt{\phi/\alpha}\right)}{2\beta_m\sqrt{\phi/\alpha}} \quad (A.5b)$$

Introducing these algebraic identities in equation (A.4) gives

$$\begin{aligned}\sum_{m=1}^{\infty}\sum_{n=1}^{\infty}\frac{e^{-\eta_n^2\frac{\alpha t}{L^2}}-e^{-\beta_m^2\frac{\phi t}{L^2}}}{\beta_m^2 C\phi-\eta_n^2 k}&=\frac{1}{2k}\sum_{n=1}^{\infty}\frac{e^{-\eta_n^2\frac{\alpha t}{L^2}}}{\eta_n^2}-\frac{1}{2k}\sum_{n=1}^{\infty}e^{-\eta_n^2\frac{\alpha t}{L^2}}\frac{\cot\left(\eta_n\sqrt{\alpha/\phi}\right)}{\eta_n\sqrt{\phi/\alpha}}\\&+\frac{1}{2k}\sum_{m=1}^{\infty}e^{-\beta_m^2\frac{\phi t}{L^2}}\frac{\tan\left(\beta_m\sqrt{\phi/\alpha}\right)}{\beta_m\sqrt{\phi/\alpha}}\end{aligned} \quad (A.6)$$

Substituting this result into equation (A.3) yields the surface temperature defined by Eq. (6). Unlike the double-summation form defined by equation (A.3), Eq. (6) converges nicely with a relatively small numbers of terms, as shown in Subsection 2.2.

References

[1] Beck J V and Arnold K J 1977 *Parameter Estimation in Engineering and Science* (New York: John Wiley and Sons, Inc.)

[2] Rossmanna C and Haemmerich D 2014 Review of temperature dependence of thermal properties, dielectric properties, and perfusion of biological tissues at hyperthermic and ablation temperatures *Crit. Rev. Biomed. Eng.* **42** 6 467-92

[3] D'Alessandro G and de Monte F 2017 Sensitivity coefficients for thermal properties measurements using a boundary condition of the 4th kind *J. Phys. Conf. Series* **796** 1 012011

[4] McMasters R L, de Monte F and Beck J V 2018 Estimating two heat-conduction parameters from two complementary transient experiments *ASME J. Heat Transfer* doi:10.1115/1.4038855

[5] Cole K D, Beck J V, Haji-Sheikh A and Litkouhi B 2011 *Heat Conduction using Green's Function 2nd Ed.* (Baca Raton: CRC Press)

[6] Maillet D, André S, Batsale J C, Degiovanni A and Moyne C 2000 *Thermal Quadrupoles – Solving the Heat Equation through Integral transforms* (Chichester: John Wiley & Sons)

[7] de Monte F, Beck J V and Amos D E 2012 Solving two-dimensional Cartesian unsteady heat conduction problems for small values of the time *Int. J. Thermal Sciences* **60** 106-113

[8] Cole K D, Beck J V, Woodbury K A and de Monte F 2014 Intrinsic verification and a heat conduction database *Int. J. Thermal Sciences* **78** 36-47

[9] D'Alessandro G and de Monte F 2018 Intrinsic verification of an exact analytical solution in transient heat conduction *Comp. Thermal Sciences* **10** 2 1-22 doi:10.1615/ComputThermalScien.2017021201

[10] Carslaw H S and Jaeger J C 1959 *Conduction of Heat in Solids 2nd Ed.* (New York: Oxford University Press)

[11] de Monte F, Beck J V and Amos D E 2011 A heat-flux based building block approach for solving heat conduction problems *Int. J. Heat Mass Transfer* **54** 2789-2800

Constructing tubular networks that occupy arbitrary regions in $\mathbb{R}^3$

F Ghasempour[1], W Jiang[1], F Mendivil[2], S D Peterson[3] and E R Vrscay[1]

[1] Department of Applied Mathematics, University of Waterloo, Waterloo, Ontario, Canada N2L 3G1

[2] Department of Mathematics and Statistics, Acadia University, Wolfville, Nova Scotia, Canada B4P 2R6

[3] Department of Mechanical and Mechatronics Engineering, University of Waterloo, Waterloo, Ontario, Canada N2L 3G1

E-mail: fghasemp@uwaterloo.ca, wenzhe.jiang0801@gmail.com, franklin.mendivil@acadiau.ca, peterson@uwaterloo.ca, ervrscay@uwaterloo.ca

Abstract. Advances in additive manufacturing have enabled industry to relax many design constraints imposed by traditional construction methods. As such, it is now possible to design and build objects or devices with complex internal structures that conform to irregular external envelopes. In applications where fluid distribution or storage is an integral part of a larger system, additive manufacturing technologies allow "left over space" in an overall device volume to now be effectively utilized. Herein we describe a general algorithmic framework for the construction of branched networks of tubes which occupy a specified, and possibly quite complicated, region $D \subset \mathbb{R}^3$ as fully as possible. Such networks can be important in a variety of industrial applications ranging from heat exchangers to storage vessels to fluid distribution networks. Depending on the specific application of interest, such a design problem can be extremely ill-posed: For a given region D, our algorithm can produce an enormous number of networks, which generally requires that a much smaller number of "best" networks be isolated by ranking the networks based upon some problem-dependent properties (*e.g.*, heat transfer rate, minimal friction losses, volume fraction, etc.). Some simple examples are presented to illustrate the algorithm and its output with comments added to illustrate its potential utility in industrial applications.

1. Introduction

Additive manufacturing, which generally builds a 3D object by adding layers of material [5], has revolutionized the design process by enabling the construction of objects with complex structures with relative ease [17]. This now allows designers to develop engineering structures that minimize material usage, have complex internal configurations and/or conform to complex external envelopes [7, 14]. In engineering systems consisting of fluid distribution or storage elements, additive manufacturing opens the possibility of exploiting left over or wasted space within a given geometric envelope to house these flow elements.

In this paper we describe a general algorithmic framework for the construction of branched tubular networks that occupy a specified, and possibly quite complicated, region $D \subset \mathbb{R}^3$ as fully as possible. Such "conformable tubular networks" (CTNs) can be important in a variety of industrial applications ranging from heat exchangers to fluid distribution or storage networks in complex geometries. Our algorithm is comprised of three fundamental modules: (i) the *analysis and decomposition* of the three-dimensional region D into subregions D_i, (ii) the *packing* of each subregion D_i with an assembly of tubes and finally (iii), the *connection* of the tubes, both within each subregion D_i as well as between neighbouring subregions.

1.1. A simple, motivating example: The "barbeque pool heater"

In [9] and [11], we considered some simple, yet nontrivial, "saddle-bag" regions D (*i.e.*, two larger rectangular regions – the "bags" – connected by a narrower rectangular channel) and produced sets of networks for these regions. One of the motivations for these constructions was the so-called "barbeque (BBQ) pool heater" problem [1] wherein water from a pool is pumped into a tubular network situated inside a barbeque via an inlet pipe. As the water travels through the network, it is heated by the barbeque flames. The heated water then exits the network through an outlet pipe and re-enters the pool. While this is possible to design and construct using conventional design and manufacturing techniques, it presents a simple framework through which to explore a more general design paradigm for encapsulating a (possibly branched) tubular network in a prescribed geometric envelope.

Figure 1 shows a particular example of a saddle-bag region representing, for example, the available space inside the barbeque (left), along with one of the many one inlet-one outlet networks produced by our algorithm (right). The network shown is a "28-8-28 saddle-bag network," where two "side bags," each of which contains 28 tubes, are connected to a middle region containing 8 tubes. In this simple example we utilize a single diameter for the tubes, which is prescribed ahead of time. For the barbque pool heater, which relies on natural convection and radiation to heat the tubes passing over the flames, the tube diameters shown in the Figure would not represent the actual vessels, but rather the desired spacing between the tubes for optimal heat transfer. The actual physical tubes would concentric with the tubes shown, but with smaller diameter to enable space for the heated air to pass around the fluid-transporting tubes.

Whether or not a given network is "good" is problem-specific. In this particular example, with the constraints that were imposed during the construction phase, hundreds of thousands of unique networks were generated. For this example, selecting the best network(s) would require an additional heat transfer analysis step. The focus of this paper is the methodology for constructing the tubular networks, not the problem-specific step of imposing additional physics and/or constraints for paring down the solution space.

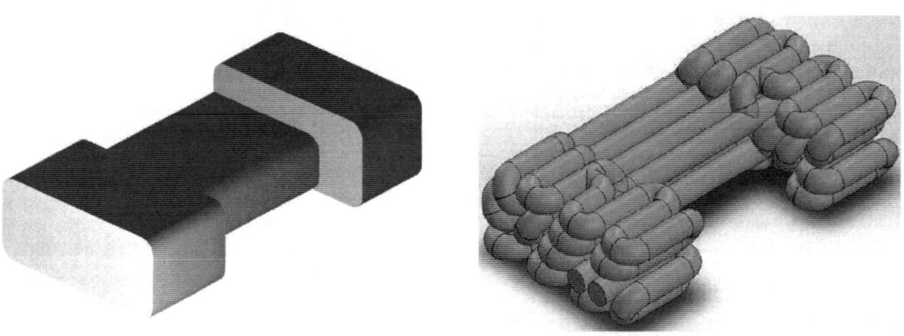

Figure 1. A "saddle-bag" region D (left) and a "28-8-28" tubular network produced by our algorithm which occupies the region.

IOP Conf. Series: Journal of Physics: Conf. Series **1047** (2018) 012003 doi:10.1088/1742-6596/1047/1/012003

2. The algorithm

This section describes the skeleton of the algorithm for constructing branched tubular networks in a given region of space. The algorithm is comprised of three phases, which will be detailed in the following subsections. It is presumed that the volume to be packed with tubes is prescribed.

Phase 1: Analysis/decomposition of region $D \in \mathbb{R}^3$ to be occupied by the network

Clearly, the simplest situation is when the 3D region D is a long block with constant cross section (*e.g.*, rectangular, triangular), in which case the tubes can be packed in a parallel manner and connected to each other at the ends. The next simplest case, a variation of the above, is when D exhibits some kind of longitudinal symmetry, *i.e.*, there exists a principal (but not necessarily straight) axis along which the cross section may vary. In this case, the construction problem is quasi-two-dimensional and the problem of packing tubes in three dimensions is reduced to circle-packing in two dimensions. A decomposition of D may then be performed by segmentation, *i.e.*, by approximating it as well as possible by a union of blocks B_i with constant cross sections. A sketch of such a generic decomposition is shown in Figure 2. Each block B_i is then packed with a parallel set of tubes as shown schematically in the figure. The "saddle-bag" region shown in Figure 1 possesses this longitudinal symmetry and is composed of three blocks.

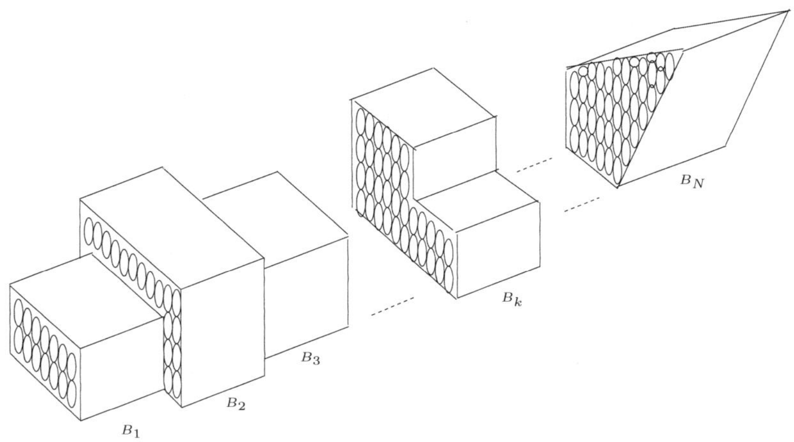

Figure 2. Generic structure of tubular network exhibiting variation along a principal direction.

In more complicated situations, it may be necessary to employ a method which can extract the fundamental morphological characteristics of region D, namely, its shape and structure. For example, if D possesses a dendritic structure in which several subprincipal axes emanate from a principal location or axis, it is desirable to determine these axes. Such information will be important for both the segmentation/discretization of the region as well as the determination of packing directions. Methods such as skeletonization [12, 13] can be employed to identify principal (of subprincipal) axes in a complex geometry, which can then segmented into a series of blocks for packing, as discussed above.

Phase 2: Packing the tubes

As mentioned in Phase 1, each block B_i comprising the envelope volume is assumed to contain a set of tubes arranged in a parallel fashion. The problem of packing each block B_i is therefore reduced to the two-dimensional problem of packing circles as efficiently as possible into its (constant) cross-sectional region. Circle packing is a well-studied problem with a long history because of its mathematical challenges as well as its important real-world applications (*e.g.*, cutting circular discs out of sheets of metal in a way that minimizes leftover material). There is a vast literature which is concerned with optimal and suboptimal methods for packing circles into regions. For a more recent review of the literature, see [8]. A significant portion of the literature has been devoted to the problem of packing as many identical circles of fixed radius into a given region, usually a regular region such as a circle, square or rectangle [2, 15, 16]. Another important set of problems is concerned with the efficient packing of square, rectangular and circular regions with circles of prescribed, but not necessarily identical, radii [4, 3].

In the the 28-8-28 "saddle-bag" network presented in Section 1, a very simple hexagonal packing of tubes of equal diameters was employed. In general, however, the cross-sectional regions are more complicated, *i.e.*, polygonal at the very best. The efficient packing of such regions will generally require tubes/circles of different sizes. Early investigations, however, showed that a more efficient packing of complicated regions could be achieved not with circles of prescribed sizes but with circles that were "tailor-made" for local regions, *i.e.*, circles designed to touch several already-packed circles as well as the boundary, if applicable. As such, the radii of these circles would have to be computed as the region was being packed. To the best of our knowledge, such algorithms did not exist in the literature, which motivated us to develop our own methods.

The starting point of our work on packing was the circle-packing algorithm by George, George and Lamar [4], hereby to be referred to as the GGL algorithm. One of the original motivations of the GGL paper was the efficient packing of pipes or bottles in a rectangular box for shipping (under the effect of gravity). Since the aims of our tubular network problem differ from those that led to the GGL algorithm, a number of differences between our algorithms and the GGL algorithm naturally arose. Here, we shall simply state that during the early stages of our research, a number of packing methods were developed and tested. A detailed discussion of these early methods is given in [11].

Our early modified-GGL methods were designed to pack regions with polygonal boundaries and worked very well for (convex and nonconvex) polygonal regions for which the lengths of the segments comprising the boundary are large compared to the circle sizes. If this condition is not satisfied – which includes the case of quite irregular boundary curves – the methods do not work very well. One could still consider employing one or more of these methods by using polygonal approximations of irregular boundary curves, but possibly at the expense of yet-unpacked regions. For this reason, we have recently developed another method of circle-packing which can be used in the case of irregular boundaries. This method, which will be described briefly below, works with a *discretization* of region D into rectangular regions of dimension Δx by Δy. The parameters Δx and Δy are chosen according to the refinement desired for the problem.

In this method, we begin once again with a Step 1 (hexagonal) packing of region D with circles of radius R. After Step 1 is completed, additional circles are packed in a stepwise fashion as follows:

(i) At the beginning of each step, let $D' \subset D$ denote the subregion of D which is not covered by packed circles and which has at least part of the boundary of D as its boundary. (This condition will exclude regions that lie between packed circles. It can be modified to accommodate such regions if they are suitably large.)

(ii) Let $\partial D'$ denote the boundary of region D'. Compute the distance of each mesh point in D' to $\partial D'$. The result is a distance function defined over D' which may be represented in terms of contours. Let $(x_{\max}, y_{\max})$ denote a point in D' at which the distance function achieves a local maximum value, to be denoted as $d_{\max}$. Pack, if possible, a circle of radius $d_{\max}$ centered at $(x_{\max}, y_{\max})$. Pack as many such circles in D' as possible. After all possible circles are packed, redefine $D' \subset D$ as the new subregion of D which remains unpacked and go to (i).

The above procedure may be stopped when circles of radii less than an acceptable radius $R_{\min}$ are packed (for example, if there is a minimum tube radius that can be manufactured, or there is a minimum radius imposed by some other exogenous constraint).

A few remarks on this method are in order.

- From a practical perspective, the computation of distances between points of region D' and its boundary must be performed over a *mesh* of interior points. The MATLAB function `bwdist` can be used.
- It is not only quite possible, but quite probable, that points of maximum distance $d_{\max}$ are not isolated but lie on a continuous curve or region. At present, we simply use a center point of such a curve or region, acknowledging that some improvements could be made in future versions of the algorithm.

This method works very well for regions with polygonal boundaries. In Figure 3 are shown some results obtained in the case of a region D with a more complicated boundary. A mesh constructed from 1086 by 326 points was employed. A prescribed radius of $R = 35$ was used for the Step hexagonal packing (top left) with zero angle of orientation. The MATLAB distance function `bwdist` was used to compute the distances of all mesh points in the unpacked region to the boundary of the region. The resulting contours of the distance function are shown in the plot at the top right. Circles are then constructed at local maxima of the distance function as shown in the plot at the bottom left. After all possible circles are plotted in this way, the distances of all mesh points in the remaining unpacked region are computed and new circles are packed (bottom right). In this example, the prescribed minimum radius was $R_{\min} = 3$. The packing procedure was repeated until no circles of radius greater than or equal to $R_{\min}$ could be packed.

Packing common cross-sectional regions first In order to facilitate connections of tubes in adjacent cross-sections, it is advantageous to look for an interior (planar) region R that is common to several, if not all, adjacent block cross-sections. These regions are packed first. In this way, it is possible that some tubes will extend continuously from one block into another possibly through all blocks comprising the network. Such tubes are clearly visible in the 28-8-28 network shown in Figure 1 (there are 8 such tubes in total). Each non-common region is then packed. The packing of each such region begins with circles that touch already-packed circles from the common region. The packing then proceeds outward to the boundaries.

Phase 3: Connecting the tubes to produce a tubular network
With reference to the general network shown in Figure 2, after all of the blocks B_i have been packed with tubes (including tubes that extend continuously from one block to the next, possibly from one end of the network to the other), it remains to connect them to produce a feasible tubular network with a specified number of inlets and outlets. As will be shown below, there are, in general, a significant number of possible connections at each end and each block-to-neighbouring-block interface which can result in an enormous number of possible networks. Herein we will restrict ourselves to a small set of elements to enable connections, detailed in the following subsection. That will be followed by a example showing how these simple connecting elements can lead to a very large number of unique networks.

Basic connection elements Currently, we have employed the following three **fundamental elements** to construct our tubular networks: (i) straight tubes, (ii) 90-degree bends and (iii) T-joints (or "Tees"). These three elements are sketched in Figure 4.

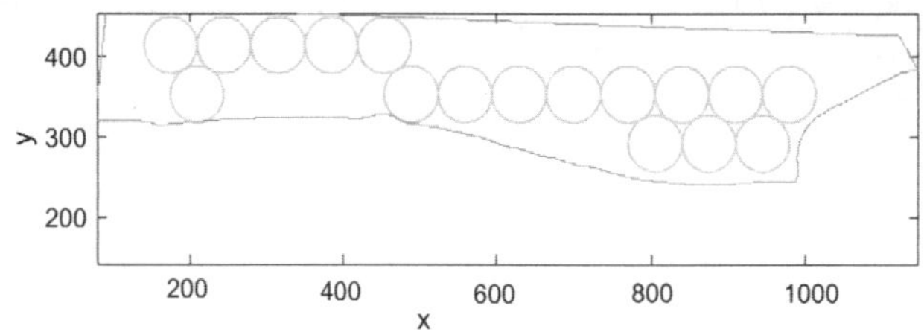

Step 1: 17 circles of radius $R = 35$.

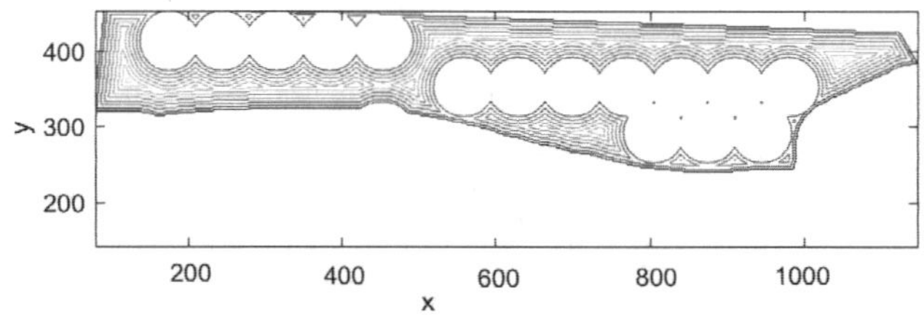

Contour plot of distance function from points in the unpacked region to its boundary.

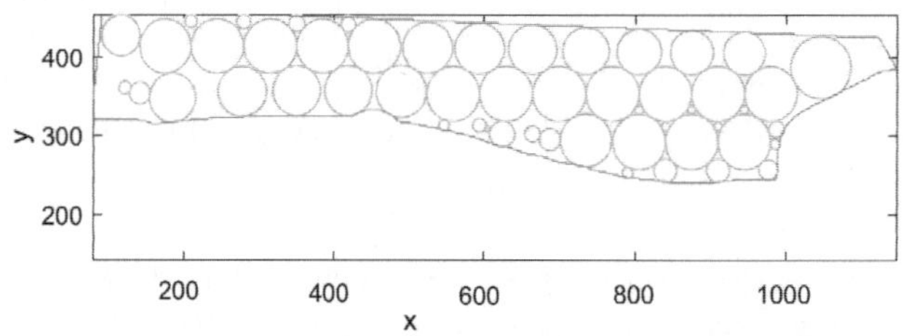

Packing the unpacked region of previous figure using distance function contours. New circles are centered at local maxima of the distance function.

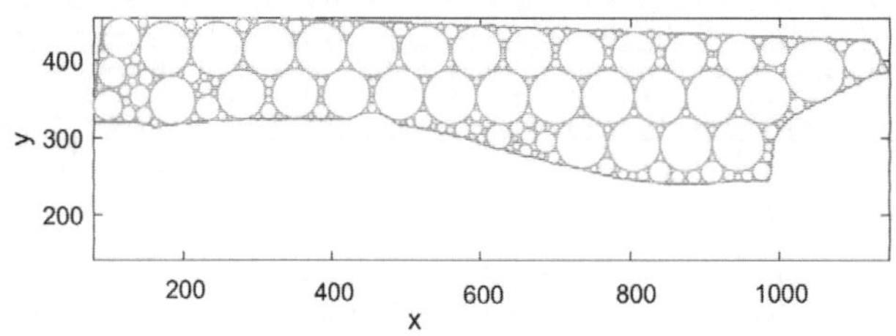

Additional packing after distance function contours are computed for unpacked region in upper plot.

Figure 3. Packing method applied to a region D with a more complicated boundary.

Note that the sketches in Figure 4 suggest that the diameters of all tubes are the same. This is not a fundamental restriction of the algorithm, although accommodating multiple tube sizes will require an augmentation to these elements. Using these elements, we currently employ four basic operations to construct our our tubular networks, shown schematically in Figure 5.

- **Endcap connection:** A 180-degree bend, formed with two 90-degree bends, which connects two neighbouring tubes at their ends. It is used at the extreme ends of the

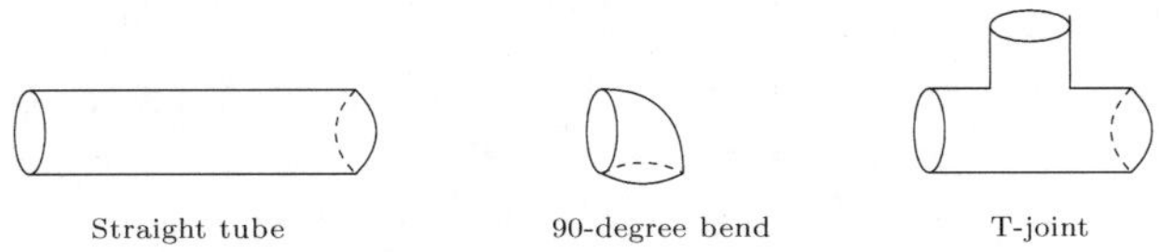

Figure 4. The three fundamental elements employed in the construction of tubular networks in this paper.

network as well as at the ends of each block B_i.

- **Simple merge:** Tube a flows into a neighbouring tube b. Formed with a 90-degree bend and a T-joint. This operation introduces branching in the tubular network. Used only at an at interface of two neighbouring blocks B_i, producing a 2-to-1 reduction in the number of tubes from one block to the next.
- **Consecutive merge:** Tube a flows into neighbouring tube b which then flows into neighbouring tube c. Used only at an interface of two neighbouring blocks B_i, producing a 3-to-1 reduction in the number of tubes.
- **Merge/shift/merge operation:** As shown below, tube a flows into neighbouring tube b. As well, tube c flows into tube d. Tube b is then shifted to the former position of c. The shift is formed with two 90-degree bends. Used only at at interface of two neighbouring blocks B_i, producing a 4-to-2 reduction in the number of tubes.

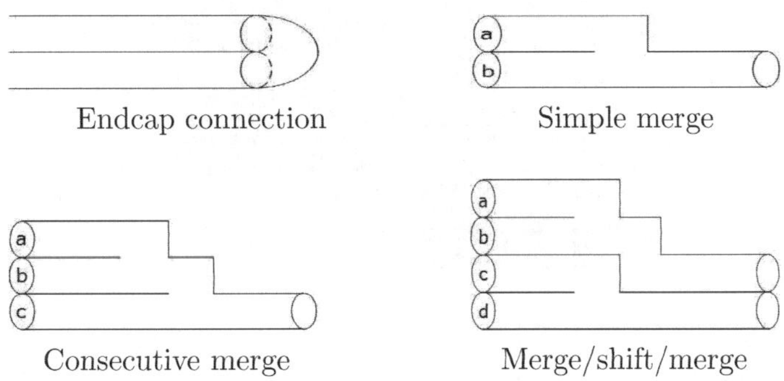

Figure 5. The four basic connection schemes used to construct the tubular networks.

The above list of connection schemes is by no means complete. Consecutive merges with higher than a 3-to-1 reduction, as well as merge/shift/merge schemes with higher reduction ratios can be considered, if necessary. Additional elements and combinations can also be introduced to accommodate specific geometries, *e.g.*, non-90-degree bends, contractions/expansions, curved tubes, manifolds connecting many tubes, etc.

The connection algorithm In general, two neighbouring blocks B_i and B_{i+1} will have a common area, to be denoted as region $\mathcal{R}_i$, as sketched in Figure 6. Circles in this region represent tubes that can run continuously from B_i to B_{i+1}. Tubes that are in block B_i but not completely in region $\mathcal{R}_i$ clearly cannot be continued into block B_{i+1}. As such, they must be either (i) turned back into block B_i via an endcap connection or (ii) merged into a neighbouring tube which lies in $\mathcal{R}_i$. There may also be the possibility of shifting a tube into a neighbouring position which extends into region $\mathcal{R}_i$ by means of a merge/shift/merge operation. The same comments apply to tubes in block B_{i+1} which cannot run into block B_i.

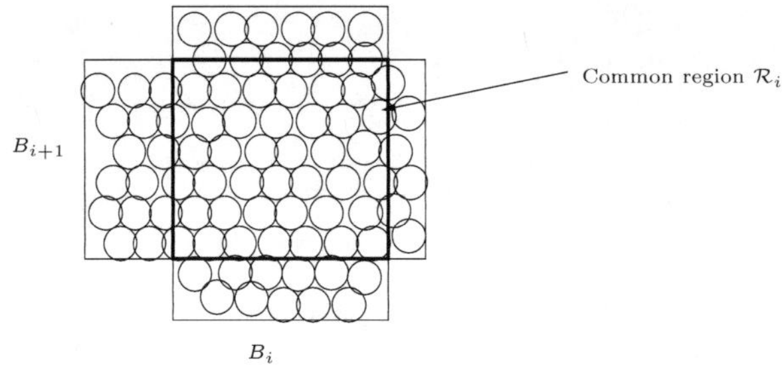

Figure 6. Two packed neighbouring blocks B_i and B_{i+1} with common region $\mathcal{R}_i$.

The 28-8-28 saddle-bag network shown in Figure 1 illustrates these connection schemes. The bags of the saddle-bag network, B_1 and B_3, contain 28 tubes whereas the interior connector region B_2 contains only 8 tubes. As such, 8 tubes run from one end, *i.e.*, B_1, through the connector region B_2, to the other end B_3. The other tubes 20 tubes in B_1 and B_3, however, must remain in their respective bags.

Before discussing the more general case, we illustrate the basic idea of our connection algorithm by means of a particular solution to a network which is composed of a much smaller number of tubes, namely, a 9-4-9 saddle-bag network. One such network is shown in Figure 7.

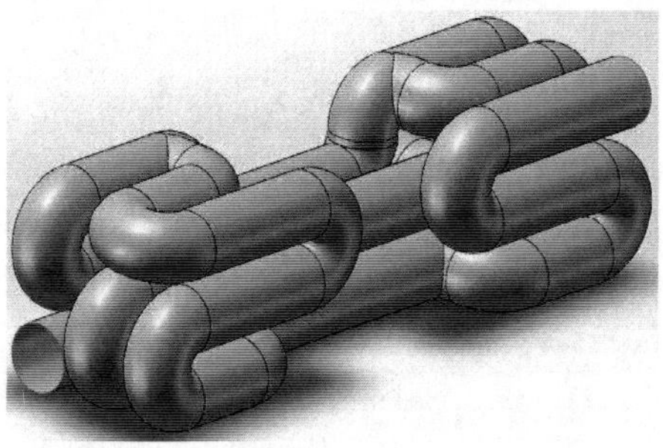

Figure 7. A 9-4-9 network produced by our algorithm.

The cross-sectional regions of the two bags as well as the connector region of a generic 9-4-9 network are shown in Figure 8.

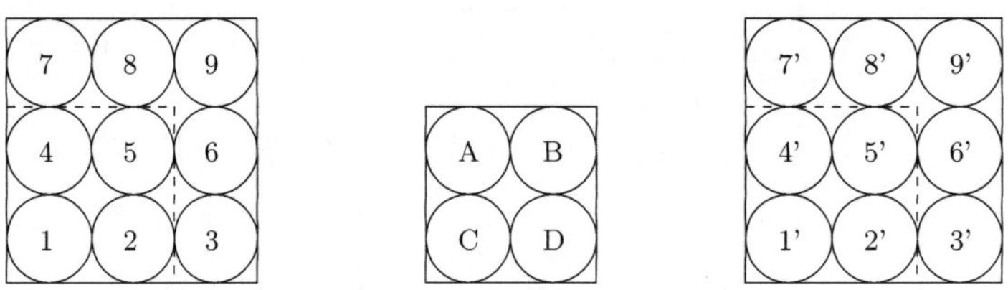

Figure 8. Cross sections of 9-4-9 saddle-bag network. **Left:** Saddle bag "L" with 9 tubes. **Middle:** Connector region with 4 tubes. **Right:** Saddle bag "R" with 9 tubes.

Using the ideas described at the beginning of this section, we present one solution for the connections at the B_1-B_2 and B_2-B_3 interfaces. We start at side L and determine how the nine tubes in the saddle bag will become Tubes A,B,C and D in the connector region.

- Tube A: Tube 7 merges into Tube 4, which then becomes Tube A. (This implies that Tube 4 is not available for any other operation.)
- Tube B: Tube 9 merges into Tube 8. Tube 8 then merges into Tube 5, which then becomes Tube B. (This implies that Tube 5 is not available for any other operation.)
  Tubes 6 and 3 must still be "removed." This means that Tube 2 will have to be moved:
- Tube C: Tube 2 merges into Tube 1, which then becomes Tube C.
- Tube D: Tube 6 merges into Tube 3, and then Tube 3 shifts to Tube 2 (after Tube 2 has merged to Tube 1).

The result of the above series of steps is a particular **merge-shift configuration** for the 9-4-9 saddle-bag network which is conveniently represented by the diagram shown in Figure 9.

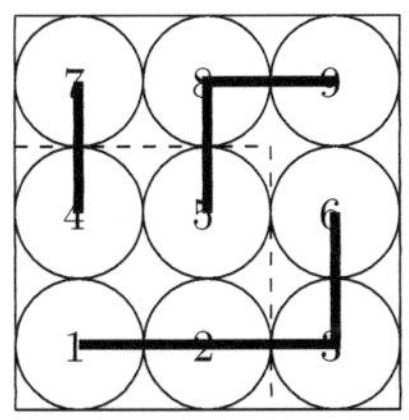

Figure 9. Diagrammatic representation of the resulting set of connections produced by the procedure outlined above.

Of course, similar shifting and merging operations must be performed on Tubes 3', 6', 7', 8' and 9' in order to move from saddle bag R into the connection region. For simplicity, we consider the same merging and shifting operations on saddle bag R as was done for saddle bag L in the steps above. Notationally, we simply replace all tube numbers with primes, *i.e.*, 3 with 3', etc. The resulting, almost-finished network may be represented by the diagram in Figure 10.

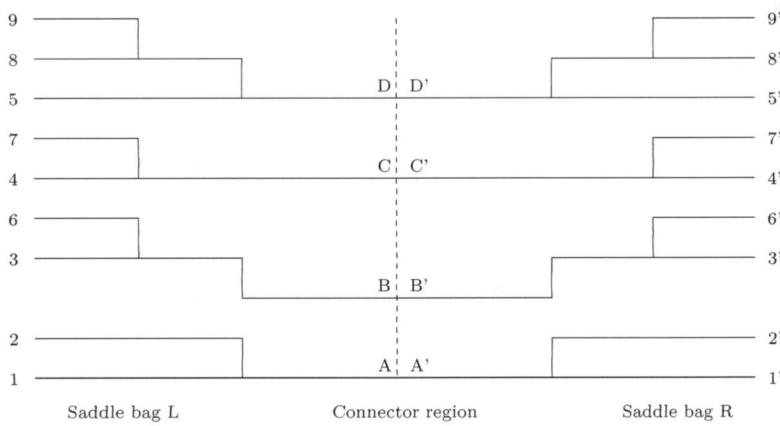

Figure 10. Graphical represntation of shift/merge connections between saddle bags and connector region.

The remaining problem is to make appropriate endcap connections between neighbouring tubes at both ends. One solution to the endcap problem for the above connection scheme is shown in Figure 11. Note that we have once again used symmetry, *i.e.*, the same set of connections on each side of the network. The net result is that Tubes 3 and 3' are, respectively input and output (or vice versa).

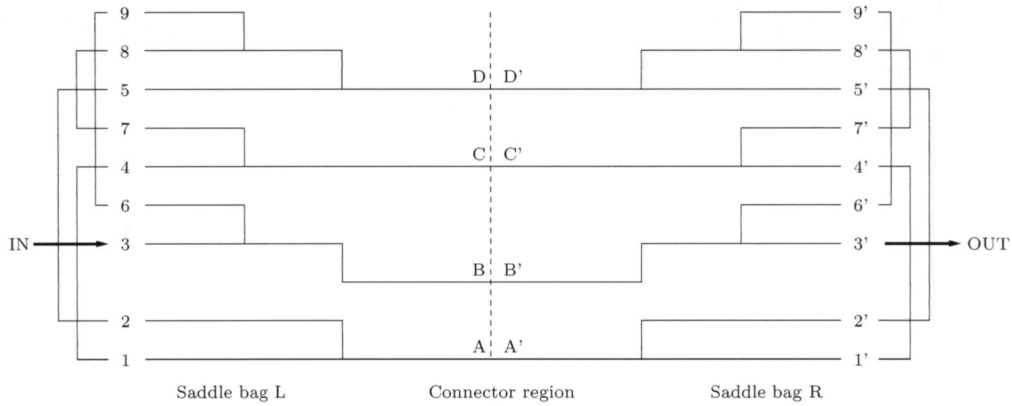

Figure 11. Graph of Figure 10 with a feasible set of endcap connections.

Here we once again emphasize that the connection scheme in Figure 11 is only one of several possible endcap schemes associated with this merge-shift configuration.

A brief description of the general connection algorithm The problem of connecting tubes at block interfaces is more complicated than that of connecting them at the extreme ends of the network. As such, we first consider the interface problem, with reference to Figure 6. For simplicity, our discussion below is restricted to the circles (which represent tubes) that are in Block B_i of the figure. Firstly, circles of B_i that lie entirely in the common region $\mathcal{R}_i$ are referred to as **common circles**. The remaining circles of B_i are referred to as **boundary circles** (BCs). It is the boundary circles which must be "processed," *i.e.*, endcapped, merged, or shifted so that they disappear when B_i is exited. The farther a circle lies from the common region, the higher the priority that it be considered for connection/shifting. This is accomplished by assigning a "layer-rank" index L_r to each circle, as follows:

- First-layer BCs, also called 1B circles, touch common circles. $L_r = 1$.
- All other BCs are called 2B circles, $L_r = 0$.
- First-layer common circles, $L_r = 2$, or 1C circles, touch at least one BC.
- Second-layer common circles, $L_r = 3$, or 2C circles, touch at least one 1C circle.
- All other common circles are not involved in this algorithm since it is not possible for them to be connected at an interface using any of the merge operations shown in Figure 5. $L_r = \infty$.

Circles are then processed in increasing order of layer rank index L_r.

There is another important factor. Since it is necessary that all BCs be "processed" and not ignored in any step, it is convenient to consider those BCs with a lower number of available connections first. As such, we define the **connectivity degree**, CD, of a BC as the number of (neighbouring) circles to which it may be connected.

Finally, it is convenient to define the **connectivity value** CV of a circle/tube, the maximum number of tubes to which it can be connected in one of the four operations shown in Figure 5:

- A 1B tube can be the middle of a consecutive merge (tube b, connected to both tubes a and c). A 1C tube can be at the end of a consecutive merge or a merge/shift/merge (tube c, connected to tubes a and b in both cases). $CV = 2$.
- 2B or 2C tubes can be connected to only one tube. $CV = 1$.

IOP Conf. Series: Journal of Physics: Conf. Series **1047** (2018) 012003 doi:10.1088/1742-6596/1047/1/012003

Tube	L_r	CV	Available tubes	CD
1	3 (2C)	1	2,4	2
2	2 (1C)	2	1,3,5	3
3	1 (1B)	2	2,6	2
4	2 (1C)	2	1,5,7	3
5	2 (1C)	2	2,4,6,7	4
6	1 (1B)	2	3,5,9	3
7	1 (1B)	2	4,8	2
8	1 (1B)	2	5,7,9	3
9	0 (2B)	1	6,8	2

Table 1. Availability list for the 9-4-9 network Saddle bag L/connector interface shown in Figure 8.

The indices associated with each tube, along with the set of tubes to which each tube can be connected, comprise an "availability list". For example, the availability list for interface between Saddle Bag L and the connector region of the 9-4-9 network shown in Figure 8 is presented in Table 1.

Starting with such an availability list, our algorithm examines all possible connections at an interface in a systematic manner. Given the different natures of the circles/tubes involved (*i.e.*, BCs vs. common circles) as well as the connection operations in Figure 5, there is a rather complicated set of rules for making connections. Furthermore, after a connection is made, the availability list is updated according to another set of rules. (For example, the CV values of the tubes involved in the connection are decreased and if the CV of any tube becomes zero, it must be removed from the availability list.) Because of space limitations, a discussion of these rules is not possible here. We therefore refer the interested reader to [11] (Section 3.4.2, p. 62) for details. Here we simply state that at each step of the algorithm, after one or more connections have been made and the availability list updated accordingly, all tubes available for connection according to the list are examined, starting with those of lowest layer-rank index L_r and proceeding with increasing L_r value. For a given tube, all tubes available for connection are examined. As such, the algorithm is naturally recursive in nature, examining all possible connections in a tree-like manner. If, at any step, all boundary circles of the interface are connected, the resulting set of connections are feasible, then the connection scheme is stored.

After all interface connection schemes have been computed, they must then be combined with each other, along with endcap connections at both ends of the network, to produce possible final tubular networks. Each of these networks must be tested for "feasibility" according to some requirements, *e.g.*, a prescribed number of inputs/outputs, no disconnected components, etc.

We now state some results for the simple 9-4-9 network problem – one inlet and one outlet – in order to give an idea of the complexity of this problem.

- With reference to the leftmost diagram in Figure 8, there are 22 feasible sets of connections at the interface of a bag and the connector region, one of which is shown in Figure 9.
- There are 5 possible sets of endcap connections at each end – four endcap connections between neighbouring tubes and one unconnected inlet/outlet.

The 9-4-9 networks are comprised of two ends and two interfaces. As such, there are $5 \times 22 \times 22 \times 5 = 133,100$ connections to examine. If we require that there must be at least one connected path between the inlet on one end and the outlet on the other end, *i.e.*, that the graph representing the tubular network be connected, then the number of such feasible networks is reduced significantly to 13,100. If, in addition, we demand that there are no closed loops or dead-ends in a network, *i.e.*, that each point in the network lies on a simple (*i.e.*, nonintersecting) path from an inlet at one end to an outlet at the other end, then the number of feasible networks is reduced to 10,066.

Note: In order to determine whether or not a network is connected/disconnected and whether or not there exist closed loops/dead ends, we must examine the incidence matrix $A(G)$ associated with the graph G that represents the network [6]. For details, see [11], Section 3.5, p. 66.

Comments on practical applications
In practice, the presented algorithm may need modification to suit problem-specific needs, particularly when selecting the "best" network from the family of generated solutions. For example, it may be desirable to minimize pressure drop across the tubular bed, or maximize volumetric capacity, or perhaps both. These outcome-based metrics can be used to develop a scheme for ranking the potential solutions. In simple cases, like the 9-4-9 example, there may not be large differences in performance between networks, but the variability in solutions is expected to increase substantially in more complex problems, as the number of solutions increases.

Concluding remarks

In this paper, we have outlined an algorithm for the construction of conformable tubular networks which occupy, as fully as possible, a given three-dimensional region D. The algorithm decomposes the region (with assumed longitudinal axis) into a series of blocks with fixed cross-sections. Circles are subsequently packed into each cross-section and connected via a set of rules to produce a tubular network.

Currently, we are focused on an extension of the work in Phase 1 to accommodate more complicated regions which may not possess a single longitudinal axis. This necessitates the use of morphological methods such as skeletonization to extract principal and subprincipal axes of more complicated regions and to then use this information to devise efficient packing strategies.

Our framework is flexible and can accommodate the specific requirements that may arise in industrial applications (*e.g.*, flow through the tubes, heat transfer, etc.) through the imposition of additional constraints either within the algorithm itself, or during the selection of viable tubular networks.

Acknowledgments

This research has been supported primarily by a Collaborative Research and Development Grant from the Natural Sciences and Engineering Research Council (Grant No. CRDPJ 453649-13). Additional support, in the form of teaching assistantships (JW), was provided by the Department of Applied Mathematics and the Faculty of Mathematics, University of Waterloo.

References

[1] http://www.redneckpoolheater.com
[2] Birgin EG and Gentil JM 2010 New and improved results for packing identical unitary radius circles within triangles, circles, rectangles and strips *Comp. Oper. Res.* **37** 1318-1327
[3] Castillo I, Kampas FJ and Pinter JD, Solving circle packing problems by global optimization: Numerical results and industrial applications 2008 *Eur. J. Oper. Res.* **191** 786-802
[4] George JA, George JM and Lamar BW 1995 Packing different-sized circles into a rectangular container *Eur. J. Oper. Res.* **84** (3) 693-712
[5] Gibson I, Rosen DW, Stuker B 2015 *Additive Manufacturing Technologies* Second Edition (NY: Springer)
[6] Gross JL and Yellen J 2005 *Graph Theory and its Applications* Second Edition (Boca Raton, FL: CRC Press)
[7] Gu DD, Meiners W, Wissenbach and Poprawe R 2012 Laser additive manufacturing of metallic components: Materials, processes and mechanisms *Int. Mat. Rev.* **57** 133-164
[8] Hifi M and M'Hallah R 2009 A literature review on circle and sphere packing problems: models and methodologies *Adv. Oper. Res.* Article ID 150624 `http://dx.doi/org/10.1155/2009/150624`

[9] Jiang W, Kettlewell B, Qiao T, Mendivil F, Peterson SD and Vrscay ER 2015 The barbeque pool heater: An algorithm to construct tubular networks the occupy arbitrary regions in $\mathbb{R}^3$ *Proc. AMMCS-CAIMS 2015 Congress* (Waterloo, ON, Canada)

[10] Jiang W, Qiao T, Mendivil F, Peterson SD and Vrscay ER 2015 Some novel circle-packing algorithms devised for the construction of tubular networks in $\mathbb{R}^3$ *Proc. AMMCS-CAIMS 2015 Congress* (Waterloo, ON, Canada)

[11] Jiang W 2015 *Construction of optimal tubular networks in arbitrary regions in* $\mathbb{R}^3$ M.Math. thesis, Dept. of Applied Mathematics, University of Waterloo

[12] Passalis G, Theoharis T, Kakadiaris IA 2007 A novel depth buffer-based shape descriptor for three-dimensional object retrieval, *The Visual Computer* **23** 5-14

[13] Saha PK, Borgefors G and Sanniti di Baja G 2016 A survey on skeletonization algorithms and their applications, *Patt. Rec. Lett.*, **76** 3-12

[14] Sames WJ, List FA, Pannala S, Dehoff RR and Babu SS 2016 The metallurgy and processing science of metal additive manufacturing, *Int. Mat. Rev.* **61** 1-46

[15] Specht E 2013 High density packings of equal circles in rectangles with variable aspect ratio *Comp. Oper. Res* **40** 58-69

[16] Szabo PG, Markot MCs, Csendes T, Specht E, Casado LG and Garcia I 2007 *New Approaches to Circle Packing in a Square with Program Codes* (NY: Springer)

[17] Vaezi M, Chianrabutra, Mellor B and Yang S 2013 Multiple material additive manufacturing Part 1: A review *Virtual and Physical Prototyping* **8** 19-50

IOP Conf. Series: Journal of Physics: Conf. Series **1047** (2018) 012004 doi:10.1088/1742-6596/1047/1/012004

Collage theorem-based approaches for solving inverse problems for differential equations: A review of recent developments

H Kunze[1], D La Torre[2,3], F Mendivil[4] and E R Vrscay[5]

[1] Department of Mathematics and Statistics, University of Guelph, Guelph, Ontario, Canada N16 2W1

[2] Dubai Business School, University of Dubai, 14143 Dubai, UAE

[3] Department of Economics, Management and Quantitative Methods, University of Milan, 20122 Milan, Italy

[4] Department of Mathematics and Statistics, Acadia University, Wolfville, Nova Scotia, Canada B4P 2R6

[5] Department of Applied Mathematics, University of Waterloo, Waterloo, Ontario, Canada N2L 3G1

E-mail: hkunze@uguelph.ca, dlatorre@ud.ac.ae, davide.latorre@unimi.it, franklin.mendivil@acadiau.ca, ervrscay@uwaterloo.ca

Abstract. In this short survey, we review the current status of fractal-based techniques and their application to the solution of inverse problems for ordinary and partial differential equations. This involves an examination of several methods which are based on the so-called Collage Theorem, a simple consequence of Banach's Fixed Point Theorem, and its extensions.

1. Introduction

In this paper, we review the current state of the art of "fractal-based methods" of solving inverse problems for ordinary and partial differential equations. This intent of this survey paper is to summarize, and for some readers introduce, a collection of so-called "Collage Methods" to the inverse problems community. The approach is different, but not necessarily better, than those employed by the majority of people working in inverse problems.

In the context of the work reviewed in this paper, we consider the direct problem to be the determination of the solution to a completely prescribed differential equation, including known initial conditions and/or boundary conditions. On the other hand, we consider the inverse problem to be estimation of values of the parameters, or perhaps a subset thereof, in a system of differential equations, based on some information about the solution, e.g., observed data values.

The literature is rich in papers studying *ad hoc* methods to address ill-posed inverse problems by minimizing a suitable approximation error along with utilizing some regularization techniques [20, 40, 44, 45, 46, 48]. Some of the approaches, for example in [48], without any rigorous justification actually perform the same minimization that is justified by the collage theorem; the references in this survey paper additionally list some publications by others who recognize that they are using the collage theorem and/or developing new related results [2, 1, 13, 14, 15, 37, 38, 41, 43, 43, 47].

IOP Conf. Series: Journal of Physics: Conf. Series **1047** (2018) 012004 doi:10.1088/1742-6596/1047/1/012004

2. Inverse problem of approximation by fixed points of contraction mappings

Most "fractal-based" methods are based on the use of contraction mappings on appropriate metric spaces. (By "appropriate", we mean appropriate to the application of concern.) It is therefore helpful if we provide some brief mathematical preliminaries.

In what follows, we shall let (X, d) denote a complete metric space. (For example, X could be a set of functions defined over an interval $[a, b] \subseteq \mathbb{R}$, e.g., the set $L^2[a, b]$ of square-integrable functions on $[a, b]$, and d the corresponding metric on that space.)

Definition 1 *(Contraction mapping) Let $T : X \to X$ be a mapping on a complete metric space (X, d). Then T is said to be contractive if there exists a constant $c \in [0, 1)$ such that $d(Tx, Ty) \le cd(x, y)$ for all $x, y \in X$.*

Generally, the smallest such $c \in [0, 1)$ for which the above inequality holds true is known as the *contraction factor* of T. We now come to what is perhaps the most famous theorem regarding contraction maps on metric spaces and certainly central to fractal-based methods.

Theorem 1 *(Banach Fixed Point Theorem [3]) Let $T : X \to X$ be a contraction mapping on X with contraction factor $c \in [0, 1)$. mapping on X. Then,*

(i) There exists a unique element $\bar{x} \in X$, the fixed point of T, for which $T\bar{x} = \bar{x}$.

(ii) Given any $x_0 \in X$, if we form the iteration sequence $x_{n+1} = T(x_n)$, then $x_n \to \bar{x}$, i.e., $d(x_n, \bar{x}) \to 0$ as $n \to \infty$. In other words, the fixed point $\bar{x}$ is globally attractive.

We consider the following general class of inverse problems:

> Let (X, d) be a complete metric space and a "target" element $x \in X$ that we wish to approximate. Given an $\epsilon > 0$, can we find a contraction mapping $T : X \to X$ with fixed point $\bar{x} \in X$ such that $d(\bar{x}, x) < \epsilon$?

Very briefly, the original motivation for this formulation was fractal image coding [16, 6, 39]. When an image x is approximated by the fixed point $\bar{x}$ of a contractive "fractal transform" T, the amount of computer memory required to store the parameters which define T is generally much less than that required to store x. Instead of storing or transmitting x, one stores or transmit T, from which $\bar{x}$, an approximation to x can be generated via iteration. The result: *fractal image compression.*

Given the complicated nature of fractal transforms, however, the determination of optimal mappings T by minimizing the approximation error $d(\bar{x}, x)$ is intractable. An enormous simplification is achieved by means of the following simple consequence of Banach's Theorem, known in the literature as the *Collage Theorem.*

Theorem 2 *("Collage Theorem" [5, 4]) Let (X, d) be a complete metric space and $T : X \to X$ a contraction mapping with contraction factor $c \in [0, 1)$. Then for any $x \in X$,*

$$d(x, \bar{x}) \le \frac{1}{1-c} d(x, Tx), \tag{1}$$

where $\bar{x}$ is the fixed point of T.

This permits a reformulation of our original inverse problem as follows,

> Given an $\epsilon > 0$, can we find a contraction mapping $T : X \to X$ such that $d(Tx, x) < \delta$?

In an effort to minimize the approximation error $d(\bar{x}, x)$, we now look for contraction maps T which minimize the so-called *collage error* $d(x, Tx)$. In other words, we look for maps T which send the target x as close as possible to itself. We refer to this approach as *collage coding* [30].

Barnsley and co-workers [5, 4] were the first to see the potential of using the Collage Theorem for the purpose of image approximation and compression. Most, if not all fractal image coding methods are based on some kind of block-based collage coding method which follows the strategy originally presented by Jacquin [18].

A collage coding approach, however, may be applied in other, "nonfractal," situations where contractive mappings are encountered, as we describe below. In a practical application, we consider a family of appropriate contraction mappings T_λ, $\lambda \in \Lambda \subset \mathbb{R}^n$, and try to find the parameter $\bar{\lambda}$ which minimizes the approximation error $d(x, \bar{x}_\lambda)$. The feasible set can be defined as $\Lambda = \{\lambda \in \mathbb{R}^n : 0 \leq c_\lambda \leq c \leq C < 1\}$ which guarantees the contractivity of T_λ for any $\lambda \in \Lambda$. (Here, C is a prescribed "cutoff.") This is perhaps the main difference between our collage coding approach and Tikhonov regularization (see [45, 46]): In the former, the constraint $\lambda \in \Lambda$ guarantees that T_λ is a contraction essentially replacing the effect of the regularization term in the Tikhonov approach. The collage-based inverse problem described above can be formulated as an optimization problem as follows,

$$\min_{\lambda \in \Lambda} d(x, T_\lambda x)\,. \tag{2}$$

In general this optimization problems is nonlinear and nonsmooth. The regularity of the objective function strictly depends on the term $d(x, T_\lambda x)$. In many cases, however, the problem in (2) can be reduced to a quadratic optimization problem. A number of algorithms can then be used to solve this problem, including, for example, penalization methods, particle swarm ant colony techniques, etc..

3. Inverse problems for DEs using the Collage Theorem

The use of the Collage Theorem to solve inverse problems for ODEs was originally proposed in [21] and developed in many subsequent works including [22, 23, 25, 26, 28]. The initial value problems (IVPs) studied in these papers had the general form,

$$\begin{cases} \dot{u} = f(t, u) \\ u(0) = u_0\,. \end{cases} \tag{3}$$

Associated with the above IVP is the following Picard integral operator,

$$(Tu)(t) = u_0 + \int_0^t f(s, u(s))\, ds. \tag{4}$$

It is well known that the solution to the IVP in (3) is a fixed point of T, i.e.,

$$Tu = u. \tag{5}$$

Consider the complete metric space $C([-\delta, \delta])$ endowed with the usual d_∞ metric and assume that $f(t, u)$ is Lipschitz in the variable u, that is there exists a $K \geq 0$ such that $|f(s, u) - f(s, v)| \leq K|u - v|$, for all $u, v \in \mathbb{R}$. For simplicity we suppose that $u \in \mathbb{R}$ but the same consideration can be developed for the case of several variables. Under these hypotheses T is Lipschitz on the space $C([-\delta, \delta] \times [-M, M])$ for some δ and $M > 0$.

In the direct or forward problem, as is well known, the above Lipschitz property of f guarantees the existence of a unique fixed point of T and therefore the solution to (3). (Standard proof: contractivity of T.) Here, however, we are concerned with the collage-based *inverse problem* associated with (3):

Given a function $u(t)$, find a Picard operator T – as defined by the function f – which maps u as close as possible to itself.

Theorem 3 *[21] The function T satisfies*

$$\|Tu - Tv\|_2 \leq c\|u - v\|_2 \tag{6}$$

for all $u, v \in C([-\delta, \delta] \times [-M, M])$ where $c = \delta K$.

Now let $\delta' > 0$ be such that $\delta' K < 1$. In order to solve the inverse problem for the Picard operator in (4) we employ the L^2 expansion of the function f. Let $\{\phi_i\}$ be a basis of functions in $L^2([-\delta', \delta'] \times [-M, M])$ and consider $f_\lambda(s, u) = \sum_{i=1}^{+\infty} \lambda_i \phi_i(s, u)$. Each sequence of coefficients $\lambda = \{\lambda_i\}_{i=1}^{+\infty}$ then defines a Picard operator T_λ. Suppose further that each function $\phi_i(s, u)$ is Lipschitz in u with constant K_i.

Theorem 4 *[21] Let $K, \lambda \in \ell^2(\mathbb{R})$. Then*

$$|f_\lambda(s, u) - f_\lambda(s, v)| \leq \|K\|_2 \|\lambda\|_2 |u - v| \tag{7}$$

for all $s \in [-\delta', \delta']$ and $u, v \in [-M, M]$ where $\|K\|_2 = \left(\sum_{i=1}^{+\infty} K_i^2\right)^{\frac{1}{2}}$ and $\|\lambda\|_2 = \left(\sum_{i=1}^{+\infty} \lambda_i^2\right)^{\frac{1}{2}}$

Given a target solution x, we now wish to minimize the collage distance $\| u - T_\lambda u \|_2$. The square of the collage distance becomes

$$\Delta^2(\lambda) = \|u - T_\lambda u\|_2^2 = \int_{-\delta}^{\delta} \left| u(t) - u_0 - \int_0^t \sum_{i=1}^{+\infty} \lambda_i \phi_i(s, u(s)) ds \right|^2 dt$$

and the inverse problem can be formulated as

$$\min_{\lambda \in \Lambda} \Delta(\lambda), \tag{8}$$

where $\Lambda = \{\lambda \in \ell^2(\mathbb{R}) : \|\lambda\|_2 \|K\|_2 < 1\}$. To solve this problem numerically, we consider the first n terms of the L^2 basis. In this case, the previous problem can be reduced to:

$$\min_{\lambda \in \tilde{\Lambda}} \tilde{\Delta}^2(\lambda) = \int_{-\delta}^{\delta} \left| u(t) - u_0 - \int_0^t \sum_{i=1}^{n} \lambda_i \phi_i(s, u(s)) ds \right|^2 dt, \tag{9}$$

where $\tilde{\Lambda} = \{\lambda \in \mathbb{R}^n : \|\lambda\|_2 \|K\|_2 < 1\}$. This is a classical quadratic optimization problem which can be solved by means of classical numerical methods.

Let $\tilde{\Delta}_{\min}^n$ be the minimum value of $\tilde{\Delta}$ over $\tilde{\Lambda}$. This is a nonincreasing sequence of numbers (depending on n). Following the method of [17], it can be shown that $\liminf_{n \to +\infty} \tilde{\Delta}_{\min}^n = 0$, i.e., the distance between the target element and the unknown solution of the differential equation can be made arbitrarily small.

In [9, 11, 25] the above approach was extended to consider the case of inverse problems for random and stochastic differential equations.

4. Inverse problems for PDEs using the Generalized Collage Theorem

We now review an extension of the Collage Theorem, the Generalized Collage Theorem, and show how it can be used to solve inverse problems for families of PDEs.

4.1. Elliptic equations
Consider the following variational equation,

$$a(u,v) = \phi(v), \qquad v \in H. \tag{10}$$

where $\phi(v)$ and $a(u,v)$ are linear and bilinear maps, respectively, both defined on an Hilbert space H. Let $\langle\cdot\rangle$ denote the inner product in H, $\|u\|^2 = \langle u,u\rangle$ and $d(u,v) = \|u-v\|$, for all $u,v \in H$. The existence and uniqueness of solutions to this kind of equation are provided by the classical Lax-Milgram representation theorem [12]: Let H be a Hilbert space and ϕ a bounded linear nonzero functional, i.e., $\phi : H \to \mathbb{R}$. Also suppose that $a(u,v)$ is a bilinear form on $H \times H$ which satisfies the following conditions:

(i) There exists a constant $M > 0$ s.t. $|a(u,v)| \leq M\|u\|\|v\|$ for all $u,v \in H$,
(ii) There exists a constant $m > 0$ s.t. $|a(u,u)| \geq m\|u\|^2$ for all $u \in H$.

Then there is a unique vector $u^* \in H$ such that $\phi(v) = a(u^*, v)$ for all $v \in H$.

The inverse problem may now be viewed as follows. Suppose that we have an observed solution u and a given (restricted) family of bilinear functionals $a_\lambda(u,v)$, $\lambda \in \mathbb{R}^n$. We now seek to find "optimal" values of λ.

Suppose that we have a given Hilbert space H, a "target" element $u \in H$ and a family of bilinear functionals a_λ. Then by the Lax-Milgram theorem, there exists a unique vector u_λ such that $\phi(v) = a_\lambda(u_\lambda, v)$ for all $v \in H$. We would like to determine if there exists a value of the parameter λ such that $u_\lambda = u$ or, more realistically, such that $\|u_\lambda - u\|$ is small enough. The following theorem will be useful for the solution of this problem.

Theorem 5 *(Generalized Collage Theorem) [27] Suppose that $a_\lambda(u,v) : \mathcal{F} \times H \times H \to \mathbb{R}$ is a family of bilinear forms for all $\lambda \in \mathcal{F}$ and $\phi : H \to \mathbb{R}$ is a given linear functional. Let u_λ denote the solution of the equation $a_\lambda(u,v) = \phi(v)$ for all $v \in H$ as guaranteed by the Lax-Milgram theorem. Given a target element $u \in H$ then*

$$\|u - u_\lambda\| \leq \frac{1}{m_\lambda} F(\lambda), \tag{11}$$

where

$$F(\lambda) = \sup_{v \in H,\ \|v\|=1} |a_\lambda(u,v) - \phi(v)|. \tag{12}$$

In order to ensure that the approximation u_λ is close to a target element $u \in H$, we can, by the Generalized Collage Theorem, try to make the term $F(\lambda)/m_\lambda$ as close to zero as possible. The appearance of the m_λ factor complicates the procedure as does the factor $1/(1-c)$ in standard collage coding, i.e., Eq. (1). If $\inf_{\lambda \in \mathcal{F}} m_\lambda \geq m > 0$ then the inverse problem can be reduced to the minimization of the function $F(\lambda)$ on the space $\mathcal{F}$, that is,

$$\min_{\lambda \in \mathcal{F}} F(\lambda). \tag{13}$$

Next sections show that, under the condition $\inf_{\lambda \in \mathcal{F}} m_\lambda \geq m > 0$, our approach is stable. Following our earlier studies of inverse problems using fixed points of contraction mappings, we shall refer to the minimization of the functional $F(\lambda)$ as a "generalized collage method."

Now let $\langle e_i \rangle \subset H$ be a basis of the Hilbert space H, not necessarily orthogonal, so that each element $v \in H$ can be written as $v = \sum_i \alpha_i e_i$. It can easily be proved that

$$\inf_{\lambda \in \mathcal{F}} \|u - u_\lambda\| \leq \frac{1}{m} \sup_{v \in H,\ \|v\|=1} \left[\sum_i \alpha_i^2\right] \inf_{\lambda \in \mathcal{F}} \left[\sum_i |a_\lambda(u, e_i) - \phi(e_i)|^2\right]. \tag{14}$$

Let $V_n =< e_1, e_2, \ldots, e_n >$ be the finite dimensional vector space generated by e_i, $V_n \subset H$. Given a target $u \in H$, let $\Pi_{V_n} u$ the projection of u on the space V_n and consider the following problem: find $u_\lambda \in V_n$ such that $\|\Pi_{V_n} u - u_\lambda\|$ is as small as possible. We have

$$\|\Pi_{V_n} u - u_\lambda\| \le \frac{M}{m} \left[\sum_i |a_\lambda(u, e_i) - \phi(e_i)|^2 \right], \tag{15}$$

where $M = \max_{v=\sum_{i=1}^n \alpha_i e_i \in V_h, \|v\|=1} \sum_{i=1}^n \alpha_i^2$, so that the problem has been reduced to the following minimization problem,

$$\inf_{\lambda \in \mathcal{F}} \|\Pi_{V_n} u - u_\lambda\| \le \frac{M}{m} \inf_{\lambda \in \mathcal{F}} \sum_{i=1}^n |a_\lambda(u, e_i) - \phi(e_i)|^2 = \frac{M}{m} (F_n(\lambda))^2. \tag{16}$$

Example 1: We consider

$$\begin{aligned} -\nabla \cdot (\kappa(x,y) \nabla u(x,y)) &= f(x,y),\ (x,y) \in \Omega = \{0 < x, y < 1\} \\ u(x,y) &= 0,\ (x,y) \in \partial\Omega. \end{aligned} \tag{17}$$

Multiply (17) by a test function $v(x,y) \in H = H_0^1([0,1]^2)$, the Hilbert space built with all L^2 functions that have a weak derivative in L^2, integrate over $\bar{\Omega}$, and apply Green's first identity, with $\hat{n}$ denoting the outward unit normal to $\partial\Omega$, to get the equation $a(u,v) = \phi(v)$, with

$$a(u,v) = \iint_\Omega \kappa \nabla v \cdot \nabla u \, dA \quad \text{and} \quad \phi(v) = \iint_\Omega f v \, dA. \tag{18}$$

Now, consider the inverse problem of recovering an estimate of the diffusivity $\kappa(x,y)$ given $f(x,y)$ and a set of values of the solution u inside Ω. Using (16), we solve the inverse problem by minimizing $F_n(\lambda)$. To produce a specific example, we set

$$\kappa_{true}(x,y) = 2 + 8xy^2(1-x) \quad \text{and} \quad u(x,y) = x(1-x)\sin(\pi y) \in H_0^1([0,1]^2).$$

We determine $f(x,y)$ from (17). We use the 49 data values $u\left(\frac{i}{8}, \frac{j}{8}\right)$, $i, j = 1, .., 7$. The grid of data points induces a finite element basis for $V_{7,7} \subset H$, within which we seek to recover an estimate of the 49 basis coefficients for κ. Minimization of $F_n(\lambda)$ produces a $\kappa(x,y)$ satisfying $\|\kappa(x,y) - \kappa_{true}(x,y)\|_2 = 0.0128$. Figure 1 presents the graphs of $\kappa_{true}(x,y)$ and the recovered $\kappa(x,y)$.

4.2. Parabolic equations

Consider the following abstract formulation of a parabolic equation,

$$\begin{cases} \langle \frac{d}{dt} u, v \rangle = \psi(v) + a(u,v) \\ u(0) = f, \end{cases} \tag{19}$$

where H is a Hilbert space, $\psi : H \to \mathbb{R}$ is a linear functional, $a : H \times H \to \mathbb{R}$ is a bilinear form and $f \in H$ is an initial condition. The inverse problem for the above equation consists of finding an approximation of the coefficients and parameters starting from a sample of observations of a target $u \in H$. To do this, we consider a family of bilinear functionals a_λ and let u_λ be the solution to

$$\begin{cases} \langle \frac{d}{dt} u_\lambda, v \rangle = \psi(v) + a_\lambda(u_\lambda, v) \\ u_0 = f. \end{cases} \tag{20}$$

We wish to determine if there exists a value of the parameter λ such that $u_\lambda = u$ or, more realistically, such that $\|u_\lambda - u\|$ is sufficiently small. To this end, Theorem 6 states that the distance between the target solution u and the solution u_λ of (20) can be reduced by minimizing a functional which depends on parameters.

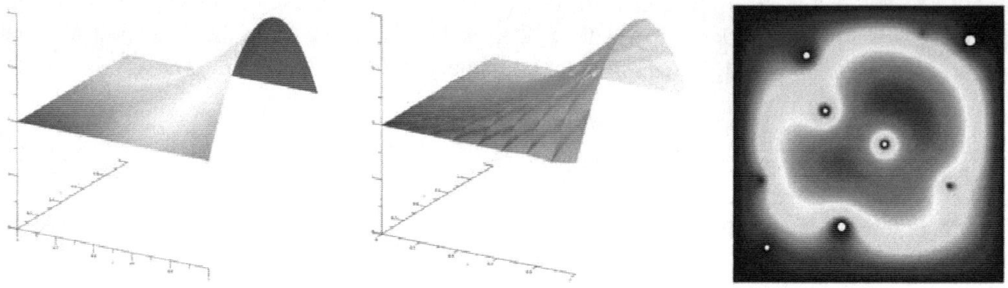

(a) $z = \kappa_{true}(x, y)$ (b) $z = \kappa(x, y)$ (c) Isotherms of $u(x, y)$

Figure 1. (a-b) The graphs of $\kappa_{true}(x, y)$ and the recovered $\kappa(x, y)$ in Example 1. (c) Isotherms of $u(x, y)$ on the perforated domain in Example 2.

Theorem 6 *[11] Let $u : [0, T] \to L^2(D)$ be the target solution which satisfies the initial condition in (19) and suppose that $\frac{d}{dt}u$ exists and belongs to H. Suppose that $a_\lambda(u, v) : \mathcal{F} \times H \times H \to \mathbb{R}$ is a family of bilinear forms for all $\lambda \in \mathcal{F}$. We have the following result:*

$$\int_0^T \|u - u_\lambda\|_H dt \leq \frac{1}{m_\lambda^2} \int_0^T \left(\sup_{\|v\|=1} \left\langle \frac{d}{dt}u, v \right\rangle - \psi(v) - a_\lambda(u, v) \right)^2 dt\,, \tag{21}$$

where u_λ is the solution of (20) s.t. $u_\lambda(0) = u(0)$ and $u_\lambda(T) = u(T)$.

Whenever $\inf_{\lambda \in \mathcal{F}} m_\lambda \geq m > 0$ then the previous result states that in order to solve the inverse problem for the parabolic equation (19) one can minimize the following functional,

$$\int_0^T \left(\sup_{\|v\|=1} \left\langle \frac{d}{dt}u, v \right\rangle - \psi(v) - a_\lambda(u, v) \right)^2 dt\,, \tag{22}$$

over all $\lambda \in \mathcal{F}$.

4.3. Hyperbolic equations

Let us now consider the following weakly-formulated hyperbolic equation,

$$\begin{cases} \langle \frac{d^2}{dt^2}u, v \rangle = \psi(v) + a(u, v) \\ u(0) = f \\ \frac{d}{dt}u(0) = g\,, \end{cases} \tag{23}$$

where $\psi : H \to \mathbb{R}$ is a linear functional, $a : H \times H \to \mathbb{R}$ is a bilinear form, and $f, g \in H$ are the initial conditions. As in previous sections, the inverse problem for the above system of equations is to reconstruct the coefficients starting from a sample of observations of a target $u \in H$. We consider a family of bilinear functionals a_λ and let u_λ be the solution to the following problem,

$$\begin{cases} \langle \frac{d}{dt}u_\lambda, v \rangle = \psi(v) + a_\lambda(u_\lambda, v) \\ u_0 = f \\ \frac{d}{dt}u(0) = g\,. \end{cases} \tag{24}$$

We wish to determine if there exists a value of the parameter λ such that $u_\lambda = u$ or, more realistically, such that $\|u_\lambda - u\|$ is sufficiently small. Theorem 7 states that the distance between the target solution u and the solution u_λ of (24) can be reduced by minimizing a functional which depends on parameters.

Theorem 7 *Let $u : [0,T] \to L^2(D)$ be the target solution which satisfies the initial condition in (23) and suppose that $\frac{d^2}{dt^2}u$ exists and belongs to H. Suppose that there exists a family of $m_\lambda > 0$ such that $a_\lambda(v,v) \geq m_\lambda \|v\|^2$ for all $v \in H$. We have the following result:*

$$\int_0^T \|u_t - (u_\lambda)_t\|^2 dt \leq \frac{1}{m_\lambda^2} \int_0^T \left(\sup_{\|v\|=1} \left\langle \frac{d^2}{dt^2} u_t, v \right\rangle - \psi(v) - a(u_t, v) \right)^2 dt \,, \tag{25}$$

where $(u_\lambda)_t$ is the solution of (24) s.t. $u(0) = (u_\lambda)(0)$ and $u(T) = (u_\lambda)(T)$.

The proof of the theorem follows the same path as that of Theorem 6.

5. Inverse Problems for DEs using a Collage Theorem for Banach spaces

The results presented in the previous two sections have been extended to a wider class of elliptic equations problems by considering not only Hilbert spaces but also reflexive Banach spaces. Let us mention that this kind of formulation arises, for instance, when the boundary constraints are weakly imposed. Details can be found in [7, 8, 31]. The following result presents an extended version of the Lax–Milgram theorem.

Let $N \geq 1$, E, $F_1, \dots, F_N$ are real vector spaces, $a_1 : E \times F_1 \longrightarrow \mathbb{R}, \dots, a_N : E \times F_N \longrightarrow \mathbb{R}$ are bilinear forms and $y_1^* : F_1 \longrightarrow \mathbb{R}, \dots, y_N^* : F_N \longrightarrow \mathbb{R}$ and consider the system,

$$x \in E \text{ such that } \begin{cases} y_1^* = a_1(x, \cdot) \\ \quad\vdots \\ y_N^* = a_N(x, \cdot) \,. \end{cases}$$

If this system admits a solution, then such a solution is unique if and only if, the corresponding homogeneous problem has one and only one solution. Given a real normed space G, we write G^* for its topological dual space.

Theorem 8 *[7, 8, 31] Suppose that E is a real reflexive Banach space, $N \geq 1$, $F_1, \dots, F_N$ are Banach spaces and that $a_1 : E \times F_1 \longrightarrow \mathbb{R}, \dots, a_N : E \times F_N \longrightarrow \mathbb{R}$ are continuous bilinear forms. Then, for all $y_1^* \in F_1^*, \dots, y_N^* \in F_N^*$ there exists a unique $x_0 \in E$ such that*

$$\begin{cases} y_1^* = a_1(x_0, \cdot) \\ \quad\vdots \\ y_N^* = a_N(x_0, \cdot) \end{cases}$$

if and only if

$$x \in E \text{ and } \left.\begin{matrix} y_1^* = a_1(x_0, \cdot) \\ \vdots \\ y_N^* = a_N(x_0, \cdot) \end{matrix}\right\} \Rightarrow x = 0$$

and there exists $\rho > 0$ satisfying

$$(y_1, \dots, y_N) \in F_1 \times \cdots \times F_N \ \Rightarrow \ \rho \sum_{k=1}^N \|y_k\| \leq \left\| \sum_{k=1}^N a_k(\cdot, y_k) \right\| .$$

Moreover, if these equivalent conditions hold and $x_0 \in E$ is the unique solution, then

$$\|x_0\| \leq \frac{1}{\rho} \max_{k=1,\dots,N} \|y_k^*\|.$$

The above result implies the following corollary which represents an extended version of the collage theorem.

Corollary 1 *[7] Let E be a real reflexive Banach space, let $N \geq 1$, let $F_1, \ldots, F_N$ be Banach spaces, let $y_1^* \in F_1^*, \ldots, y_N^* \in F_N^*$ and let Λ be a nonempty set such that for all $\lambda \in \Lambda$ there exist N continuous bilinear forms $a_{1\lambda} : E \times F_1 \longrightarrow \mathbb{R}, \ldots, a_{N\lambda} : E \times F_N \longrightarrow \mathbb{R}$ and $\rho_\lambda > 0$ with*

$$x \in E \text{ and } \left.\begin{array}{rl} y_1^* &= a_{1\lambda}(x_0, \cdot) \\ & \vdots \\ y_N^* &= a_{N\lambda}(x_0, \cdot) \end{array}\right\} \Rightarrow x = 0$$

and

$$(y_1, \ldots, y_N) \in F_1 \times \cdots \times F_N \Rightarrow \rho_\lambda \sum_{k=1}^{N} \|y_k\| \leq \left\| \sum_{k=1}^{N} a_{k\lambda}(\cdot, y_k) \right\|.$$

Let us also suppose that for all $\lambda \in \Lambda$, $x_\lambda \in E$ is the unique solution of the variational system,

$$x \in E \text{ and } \left\{\begin{array}{rl} y_1^* &= a_{1\lambda}(x, \cdot) \\ & \vdots \\ y_N^* &= a_{N\lambda}(x, \cdot) \end{array}\right. .$$

Then for each $x_0 \in E$ and for all $\lambda \in \Lambda$ the inequality,

$$\|x_\lambda - x_0\| \leq \frac{1}{\rho_\lambda} \max_{k=1,\ldots,N} \|y_k^* - a_{k\lambda}(x_0, \cdot)\|,$$

is valid.

Let us observe that if one wants to approximate the solution x_0 in the sense of the collage distance, that is, minimize $\{\|x_\lambda - x_0\| : \lambda \in \Lambda\}$, then according to Corollary 1, it suffices to minimize

$$\left\{ \frac{1}{\rho_\lambda} \max_{k=1,\ldots,N} \|y_k^* - a_{k\lambda}(x_0, \cdot)\| : \lambda \in \Lambda \right\}.$$

If $\rho := \inf_{\lambda \in \Lambda} \rho_\lambda > 0$, then the approximation problem is reduced to

$$\left\{ \max_{k=1,\ldots,N} \|y_k^* - a_{k\lambda}(x_0, \cdot)\| : \lambda \in \Lambda \right\}.$$

Some more details about the implementation of the numerical scheme and more numerical examples that demonstrate the validity of this approach can be found in [7, 8, 31].

6. Inverse Problems for DEs on perforated domains using the Collage Theorem

In this section we review one of the latest applications of the Collage Theorem to the solution of inverse problems, namely, on perforated or porous media. The results recalled in this section can be found with more details and applications in [32, 33]. In addition, [34] provides a numerical exploration for systems of DEs on perforated domains. When a differential equation is formulated over a porous medium, the term "porous" implies that the state equation is written only in the solid portion (or matrix) while boundary conditions should be imposed on the entire boundary of the matrix, including the boundary of the pore space (or holes).

Given a compact and convex set Ω, we denote by Ω_B the collection of circular holes $\cup_{j=1}^{m} B(x_j, \varepsilon_j)$ where $x_j \in \Omega$, ε_j are strictly positive numbers, and the holes $B(x_j, \varepsilon_j)$ are assumed

IOP Conf. Series: Journal of Physics: Conf. Series **1047** (2018) 012004 doi:10.1088/1742-6596/1047/1/012004

to be nonoverlapping and to lie strictly in the interior of Ω. Let $\varepsilon = \max_j \varepsilon_j$. and denote by Ω_ε the closure of the set $\Omega \backslash \Omega_B$. In this section, we set $H = H_0^1(\Omega)$ and $H_\varepsilon = H_0^1(\Omega_\varepsilon)$.

In [32, 33] we have considered the linear system

$$\text{Find } u \in H \text{ that satisfies } a_\lambda(u, v) = \varphi(v) \quad \forall v \in H. \tag{P}$$

where $\lambda \in \Lambda$ denotes some parameters of the functionals and the corresponding system on the domain with holes,

$$\text{Find } u \in H_\epsilon \text{ that satisfies } a_\lambda(u, v) = \varphi(v) \ \forall v \in H_\epsilon. \tag{P_ε}$$

Our goal is to address the following inverse problem: Given observational data for a solution to (P_ε), estimate λ. Our approach is to use the data in (P) to estimate λ by establishing connections between the parameters λ in (P) and (P_ε) for ε small.

Since any function in $H_0^1(\Omega_\varepsilon)$ can be extended to be zero-valued over the holes, $H_0^1(\Omega_\varepsilon)$ can be embedded in $H_0^1(\Omega)$. Given u, letting $P_\varepsilon u$ be the projection of $u \in H_0^1(\Omega_\varepsilon)$ onto $H_0^1(\Omega)$, one can prove

$$\|u - P_\varepsilon u\|_{H_0^1(\Omega)} \to 0 \text{ whenever } \varepsilon \to 0.$$

When Neumann boundary conditions are considered, it is still possible to extend a function in $H_0^1(\Omega_\varepsilon)$ to a function of $H_0^1(\Omega)$.

Now supposing that there exist positive constants m, M, and μ such that

$$\begin{aligned} a_{\lambda,\epsilon}(u, u) &\geq m\|u\|^2 \ \ \forall u \in H_\varepsilon \ \ \text{(coercivity)} \\ a_{\lambda,\epsilon}(u, v) &\leq M\|u\|\|v\| \ \ \forall u, v \in H_\varepsilon \\ \phi_{\lambda,\epsilon}(u) &\leq \mu\|u\| \ \ \forall u \in H_\varepsilon \end{aligned}$$

means that problem (P) has a unique solution u^λ for each $\lambda \in \Lambda$ and problem (P_ε) has a unique solution u_ε^λ and for each positive ε and each $\lambda \in \Lambda$, by the Lax-Milgram type theorem in [31]. Now, we can establish relationships between (P) and (P_ε). For each $u \in (H_0^1(\Omega_\varepsilon))$, define

$$F_\varepsilon(u, \lambda) = \sup_{v \in H_\varepsilon, \|v\|=1} \|a_\lambda^\varepsilon(u, v) - \phi_\lambda^\varepsilon(v)\| . \tag{26}$$

The following four results were proved in [32].

Proposition 1 *The function $F(u, \lambda)$ is Lipschitz with Lipschitz constant equal to M.*

Proposition 2 *The following inequality holds:*

$$\|P_\varepsilon u - u_\varepsilon^\lambda\|_{H_\varepsilon} \leq \frac{F(u, \lambda)}{m} + \frac{M}{m}\|P_\varepsilon u - u\|_H .$$

Proposition 3 *The exists a constant $C(\varepsilon, u)$, which depends only on ε and u, such that the following inequality holds:*

$$F(P_\varepsilon u, \lambda) \leq F_\varepsilon(P_\varepsilon u, \lambda) + C(\varepsilon, u) \sup_{\|v\|=1} \|P_\varepsilon v - v\|_{H_\varepsilon} .$$

The nature of C in Proposition 3 is changed slightly compared to [32], but the same proof works. Note that the constant $C(\epsilon, u) = M\|P_\varepsilon u\|_{H_\varepsilon} + \mu$ converges to $C(u) = M\|u\| + \mu$ whenever $\varepsilon \to 0$. We have also corrected an imprecision in the statement of Proposition 4 from [32], with the same proof working.

Proposition 4 *Let us suppose that, for each fixed* $u \in H_0^1(\Omega_\varepsilon)$, F *is lower continuous w.r.t.* $\lambda \in \Lambda$. *If* $\lambda_\varepsilon = \arg\min_{\lambda\in\Lambda} F_\varepsilon(P_\varepsilon u, \lambda)$, *and* $\lambda_\varepsilon \to \lambda^* \in \Lambda$ *then* $\lambda^* = \arg\min_{\lambda\in\Lambda} F(u, \lambda)$.

Example 2: We consider the model problem (17), replacing the domain Ω by a domain Ω_ε that has a number of holes. Using the same $\kappa_{true}(x, y) = 2 + 8x^2y - 8x^2y^2$ and $f(x, y)$ as in Example 1, we solve the problem numerically on Ω_ε, using homogeneous Dirichlet boundary conditions on the interior holes. The isotherms of the solution are depicted in Figure 1(c). As in the earlier example, we sample this solution at 49 uniformly-distributed points inside Ω_ε; if a point lies inside a hole, we obtain no data for that point. Using this data points, we then solve the inverse problem on the region with no holes, Ω, appealing to Proposition 4. When we seek a κ of the form $\kappa(x, y) = \lambda_0 + \lambda_1 x^2 y + \lambda_2 x^2 y^2$, we obtain $(\lambda_0, \lambda_1, \lambda_2) = (2.704, 7.301, -7.934)$, with L^2 error 0.479. If we shrink the holes, use Neumann boundary conditions on them, and/or use more data points, the estimation improves.

References

[1] Aida-zade K R and Abdullayev V M 2016 Solution To A Class Of Inverse Problems For A System Of Loaded Ordinary Differential Equations With Integral Conditions *Journal of Inverse and Ill-posed Problems* **24** 5 pp 543558

[2] Aida-zade K R and Abdullayev V M 2014 Numerical Approach to Parametric Identification of Dynamical Systems *Journal of Automation and Information Sciences* **46** 3 pp 30-46

[3] Banach S 1922 Sur les opérations dans les ensembles abstraits et leurs applications aux équations intégrales *Fund Math* 3 pp 33-181

[4] Barnsley M 1989 *Fractals Everywhere* (New York: Academic Press)

[5] Barnsley M F, Ervin V, Hardin D and Lancaster J 1985 Solution of an inverse problem for fractals and other sets *Proc Nat Acad Sci USA* 83 pp 1975-1977

[6] Barnsley M and Hurd L 1995 *Fractal image compression* (AK Peters Ltd)

[7] Berenguer M I, Kunze H, La Torre D and Ruiz Galán M 2015 Galerkin schemes and inverse boundary value problems in reflexive Banach spaces *Journal of Computational and Applied Mathematics* 275 pp 100-112

[8] Berenguer M I, Kunze H, La Torre D and Ruiz Galán M 2016 Galerkin method for constrained variational equations and a collage-based approach to related inverse problems *Journal of Computational and Applied Mathematics* 292 pp 67-75

[9] Capasso V, Kunze H, La Torre D and Vrscay E R 2014 Solving inverse problems for differential equations by a "generalized collage" method and application to a mean field stochastic model *Nonlinear Analysis: Real World Applications* 15 pp 276-289

[10] Capasso V, Kunze H, La Torre D and Vrscay E R 2013 Solving inverse problems for biological models using the collage method for differential equations *Journal of Mathematical Biology* **67** 1 pp 25-38

[11] Capasso V, Kunze H, La Torre D and Vrscay E R 2009 Parameter identification for deterministic and stochastic differential equations *Advances in Nonlinear Analysis: Theory, Methods and Applications* ed S.Sivasundaram (Cambridge Scientific Publisher) pp 71-84

[12] Ciarlet P G 2013 *Linear and Nonlinear Functional Analysis with Applications* (SIAM Press)

[13] Deng X, Wang B and Long G 2008 The Picard contraction mapping method for the parameter inversion of reaction-diffusion systems *Computers and Mathematics with Applications* **56** 9 pp 2347-2355

[14] Deng X and Liao Q 2009 Parameter Estimation for Partial Differential Equations by Collage-Based Numerical Approximation *Mathematical Problems in Engineering* 510934.

[15] Drew C 2017 Peak Oil: A Summary of Models and Predictions (Université du Québec 1.à Montréal: Institute des Sciences Mathematiques)

[16] Fisher Y 1996 *Fractal image compression: Theory and applications* (New York: Springer-Verlag)

[17] Forte B and Vrscay E R 1998 *Inverse problem methods for generalized fractal transforms Fractal Image Encoding and Analysis* (NATO ASI Series F) 159 ed Y Fisher (New York: Springer Verlag)

[18] Jacquin A 1992 Image coding based on a fractal theory of iterated contractive image transformations (IEEE Trans Image Proc) 1 pp 18-30

[19] Keller J B 1976 Inverse Problems *The American Mathematical Monthly* **83** 2 pp 107-118

[20] Kirsch A 2011 *An introduction to the mathematical theory of inverse problems* (Springer)

[21] Kunze H and Vrscay E R 1999 Solving inverse problems for ordinary differential equations using the Picard contraction mapping *Inverse Problems* 15 pp 745-770

[22] Kunze H and Gomes S 2003 Solving An Inverse Problem for Urison-type Integral Equations Using Banach's Fixed Point Theorem *Inverse Problems* 19 pp 411-418

[23] Kunze H, Hicken J and Vrscay E R 2004 Inverse Problems for ODEs Using Contraction Maps: Suboptimality of the "Collage Method" *Inverse Problems* 20 pp 977-991

[24] Kunze H and Crabtree D 2005 Using Collage Coding to Solve Inverse Problems in Partial Differential Equations *Maplesoft Conference Proceedings*

[25] Kunze H, La Torre D, Vrscay E R 2007 Random fixed point equations and inverse problems using collage method for contraction mappings *Journal of Mathematical Analysis and Applications* 334 pp 1116-1129

[26] Kunze H, La Torre D, Vrscay E R 2009 Inverse problems for random differential equations using the collage method for random contraction mappings *Journal of Computational and Applied Mathematics* **223** 2 pp 853-861

[27] Kunze H, La Torre D and Vrscay E R 2009 A generalized collage method based upon the Lax-Milgram functional for solving boundary value inverse problems *Nonlinear Analysis* **71** 12 pp e1337-e1343

[28] Kunze H, La Torre D and Vrscay E R 2010 Solving inverse problems for variational equations using the "generalized collage methods," with applications to boundary value problems *Nonlinear Analysis Real World Applications* **11** 5 pp 3734-3743

[29] Kunze H, La Torre D and Vrscay E R 2012 Solving inverse problems for DEs using the collage theorem and entropy maximization *Applied Mathematics Letters* 25 pp 2306-2311

[30] Kunze H, La Torre D, Mendivil F and Vrscay E R 2012 *Fractal-based methods in analysis* (Springer)

[31] Kunze H, La Torre D, Levere K and Ruiz Gálan M 2015 Inverse problems via the "generalized collage theorem" for vector-valued Lax-Milgram-based variational problems *Mathematical Problems in Engineering* 764643

[32] Kunze H and La Torre D 2015 Collage-type approach to inverse problems for elliptic PDEs on perforated domains *Electronic Journal of Differential Equations* 48 pp 1-11

[33] Kunze H and La Torre D 2016 Computational Aspects of Solving Inverse Problems for Elliptic PDEs on Perforated Domains Using the Collage Method *Mathematical and Computational Approaches in Advancing Modern Science and Engineering* ed J. Bélair and others (Springer) pp 113-120.

[34] Kunze H and La Torre D 2017 An Inverse Problem for a 2-D System of Steady-State Reaction-Diffusion Equations on a Perforated Domain *AIP Conference Proceedings* 1798 020089

[35] La Torre D and Vrscay E R 2009 A generalized fractal transform for measure-valued images *Nonlinear Analysis, Theory, Methods and Applications* 71 pp e1598-e1607

[36] Levere K, Kunze H, La Torre D 2013 A collage-based approach to solving inverse problems for second-order nonlinear parabolic PDEs *Journal of Mathematical Analysis and Applications* **406** 1 pp 120-133

[37] Liu F and Burrage K 2011 Novel techniques in parameter estimation for fractional dynamical models arising from biological systems *Computers & Mathematics with Applications* **62** 3 pp 822-833

[38] Liu F, Burrage K, and Hamilton N A 2013 Some Novel Techniques Of Parameter Estimation For Dynamical Models In Biological Systems *IMA Journal of Applied Mathematics* **78** 2 pp 235260

[39] Lu N 1997 *Fractal imaging* (Morgan Kaufmann Publishers Inc.)

[40] Moura Neto F D and da Silva Neto A J 2013 An Introduction to Inverse Problems with Applications (New York: Springer)

[41] Quinn J R 2011 Applications of the Contraction Mapping Principle *Fractal Geometry and Dynamical Systems in Pure and Applied Mathematics: Fractals in Applied Mathematics* ed D Carfi *AMS Contemporary Mathematics* 601 pp 345-358

[42] Şoltuz S M 2009 Solving Inverse Problems Via Hemicontractive Maps *Nonlinear Analysis: Theory, Methods & Applications* **71** 78 pp 2387-2390

[43] Şoltuz S M 2009 Extending The Collage Theorem To Contractive Like Operators *Revue D'Analyse Numerique Et De Theorie De L'Approximation* **38** 2 pp 177-181

[44] Tarantola A 2005 Inverse Problem Theory and Methods for Model Parameter Estimation (Philadelphia: SIAM)

[45] Tychonoff A N 1963 Solution of incorrectly formulated problems and the regularization method *Doklady Akademii Nauk SSSR* 151 pp 501-504.

[46] Tychonoff A N and Arsenin N Y 1977 Solution of Ill-posed Problems (Washington: Winston & Sons)

[47] Vanhems A 2006 Nonparametric Study Of Solutions Of Differential Equations *Econometric Theory* **22** 1 pp 127-157

[48] Vogel C R 2002 Computational Methods for Inverse Problems (New York: SIAM)

IOP Conf. Series: Journal of Physics: Conf. Series **1047** (2018) 012005 doi:10.1088/1742-6596/1047/1/012005

An inverse problem for a system of steady-state reaction-diffusion equations on a porous domain using a collage-based approach

H Kunze[1], **D La Torre**[2,3]

[1] Department of Mathematics and Statistics, University of Guelph, Guelph, Ontario, Canada N16 2W1
[2] Dubai Business School, University of Dubai, 14143 Dubai, UAE
[3] Department of Economics, Management and Quantitative Methods, University of Milan, Milan, Italy

E-mail: hkunze@uguelph.ca, dlatorre@ud.ac.ae, davide.latorre@unimi.it

Abstract. We consider an inverse problem for a system of steady-state reaction-diffusions acting on a perforated domain. We establish several results that connect the parameter values for the problem on the perforated domain with the corresponding problem on the related unperforated of solid domain. This opens the possibility of estimating a solution to the inverse problem on the perforated domain by instead working with the easier-to-solve inverse problem on the solid domain. We illustrate the results with an example.

1. Introduction

In the context of the subject work, the "direct problem" involves determining the solution to a completely prescribed two-dimensional system of steady-state reaction-diffusion equations, including known boundary conditions. On the other hand, the "inverse problem" involves estimating the values of (perhaps only some of the) parameters of the system given some information about the solution, perhaps observational data values.

In the next section, we present a summary of the collage method for solving inverse inverse problems for variational equations [5] and related systems [4]. Although the frameworks for the single equation and system cases have significant theoretical differences, both settings lead to a minimization problem for a function of the parameters λ one desires to estimate. Letting u denote the observed solution, perhaps an interpolation of observational data values, we must solve

$$\min_{\lambda\in\Lambda} F(u,\lambda).$$

In fact, in the case that the parameters λ appear linearly in the variational equation/system, upon approximating the problem in an appropriate finite-dimensional subspace, the function F is a quadratic function of the parameters and once F is built its minimization is quite straightforward.

In this paper, we wish to consider inverse problems on perforated or porous domains. A porous medium or perforated domain is a material characterized by a partitioning of the total volume into a solid portion often called the "matrix" and a pore space usually referred to as "holes" that can be either materials different from that of the matrix or real physical holes. When formulating differential equations over porous media, the term "porous" implies that the state equation is written in the matrix only, while boundary conditions should be imposed on the whole boundary of the matrix, including the boundary of the holes. Porous media can be found in many areas of applied sciences and engineering including petroleum engineering, chemical engineering, civil engineering, aerospace engineering, soil science, geology, material science, and many more areas.

Since porosity in materials can take different forms and appear in varying degrees, solving differential equations over porous media is often a complicated task and the holes' size and their distribution play an important role in its characterization. Furthermore numerical simulations over perforated domains need a very fine discretization mesh which often requires a significant computational time. The mathematical theory of differential equations on perforated domains is usually based on the theory of "homogenization" in which heterogeneous material is replaced by a fictitious homogeneous one. Of course this implies the need of convergence results linking together the model on a perforated domain and on the associated homogeneous one. In the case of porous media, or heterogeneous media in general, characterizing the properties of the material is a tricky process and can be done on different levels, mainly the microscopic and macroscopic scales, where the microscopic scale describes the heterogeneities and the macroscopic scale describes the global behavior of the composite.

In [3], two related problems for a single steady-state reaction-diffusion, problem (P_ε) on a perforated domain Ω_ε and problem (P) on the related solid domain Ω are considered:

$$\begin{cases} \nabla \cdot (K^\lambda(x,y)\nabla u(x,y)) = f^\lambda(x,y), & \text{in } \Omega_\varepsilon, \\ u(x,y) = 0, & \text{on } \partial\Omega_\varepsilon, \end{cases} \tag{P_ε}$$

and

$$\begin{cases} \nabla \cdot (K^\lambda(x,y)\nabla u(x,y)) = f^\lambda(x,y), & \text{in } \Omega, \\ u(x,y) = 0, & \text{on } \partial\Omega. \end{cases} \tag{P}$$

The inverse problem of interest for (P_ε) is to estimate λ given observational data for a solution. The paper establishes that a relationship between parameter values λ in the two problems: one can use the data from the solution to (P_ε) in the inverse problem for (P) to estimate λ, with the connection strengthening as ε decreases.

In many applications to material science it happens that data are collected without a priori knowledge of the geometry of the domain: In these situations is much simpler to assume the hypothesis of solid domain and use it to determine an estimation of the unknown parameters of the model.

In the third section, we build on these preceding results in developing a solution approach for the inverse problem for a system of steady-state reaction-diffusion equations on a perforated domain. Following that, in the final section we present an example to illustrate the method.

2. The Collage Method for Variational Equations

Consider the variational equation associated with a single elliptic equation,

$$a(u,v) = \phi(v), \qquad v \in H, \tag{1}$$

where ϕ and a are linear and bilinear maps, respectively, both defined on a Hilbert space H. We denote by $\langle\cdot,\cdot\rangle$ the inner product in H, $\|u\|^2 = \langle u,u\rangle$ and $d(u,v) = \|u-v\|$, for all $u, v \in H$. Typically $H = H_0^1(\Omega)$, that is the space of all $L^2(\Omega)$ functions that possess a weak derivative in $L^2(\Omega)$. The existence and uniqueness of solutions to this kind of equation are provided by the classical Lax-Milgram representation theorem, which requires that the bilinear form satisfy coercivity and boundedness properties and the linear map be bounded.

The following theorem proves particularly useful for treating a related inverse problem of approximating a target element u by a solution of a family of operators a^λ and ϕ^λ, satisfying (1) for each $\lambda \in \Lambda$.

Theorem 1. *(Generalized Collage Theorem) [5] For all $\lambda \in \Lambda$, suppose that $a^\lambda : \Lambda \times H \times H \to \mathbb{R}$ is a family of bilinear forms and $\phi^\lambda : \Lambda \times H \to \mathbb{R}$ is a family of linear functionals, satisfying*

$$\begin{aligned} a^\lambda(u,u) &\geq m^\lambda \|u\|^2 \ \forall u \in H \quad (\textit{coercivity}) \\ a^\lambda(u,v) &\leq M^\lambda \|u\|\|v\| \ \forall u, v \in H \\ \phi^\lambda(u) &\leq \mu^\lambda \|u\| \ \forall u \in H, \end{aligned}$$

for some positive constants m^λ, M^λ, and μ^λ. Let u^λ denote the solution of the equation $a^\lambda(u,v) = \phi^\lambda(v)$ for all $v \in H$ as guaranteed by the Lax-Milgram theorem. Then, given a target element $u \in H$,

$$\|u - u^\lambda\| \leq \frac{1}{m^\lambda} F(u, \lambda), \tag{2}$$

where

$$F(u,\lambda) = \left\| a^\lambda(u,\cdot) - \left(\phi^\lambda\right)^* \right\|. \tag{3}$$

If $\inf_{\lambda \in \Lambda} m^\lambda \geq m > 0$ then the inverse problem can be reduced to the minimization of the function $F(u, \lambda)$ on Λ, that is,

$$\min_{\lambda \in \Lambda} F(u, \lambda), \tag{4}$$

and we refer to this minimization as a "generalized collage method." To finite dimensionalize the problem, let $V_n = \langle e_1, e_2, \ldots, e_n \rangle$ be the finite dimensional vector space generated by e_i, so that $V_n \subset H$. Given a target $u \in H$, let $P_n u$ be the projection of u on the space V_n. We approximate the true error by $\|P_n u - u^\lambda\|$, noting that u^λ in this expression is in general not the same as the u^λ in the infinite dimensional problem. We can establish that

$$\|P_n u - u^\lambda\| \leq \frac{R}{m} \sum_i \left| a^\lambda(u, e_i) - \phi(e_i) \right|^2,$$

where R is a constant, leading to the minimization problem

$$\inf_{\lambda \in \Lambda} F_n(u, \lambda), \text{ where } F_n(u,\lambda) = \sum_{i=1}^{n} \left| a^\lambda(u, e_i) - \phi^\lambda(e_i) \right|^2. \tag{5}$$

This approach has been extended in [4] to the case of linear systems,

$$\left\{ \begin{aligned} a_1^\lambda(u, v_1) &= \phi_1(v_1) \\ &\ \vdots \\ a_N^\lambda(u, v_N) &= \phi_N(v_N) \end{aligned} \right. , \ \forall\, v_1, \ldots, v_N \in H,$$

where $u = (u_1, \ldots, u_N)$, to arrive at

$$F(u,\lambda) = \max_{1 \leq k \leq N} \left\| a_k^\lambda(u,\cdot) - \left(\phi_k^\lambda\right)^* \right\| = \max_{1 \leq k \leq N} \sup_{\|v\|=1, v \in H} \left| a_k^\lambda(u,v) - \phi_k^\lambda(v) \right|. \tag{6}$$

In this setting,

$$\|u - u^\lambda\| \leq \frac{1}{m^\lambda} F(u, \lambda),$$

where $m^\lambda = \min_k m_k^\lambda$.

3. A System of Steady-State Reaction-Diffusion Equations on Perforated and Unperforated Domains

Given a compact and convex set Ω, we denote by Ω_B the collection of circular holes $\cup_{j=1}^m B(x_j, \varepsilon_j)$ where $x_j \in \Omega$, ε_j are strictly positive numbers, and the holes $B(x_j, \varepsilon_j)$ are nonoverlapping and lie strictly in the interior of Ω. We let $\varepsilon = \max_j \varepsilon_j$. We denote by Ω_ε the closure of the set $\Omega \backslash \Omega_B$. In this section, we will set $H = H_0^1(\Omega)$ and $H_\varepsilon = H_0^1(\Omega_\varepsilon)$. We consider the linear system

$$(P)\text{: Find } u = (u_1, ..., u_N) \in H^N \text{ that satisfies } \begin{cases} a_1^\lambda(u, \cdot) & = & \left(\varphi_1^\lambda\right)^* \\ & \cdots & \\ a_N^\lambda(u, \cdot) & = & \left(\varphi_N^\lambda\right)^* \end{cases}$$

where $\lambda \in \Lambda$ denotes some parameters of the functionals, and, on the domain with holes, the corresponding system

$$(P_\varepsilon)\text{: Find } u = (u_1, ..., u_N) \in (H_\epsilon)^N \text{ that satisfies } \begin{cases} a_{1,\varepsilon}^\lambda(u, \cdot) & = & \left(\varphi_{1,\varepsilon}^\lambda\right)^* \\ & \cdots & \\ a_{N,\varepsilon}^\lambda(u, \cdot) & = & \left(\varphi_{N,\varepsilon}^\lambda\right)^* \end{cases}$$

Our goal is to address the inverse problem: Given observational data for a solution to (P_ε), estimate λ. Our approach is to use the data in (P) to estimate λ by establishing connections between the parameters λ in (P) and (P_ε) for ε small.

As any function in $H_0^1(\Omega_\varepsilon)$ can be extended to be zero over the holes, it is trivial to prove that $H_0^1(\Omega_\varepsilon)$ can be embedded in $H_0^1(\Omega)$. Let $u = (u_1, ..., u_N)$ and $P_\varepsilon u_k$ be the projection of $u_k \in H_0^1(\Omega_\varepsilon)$ onto $H_0^1(\Omega)$, $k = 1, ..., N$. It is easy to prove that

$$\|u_k - P_\varepsilon u_k\|_{H_0^1(\Omega)} \to 0 \text{ whenever } \varepsilon \to 0.$$

When Neumann boundary conditions are considered, it is still possible to extend a function in $H_0^1(\Omega_\varepsilon)$ to a function of $H_0^1(\Omega)$: these extension conditions are well studied (see [7]) and they typically hold when the domain Ω has a particular structure. In any case, it holds for a wide class of disperse media, that is, media consisting of two media that do not mix.

We assume that there exist three strictly positive constants m, M, and μ such that

$$\begin{aligned} a_k^{\lambda,\epsilon}(u, u) &\geq m\|u\|^2 \ \forall u \in H_\varepsilon \\ a_k^{\lambda,\epsilon}(u, v) &\leq M\|u\|\|v\| \ \forall u, v \in H_\varepsilon \\ \phi_k^{\lambda,\epsilon}(u) &\leq \mu\|u\| \ \forall u \in H_\varepsilon, \end{aligned}$$

Then by the Lax-Milgram type theorem in [4], problem (P) has a unique solution u^λ for each $\lambda \in \Lambda$ and problem (P_ε) has a unique solution u_ε^λ and for each positive ε and each $\lambda \in \Lambda$.

In the following results, we establish relationships between (P) and (P_ε). For each $u \in (H_0^1(\Omega_\varepsilon))^N$, let us introduce the function

$$F_\varepsilon(u, \lambda) = \max_{1\leq k\leq N} \left\| a_k^{\lambda,\varepsilon}(u, \cdot) - \left(y_k^{\lambda,\varepsilon}\right)^* \right\|. \tag{7}$$

associated with problem (P_ε).

Proposition 1. *The function $F(u, \lambda)$ is Lipschitz with Lipschitz constant equal to M.*

Proof. Computing, we have:

$$\begin{aligned}
F(u,\lambda) &= \max_{1\le k\le N} \left\|a_k^\lambda(u,\cdot) - (\varphi_k^\lambda)^*\right\| \\
&= \max_{1\le k\le N} \sup_{\|v\|=1} |a_k^\lambda(u,v) - \varphi_k^\lambda(v)| \\
&= \max_{1\le k\le N} \sup_{\|v\|=1} |a_k^\lambda(u,v) - a_k^\lambda(w,v) + a_k^\lambda(w,v) - \varphi_k^\lambda(v)| \\
&\le \max_{1\le k\le N} \sup_{\|v\|=1} |a_k^\lambda(u-w,v)| + |a_k^\lambda(w,v) - \varphi_k^\lambda(v)| \\
&\le M\|u-w\| + F(w,\lambda)
\end{aligned}$$

which easily implies that

$$|F(u,\lambda) - F(w,\lambda)| \le M\|u-w\|.$$

□

In what follows, for each $u \in (H_0^1(\Omega_\varepsilon))^N$ let us $P_\varepsilon u = (P_\epsilon u_1, \ldots, P_\varepsilon u_N)$.

Proposition 2. *The following inequality holds:*

$$\|P_\varepsilon u - u_\varepsilon^\lambda\| \le \frac{F(u,\lambda)}{m} + \frac{M}{m}\|P_\varepsilon u - u\|$$

Proof. Computing, we have:

$$\begin{aligned}
\|P_\varepsilon u - u_\varepsilon^\lambda\| &\le \tfrac{1}{m} F_\varepsilon(P_\varepsilon u,\lambda) \\
&\le \tfrac{1}{m} F(P_\varepsilon u,\lambda) \\
&\le \tfrac{1}{m}[F(u,\lambda) + M\|P_\varepsilon u - u\|]
\end{aligned}$$

□

Proposition 3. *The exists a constant $C(\varepsilon,u)$, which depends only on ε and u, such that the following inequality holds:*

$$F(P_\varepsilon u,\lambda) \le F_\varepsilon(P_\varepsilon u,\lambda) + C(\varepsilon,u) \sup_{\|v\|=1} \|P_\varepsilon v - v\|$$

Proof. Easy calculations imply that:

$$\begin{aligned}
F(P_\varepsilon u,\lambda) &= \max_{k=1\ldots N} \sup_{\|v\|=1} |a_k^\lambda(P_\varepsilon u,v) - \phi_k^\lambda(v)| \\
&\le \max_{k=1\ldots N} \sup_{\|v\|=1} |a_k^\lambda(P_\varepsilon u,v) - a_k^\lambda(P_\varepsilon u,P_\varepsilon v)| \\
&+ \max_{k=1\ldots N} \sup_{\|v\|=1, v\in H_\varepsilon^N} |a_k^\lambda(P_\varepsilon u,v) - \phi_k^\lambda(v)| \\
&+ \max_{k=1\ldots N} \sup_{\|v\|=1} |\phi_k^\lambda(P_\varepsilon v) - \phi_k^\lambda(v)| \\
&\le \sup_{\|v\|=1} M\|P_\varepsilon u\|\|P_\varepsilon v - v\| \\
&+ F_\varepsilon(P_\varepsilon u,\lambda) + \mu \sup_{\|v\|=1} \|P_\varepsilon v - v\| \\
&\le (M\|P_\varepsilon u\| + \mu) \sup_{\|v\|=1} \|P_\varepsilon v - v\| + F_\varepsilon(P_\varepsilon u,\lambda) \\
&= C(\varepsilon,u) \sup_{\|v\|=1} \|v - P_\varepsilon v\| + F_\varepsilon(P_\varepsilon u,\lambda)
\end{aligned}$$

□

Let us notice that the constant $C(\epsilon,u) = M\|P_\varepsilon u\| + \mu$ converges to $C(u) = M\|u\| + \mu$ whenever ε tends to zero. The following proposition states a convergence theorem for the sequence of minimizers.

IOP Conf. Series: Journal of Physics: Conf. Series **1047** (2018) 012005 doi:10.1088/1742-6596/1047/1/012005

Proposition 4. *Let us suppose that, for each fixed* $u \in (H_0^1(\Omega_\varepsilon))^N$, *F is lower continuous w.r.t.* $\lambda \in \Lambda$. *If* $\lambda_{\varepsilon_n} = \arg\min_{\lambda\in\Lambda} F_{\varepsilon_n}(P_{\varepsilon_n}u, \lambda)$, *and* $\lambda_{\varepsilon_n} \to \lambda^* \in \Lambda$ *then* $\lambda^* = \arg\min_{\lambda\in\Lambda} F(u, \lambda)$.

Proof. Computing, we have:

$$\begin{aligned}
F(u,\lambda^*) &\leq \liminf_{\varepsilon_n\to 0} F(u,\lambda_{\varepsilon_n}) \\
&\leq \liminf_{\varepsilon_n\to 0} F(P_{\varepsilon_n}u,\lambda_{\varepsilon_n}) + M\|P_{\varepsilon_n}u-u\| \\
&\leq \liminf_{\varepsilon_n\to 0} F_{\varepsilon_n}(P_{\varepsilon_n}u,\lambda_{\varepsilon_n}) + C(\varepsilon_n,u)\sup_{\|v\|=1}\|P_{\varepsilon_n}v-v\| + M\|P_{\varepsilon_n}u-u\| \\
&\leq \liminf_{\varepsilon_n\to 0} F_{\varepsilon_n}(P_{\varepsilon_n}u,\lambda_{\varepsilon_n}) + \tilde{C}(\varepsilon_n,u)\varepsilon_n \\
&\leq \liminf_{\varepsilon_n\to 0} F_{\varepsilon_n}(P_{\varepsilon_n}u,\lambda) + \tilde{C}(\varepsilon_n,u)\varepsilon_n \\
&\leq \liminf_{\varepsilon_n\to 0} F(P_{\varepsilon_n}u,\lambda) + \tilde{C}(\varepsilon_n,u)\varepsilon_n \\
&\leq \liminf_{\varepsilon_n\to 0} F(u,\lambda) + \tilde{C}(\varepsilon_n,u)\varepsilon_n + M\|P_{\varepsilon_n}u-u\| = F(u,\lambda)
\end{aligned}$$

for all $\lambda \in \Lambda$, which implies that λ^* is a global minimizer. □

4. Examples

We consider the problem

$$-\nabla\cdot(K_1(x,y)\nabla u_1) + b_1u_2 = f_1(x,y) \tag{8}$$
$$-\nabla\cdot(K_2(x,y)\nabla u_2) + b_2u_1 = f_2(x,y) \tag{9}$$

on Ω_ε, the perforated unit square, with N arbitrarily placed holes, and with $u_k = 0$ on the boundary of the square and $\frac{\partial u_k}{\partial n} = 0$ on the boundaries of the holes, $k = 1, 2$.

We consider the following inverse problem: given a target solution $u = (u_1, u_2)$, f_1, and f_2 on Ω_ε, estimate K_k and b_k, $k = 1, 2$.

Following Proposition 4, we solve the associated inverse problem on $\Omega = [0, 1]^2$, the unit square with no holes. We minimize the objective function $F(u, \lambda)$, where λ is the vector of parameters defining K_i and b_i. The functions a_k^λ and ϕ_k^λ in F are straightforward to construct: multiply (8) by $v_1 \in H_0^1(\Omega)$ and (8) by $v_2 \in H_0^1(\Omega)$ and integrate over Ω. Using Green's Identity on the left, we obtain

$$a_k^\lambda(u, v_k) = \phi_k^\lambda(v_k),\ k = 1, 2,$$

where

$$\begin{aligned}
a_k^\lambda(u,v_k) &= \int_\Omega K_k\nabla u_k\cdot\nabla v_k\,dA \\
\phi_k^\lambda(v_k) &= \int_\Omega f_k v_k\,dA.
\end{aligned}$$

To finite dimensionalize the problem, we introduce n^2 nodes, positioned at the points $(x_i, y_i) = \left(\frac{i}{n+1}, \frac{j}{n+1}\right)$, $i, j = 1, 2, \ldots, n$, and we define the triangles

$$D_n^{ij^+} = \left\{(x,y)\ :\ x_{i-1}\leq x\leq x_i,\ \frac{y_i-y_{i-1}}{x_i-x_{i-1}}x + y_{i-1}\leq y\leq y_i\right\},\ i,j = 1,\ldots,n+1,$$

and

$$D_n^{ij^-} = \left\{(x,y)\ :\ x_{i-1}\leq x\leq x_i,\ y_{i-1}\leq y\leq \frac{y_i-y_{i-1}}{x_i-x_{i-1}}x + y_{i-1}\right\},\ i,j = 1,\ldots,n+1.$$

IOP Conf. Series: Journal of Physics: Conf. Series **1047** (2018) 012005 doi:10.1088/1742-6596/1047/1/012005

We will work with functions v_k in

$$V_n^1 = \{v \in C(\Omega) \,:\, v \text{ is piecewise linear on } D_n^{ij^+} \text{ and } D_n^{ij^-},\ i,j = 1,\ldots,n+1, \text{ and } v = 0 \text{ on } \partial\Omega\}.$$

A basis for V_n^1 consists of the n^2 functions $e_{ij}(x,y)$ satisfying

$$e_{ij} \text{ is piecewise linear and } e_{ij}(x_k, y_l) = \begin{cases} 1 & \text{if } (k,l) = (i,j) \\ 0 & \text{if } (k,l) \neq (i,j) \end{cases},\ i,j = 1,\ldots,n.$$

Similar to [5] and [4], the problem we solve is

$$\min_{\lambda} \sum_{k=1}^{2} \sum_{i,j=1}^{n} \left(a_k^{\lambda}(u, e_{ij}) - \phi_k^{\lambda}(e_{ij})\right)^2. \tag{10}$$

In the examples that follow, we choose values for K_i, f_i, and b_i, $i = 1,2$, solve the problem on the perforated domain, and then sample the numerical solutions at the n^2 nodes to produce observational data. If a node lies inside one of the holes, we discard it. We produce a target solution $u = (u_1, u_2)$ by performing a least squares fit of the polynomial function (of degree $L + 4$)

$$w(x,y) = xy(1-x)(1-y) \sum_{j=0}^{L} \sum_{i=0}^{k} a_{ij} x^i y^{j-i},$$

which satisfies the homogenous Dirichlet boundary conditions on $\partial\Omega$, to each component. We then solve the inverse problem, as outlined above, to obtain estimates of K_i and b_i.

Example 1: We set $K_1(x,y) = 1 + 2x + \frac{1}{2}y$, $K_2(x,y) = 1 + \frac{1}{4}x + 2y$, $b_1 = 2$, $b_2 = 1$, $f_1(x,y) = x^2 + y^2$, and $f_2(x,y) = x + 2y$. We arbitrarily distribute 10 holes of various sizes inside Ω. The domain Ω_ε and the level curves of the solution are presented in Figure 1. In the figure, the holes are numbered by increasing

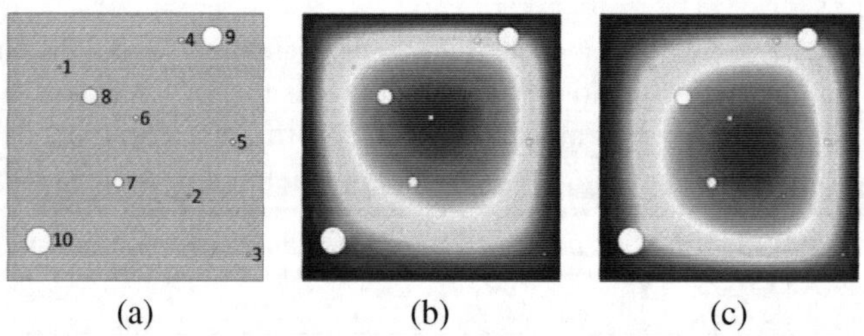

Figure 1. (a) The unit square with 10 assorted holes. (b-c) Isotherms for numerical solutions u_1 and u_2, respectively.

radius. In solving the problem, we set $n = 9$, $L = 8$,

$$K_1(x,y) = \lambda_1 + \lambda_2 x + \lambda_3 y,\ K_1(x,y) = \lambda_4 + \lambda_5 x + \lambda_6 y,\ b_1 = \lambda_7, \text{ and } b_2 = \lambda_8.$$

Table 1 presents the results for 11 different choices of domain: we solve the problem with N holes, $N = 0,\ldots,10$, each time using the first N holes, as numbered in Figure 1(a). It is instructive to visualize the solutions as the number of holes changes; in Figure 2, we present the isotherms for several cases. When comparing rows, it is important to note that the numerical solutions, and hence the target solutions, change for each run. We see that most of the cases, the parameter values are recovered well. As the number of holes grows, which corresponds to the larger holes being part of the problem, the recovered values of the adsorption coefficients b_1 and b_2 become poorer.

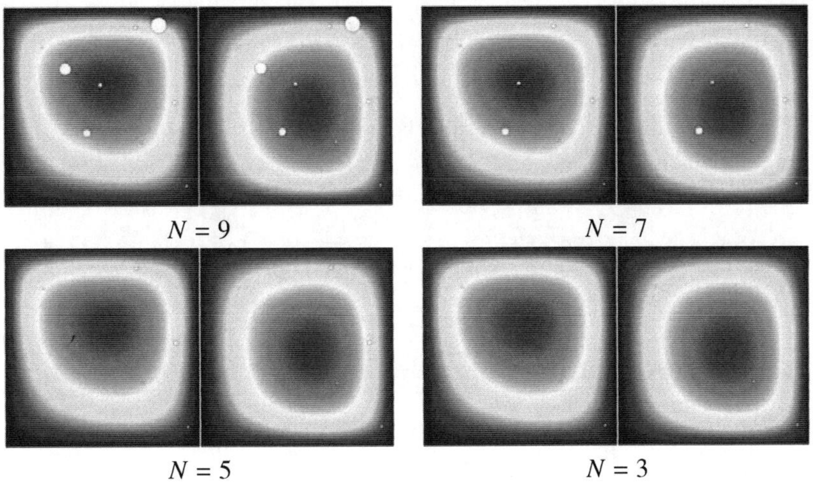

Figure 2. Isotherms for numerical solutions on the domain with N holes.

# holes	λ_1	λ_2	λ_3	λ_4	λ_5	λ_6	λ_7	λ_8
10	0.9816	1.9292	0.4962	0.8718	0.3178	1.8689	2.4295	5.1526
9	1.0600	1.8997	0.4332	0.9746	0.2316	1.8578	2.1646	3.8523
8	1.0064	1.9819	0.5075	0.9927	0.2374	1.9967	2.0416	1.6407
7	1.0044	1.9972	0.4968	1.0178	0.2305	2.0013	1.9665	0.7470
6	1.0035	1.9995	0.4979	1.0089	0.2400	2.0039	1.9366	0.7560
5	1.0071	2.0002	0.4946	1.0099	0.2407	2.0035	1.8871	0.6711
4	0.9953	1.9999	0.5040	1.0041	0.2439	2.0012	2.0069	0.8857
3	0.9882	2.0055	0.5111	1.0006	0.2472	2.0085	2.0375	0.8906
2	0.9958	2.0067	0.5040	1.0029	0.2477	2.0075	1.9835	0.8333
1	0.9951	2.0060	0.5052	1.0034	0.2465	2.0083	1.9861	0.8213
0	0.9907	2.0084	0.5065	0.9985	0.2508	2.0056	2.0176	0.9427
true values	1.0000	2.0000	0.5000	1.0000	0.2500	2.0000	2.0000	1.0000

Table 1. Recovered parameter values, to 4 decimal places, for Example 1.

Example 2: We use the same true parameter values and position 10 circular holes in Ω, as in Example 1. This time, however, all of the holes have the same radius. As in Example 1, we set $n = 9$, $L = 8$, and recover λ_i, $i = 1, \ldots, 8$. Table 2 presents the results for 4 different radius values. As the radius decreases, the accuracy of the recovery improves.

radius	λ_1	λ_2	λ_3	λ_4	λ_5	λ_6	λ_7	λ_8
0.02	1.0403	1.9089	0.5002	1.0571	0.1616	1.9873	1.9242	0.5144
0.01	1.0124	1.9832	0.4972	1.0198	0.2241	2.0028	1.9278	0.6748
0.005	0.9961	2.0024	0.5045	1.0049	0.2433	2.0061	1.9906	0.8389
0.001	0.9908	2.0079	0.5065	0.9987	0.2505	2.0055	2.0195	0.9435
true values	1.0000	2.0000	0.5000	1.0000	0.2500	2.0000	2.0000	1.0000

Table 2. Recovered parameter values, to 4 decimal places, for Example 2.

Figure 3 shows the numerical solutions' isotherms for the different cases.

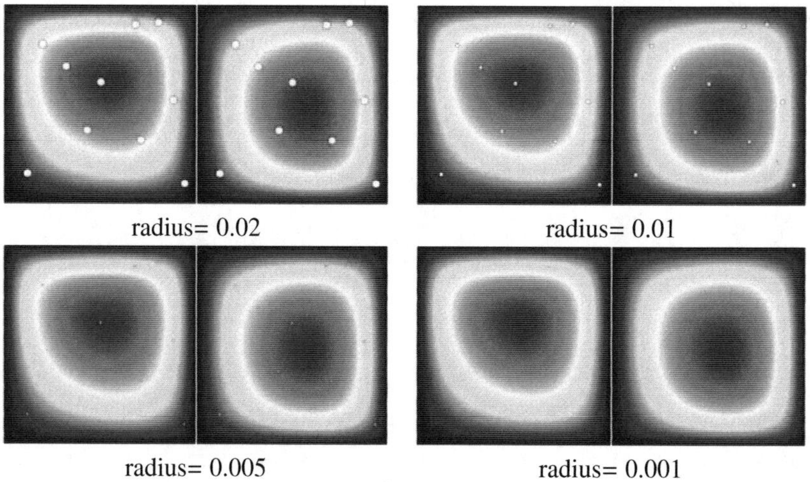

Figure 3. Isotherms for numerical solutions with different radius.

References

[1] Engl H W and Grever W 1994 Using the L-curve for determining optimal regularization parameters *Numer Math* **69** 1 pp 25-31

[2] Kunze H, Hicken J and Vrscay E R 2004 Inverse Problems for ODEs Using Contraction Maps: Suboptimality of the "Collage Method" *Inverse Problems* 20 pp 977-991

[3] Kunze K and La Torre D 2015 Collage-type approach to inverse problems for elliptic PDEs on perforated domains *Electronic Journal of Differential Equations* 48 pp 1-11

[4] Kunze H, La Torre D, Levere K, and Ruiz Gálan M 2015 Inverse problems via the "generalized collage theorem" for vector-valued Lax-Milgram-based variational problems *Mathematical Problems in Engineering* 764643

[5] Kunze H, La Torre D and Vrscay E R 2009 A generalized collage method based upon the Lax-Milgram functional for solving boundary value inverse problems *Nonlinear Analysis* **71** 12 pp 1337-1343

[6] Kunze H and Vrscay E R 1999 Solving inverse problems for ordinary differential equations using the Picard contraction mapping *Inverse Problems* 15 pp 745-770

[7] Marchenko V A and Khruslov E Y 2006 Homogenization of Partial Differential Equations (Boston: Birkhauser)

IOP Conf. Series: Journal of Physics: Conf. Series **1047** (2018) 012006 doi:10.1088/1742-6596/1047/1/012006

Indirect measurement of temperature inside a furnace, ARX model identification

T Loussouarn[1,2], D Maillet[1], B Remy[1], V Schick[1] and D Dan[2]

[1] LEMTA, Universite de Lorraine and CNRS, Vandoeuvre-Les-Nancy, 54504, France
[2] Safran Aircraft Engines, Evry, 91019, France

E-mail: thomas.loussouarn@univ-lorraine.fr

Abstract. Thermal processing of ceramic parts at high temperature can be implemented in induction furnaces under vacuum conditions. In production, there is no direct access to these temperatures because of the high temperature level (higher than $1000°C$) and also because the load cannot be equipped with temperature sensors. This paper deals with the indirect estimation of load temperatures. The estimation has to be made using transient point temperature measurements in the heating element. Indirect measurement of its temperature requires the calibration of a reference model that relates load temperatures to temperature in the susceptor first. A simplified reference furnace model, based on experimentations and thermal properties of its components, is derived. It is the reference model used to test the capacity of an identified AutoRegressive with eXogeneous variable model (ARX) to reproduce its output. This paper underlines the benefits of using ARX models instead of an analytical model as well as their close connection with convolutive products.

1. Introduction

Thermal processing of ceramic parts at high temperature can be implemented in induction furnaces under vacuum conditions. The furnaces used for this process are equipped with few sensors. The only information required for the control of the furnace during a manufacturing cycle are the control parameters. In our simplified configuration, the furnace has only one heating element. It is possible to study the complete system, from electrical power to temperature in the load and in the heating element.

The aim of this article is to get a reduced model of an induction furnace working under vacuum conditions, starting from the electrical power input and temperature outputs of the simulations that can be given by a detailed model, a Finite Element (FE) code. We have already studied the system with both the detailed Finite Element model and a 0D physical reduced model (white box method). This work has been published in a previous article [1]. The present article is the continuation, as we found that the 0D model was biased, because it could not replicate the outputs of the FE detailed model when one tried to estimate its parameters in a non-linear least square procedure, with some of their estimates taking non-physical values.

So, we have decided to investigate the capacity of ARX models to reproduce the outputs of a reference model that is not the FE model anymore but a 0.5 D model, whose structure will be given further on. The 0.5 D model is simpler than the FE model and it is also more pertinent than the 0D model in the heat transfer configuration of the studied furnace. The literature about physical detailed or reduced modeling of such furnaces is scarce. One can find a description of a detailed model of a vacuum furnace in [2]. 0D and 1D modeling of a furnace, including radiation, is presented in [3], while a focus of different identification methods for a furnace equipped with burners can be found in [4].

ARX models have been used for very different applications. They have been used to predict temperatures of a slab in a slab reheating furnace [5], temperatures in aircraft powerplant systems[6] or temperatures in a building [7].

ARX (reduced) models have many advantages. We show in this article that ARX models (which are of the grey box type) are related to convolutive models, which have a structure that corresponds to the forced solution of a linear partial differential equation mathematical system with time independent coefficients. Convolutive models represent here transient heat diffusion in each solid subsystem of the furnace and linearized radiation between its surfaces (see: [8]). These convolutive models are based on transfer functions of the corresponding system and are definitely of the white box type since they characterize its forced response, even if explicit analytical forms are not available if the geometry is not 1D. Convolutive models can also be used in experimental inverse input problems [9].

ARX models have also other nice properties: they can model some non-zero initial conditions (relaxation of a non steady temperature field, before initial time here) and they can also accommodate some slight non-linearities. ARX models are useful also for the finality of our project because it also takes in account the bias of experimental measurements as they will be calibrated in production conditions to be used later on in production.

2. Presentation of the furnace

2.1. The furnace configuration

In order to understand the global behaviour of the system, a simplified configuration has been chosen, see figure 1. The furnace operates under vacuum conditions and it has only one inductor. We assume in the whole study that the surface temperature of the inductor is constant and equal to 473 K. The heating element, also known as a susceptor, is heated up by induction phenomena. Thermal radiation exchanges occur between insulator and inductor. Thermocouples are not represented in the figure. They will be considered as measurement points inside the susceptor here. In the heating module, the base of the cylindrical load, a mould here, rests on a cooling copper base, which is called the hearth. The surface temperature of the hearth is 293 K. Thermal contact between load and hearth is modeled by a low exchange coefficient h_h corresponding to poor contact conditions. The load is heated up by radiation from the susceptor. A measurement campaign has been made. It has shown that radiation between load and susceptor can be linearized for the operating conditions considered. The upper and lower closings of the heating chamber are considered as perfect insulators. The system has a rotational symmetry relative to the vertical axis. That is why it will be represented in 2D axisymmetric coordinates here.

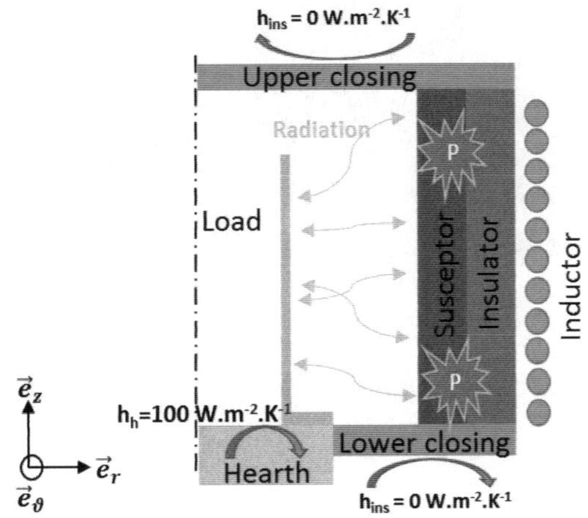

Figure 1. Heating module configuration studied with the load inserted

2.2. Simplified modeling of heat transfer in the heating module

The present modeling corresponds to the system made up of the load and of the heating element, see a schematic representation in figure 2. The susceptor is considered as a 0D lumped body at temperature $T(t)$ because of a low Biot number (in a first approximation $Bi \approx 0.014$ and a previous detailed modeling has shown that the susceptor has a uniform temperature. The susceptor is heated up by a volumetric power density (ηP) where P is the electrical power consumed by the device and η is the induction yield. It exchanges heat with the load through radiation transfer that is linearized here to be written as $(hS_{sus}(T(t) - \tilde{T}_m(t)))$ with h the linearized radiative exchange coefficient, S_{sus} the cylindrical inner surface of the susceptor in front of the load at a spatial mean temperature $\tilde{T}_m(t)$. It losses heat towards the outside $(\frac{1}{R_{out}}(T(t) - T_{out}))$, where R_{out} represents the conductive resistance of the insulator and the linearized radiation to the outside environment at a temperature T_{out}. The temperature of the susceptor T is driven by the heat equation (1) :

$$\rho CV \frac{dT}{dt}(t) = \eta P - hS_{sus}(T(t) - \tilde{T}_m(t)) - \frac{1}{R_{out}}(T(t) - T_{out}) \quad (1)$$

with ρ the specific mass, C the specific heat and V the volume of the susceptor. $\tilde{T}_m$, the spatial mean temperature of the load has been defined as:

$$\tilde{T}_m(t) = \frac{1}{H} \int_0^H T_m(t, z)\, dz \quad (2)$$

The load is considered as a thermal fin. We make the assumption that there is no exchange through the inner surface because of the symmetry of the system. There is also no exchange through the upper surface of the load because of the small fin cross-section. The temperature in the load is uniform at each elevation (fin approximation), so the temperature field in the load is defined by $T_m(t, z)$. Only linearized radiative transfer with the susceptor and thermal contact with the cooled hearth are considered.The heat equation of the load (3) is therefore:

$$\frac{\partial^2 T_m(t, z)}{\partial z^2} + \frac{hm}{\lambda_m S_m}(T(t) - T_m(t, z)) = \frac{1}{a_m} \frac{\partial T_m(t, z)}{\partial t} \quad (3)$$

with λ_m the thermal conductivity and a_m the thermal diffusivity of the mould, and

$$m = 2\pi r_{out} \text{ and } S_m = \pi(r_{out}^2 - r_{in}^2)$$

and the boundary conditions are

$$\lambda_m \frac{\partial T_m}{\partial z}(t, z = 0) = h_h(T_m(t, z = 0) - T_h) \quad (4a)$$

$$\frac{\partial T_m}{\partial z}(t, z = H) = 0 \quad (4b)$$

The initial temperature in the load is not uniform:

$$T_m(t = 0, z) = T_{m,0}(z) \quad (5)$$

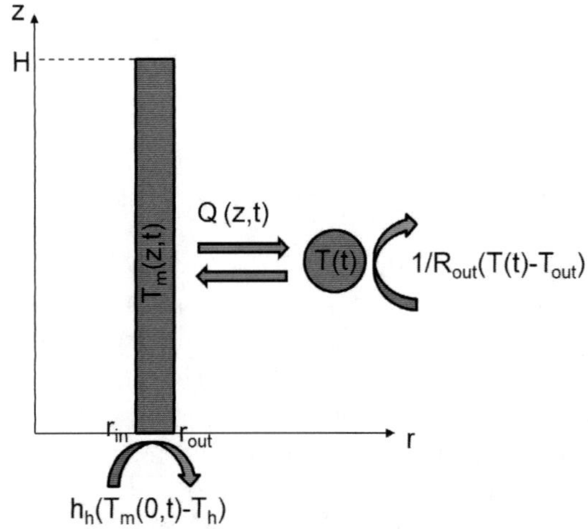

Figure 2. Schematic representation of the modeling of the heating module where $H = 0.276m$, $r_{in} = 0.03m$ and $r_{out} = 0.037m$. $T(t)$ is the temperature of the susceptor considered as a 0D solid, $T_m(t, z)$ is the temperature field in the load, considered as a 1D thermal fin, R_{out} is thermal resistance between the susceptor and outside environment at temperature T_{out}. It integrates the resistance of the insulator. h_h is the unit area conductance between the basis of the load and the hearth at temperature T_h. $Q(z, t)$ represents the radiative exchange between load and susceptor that is linearized through a uniform radiative heat transfer coefficient.

Thanks to this modeling, the temperature of the load can be modeled at each vertical location. It allows the construction of virtual sensors that explains the temperature at each height of the load by the temperature of the susceptor.

2.3. Steady state solutions and upgrading of the transient energy equations

2.3.1. Steady state solution The two heat equations have been given above with their boundary conditions and with their respective initial conditions. All the variables are noted with a subscript $\cdot_{ss}$ for the steady state resolution problem. The heat equation (1) of the susceptor becomes:

$$0 = \eta P_{ss} - hS_{sus}(T_{ss} - \tilde{T}_{m,ss}) - \frac{1}{R_{out}}(T_{ss} - T_{out}) \quad (6)$$

and the heat equation for the load (3) becomes:

$$\frac{d^2T_{m,ss}(z)}{dz^2} + \frac{hm}{\lambda_m S_m}(T_{ss} - T_{m,ss}(z)) = 0 \tag{7}$$

So, the steady state system can be solved with the help of boundary conditions (4a) and (4b) and the temperature ($T_{m,ss}$, $\tilde{T}_{m,ss}$ and T_{ss}) solutions can be found. Finally, the temperature of the load is:

$$T_{m,ss}(z) = T_{ss} - h_h(T_{ss} - T_h)\frac{\cosh(\alpha(H-z))}{h_h\cosh(\alpha H) + \lambda_m\alpha\sinh(\alpha H)} \tag{8}$$

with

$$\alpha = \sqrt{\frac{hm}{\lambda_m S_m}}$$

The expression of the mean spatial temperature of the load is:

$$\tilde{T}_{m,ss} = T_{ss} - \frac{1}{H\alpha}(T_{ss} - T_h)\frac{\tanh(\alpha H)}{1 + \frac{\lambda_m\alpha}{h_h}\tanh(\alpha H)} \tag{9}$$

And finally, we get the value of the steady state temperature of the susceptor :

$$T_{ss} = \frac{\eta P_{ss} + \frac{T_{out}}{R_{out}} + \frac{hS_m}{H\alpha}\frac{\tanh(\alpha H)}{1+\frac{\lambda_m\alpha}{h_h}\tanh(\alpha H)}T_h}{\frac{1}{R_{out}} + \frac{hS_m}{H\alpha}\frac{\tanh(\alpha H)}{1+\frac{\lambda_m\alpha}{h_h}\tanh(\alpha H)}} \tag{10}$$

2.3.2. Rewriting of the transient equations The system is considered to be in steady state at initial time. So, the initial state equations can be subtracted from the transient equations. Our objective is getting variables with a zero value at initial time. For this purpose, we subtract equation (6) from equation (1):

$$\rho CV\frac{d\theta}{dt}(t) = \eta P' - hS_{sus}(\theta(t) - \tilde{\theta}_m(t)) - \frac{1}{R_{out}}\theta(t) \tag{11}$$

with

$\theta(t) = T(t) - T_{ss}$, $\tilde{\theta}_m(t) = \tilde{T}_m(t) - \tilde{T}_{m,ss}$, $\theta_m(z,t) = T_m(z,t) - T_{m,ss}(z)$ and $P'(t) = P(t) - P_{ss}$
The heat equation of the load becomes

$$\frac{\partial^2\theta_m(z,t)}{\partial z^2} + \frac{hm}{\lambda_m S_m}(\theta(t) - \theta_m(z,t)) = \frac{1}{a_m}\frac{\partial\theta_m(z,t)}{\partial t} \tag{12}$$

and the associated boundary conditions are

$$\lambda_m\frac{\partial\theta_m}{\partial z}(z=0,t) = h_h\theta_m(z=0,t) \tag{13a}$$

$$\frac{\partial\theta_m}{\partial z}(z=H,t) = 0 \tag{13b}$$

The two heat equations are solved through integral transformations in the Laplace domain. Equation (11) is transformed in the Laplace domain as

$$(\rho CVs + hS_{sus} + \frac{1}{R_{out}})\overline{\theta} = \eta\overline{P'} + hS_{sus}\overline{\tilde{\theta}_m} \tag{14}$$

where the upper bar notation designates the Laplace transform of the corresponding quantity and where s is the Laplace parameter. Equation (12) is transformed in Laplace domain as

$$\frac{\partial^2\overline{\theta_m}}{\partial z^2}(z,s) - (\frac{s}{a_m} + \frac{hm}{\lambda_m S_m})\overline{\theta_m}(z,s) + \frac{hm}{\lambda_m S_m}\overline{\theta}(s) = 0 \tag{15}$$

IOP Conf. Series: Journal of Physics: Conf. Series **1047** (2018) 012006 doi:10.1088/1742-6596/1047/1/012006

2.4. Output from the physical model

In the Laplace domain, the relative load temperatures $(\overline{\theta_m}, \overline{\tilde{\theta}_m})$ and relative susceptor temperature $(\overline{\theta})$ are function of the relative input $\overline{P'}$. The heat equation of the load (15) is solved through the thermal quadrupoles method [10]. The temperature of the load is then expressed as a function of the susceptor temperature with a transfer function W_z called a transmissivity here:

$$\overline{\theta_m}(z,s) = \overline{W_z}(s)\overline{\theta}(s) \tag{16}$$

where

$$\overline{W_z}(s) = \frac{hm}{\lambda_m S_m \gamma^2}[1 - \frac{h_h \cosh(\gamma(H-z))}{h_h \cosh(\alpha H) + \lambda_m \gamma \sinh(\gamma H)}]$$

and

$$\gamma = \sqrt{\frac{s}{a_m} + \frac{hm}{\lambda_m S_m}}$$

The mean temperature of the load is obtained by integration of the previous expression of $\overline{W_z}$:

$$\overline{\tilde{\theta}_m}(s) = \overline{\tilde{W}}(s)\overline{\theta}(s) = \frac{hm}{\lambda_m S_m \gamma^2}[1 - \frac{1}{H\gamma}\frac{\tanh(\gamma H)}{1 + \frac{\lambda_m \gamma}{h_h}\tanh(\gamma H)}]\overline{\theta}(s) \tag{17}$$

Finally, equations (14) and (17) are combined to get the expression of the susceptor temperature that is a function of the relative heating power and of impedance $\overline{Z}$

$$\overline{\theta}(s) = \overline{Z}(s)\overline{P'}(s) = \frac{1}{\rho C V s + \frac{1}{R_{out}} + h S_{sus}(1 - \overline{\tilde{W}}(s))}\overline{P'}(s) \tag{18}$$

The load temperature is a function of the relative heating power and of the associated impedance $\overline{Z_m}(z,s)$.

$$\overline{\theta_m}(z,s) = \overline{Z_m}(z,s)\overline{P'} \tag{19}$$

Equations (16) and (18) yield the following relationship between these impedances and the transmissivity:

$$\overline{W_z}(s) = \frac{\overline{Z_m}(z,s)}{\overline{Z}(s)} \tag{20}$$

Our aim is to obtain the impulse responses, that is the original of the transfer functions in the time domain, so we apply an inverse Laplace transformation to the transmissivity $\overline{W_z}(s)$ and to impedances $\overline{Z}(s)$ and $\overline{Z_m}(z,s)$ using a numerical inversion algorithm [11]. Since a product in the Laplace domain corresponds to a convolution product in the time domain, the temperature variation of the susceptor or at each point in the load in the time domain is a convolution product between the corresponding impulse response (a time impedance or a time transmissivity) and the relative power or the variation of temperature of the susceptor :

$$\theta_m(z,t) = W_z(t) * \theta(t), \qquad \theta(t) = Z(t) * P', \qquad \theta_m(z,t) = Z_m(z,t) * P' \qquad \text{(21a,21b,21c)}$$

where $*$ stands for this convolution product. This can be written in continuous time or in discrete time, for $u(t)$ the input, $H(t)$ the impulse response and $y(t)$ the output:

$$y(t) = H(t) * u(t) = \int_0^t H(t-v)\,u(v)\,dv \Rightarrow y_i = y(t_i) = \Delta t \sum_{j=1}^{i} H_{i-j+1}\,u_j \tag{22}$$

with $H_j = H(t_j)$, $u_j = u(t_j)$ and $t_j = j\Delta t$ for $j = 1$ to i.

2.5. ARX modeling

An AutoRegressive with eXogenous input model (ARX) has the structure presented in Eq. (23). The solution $y(i)$ is a combination of the previous outputs, its previous input values and of a white noise $\epsilon(i)$. This type of ARX model, which belongs to what is called a "grey box" model, has been extensively studied by Ljung [12].

$$y(i) = -\sum_{j=1}^{n_a} a_j y(i-j) + \sum_{j=1}^{n_b} b_j e(i-j-n_d) + \epsilon(i) \tag{23}$$

Here $y(i)$ and $e(i)$ are respectively the output and the input at time t_i. The order of this model is defined by the triplet (n_a, n_b, n_d). In this paper, we will not delay the input, so $n_d = 0$ and there will be no stochastic zero mean input so $\epsilon(i) = 0$. The output y and the input e correspond respectively to the load temperature θ_m and to the susceptor temperature θ. Equation (23) can also be written in terms of matrices with $\mathbf{A}$ and $\mathbf{B}$ lower triangular matrices of size (n, n), where n is the number of observations, as

$$\mathbf{Ay} = \mathbf{Be} \tag{24}$$

where

$$\mathbf{A} = \begin{pmatrix} 1 & 0 & \cdots & \cdots & \cdots & \cdots & 0 \\ a_1 & \ddots & & & & & \vdots \\ \vdots & & \ddots & & & & \vdots \\ a_{n_a} & & & 1 & & & \vdots \\ 0 & & & a_1 & \ddots & & \vdots \\ \vdots & & & & & \ddots & \vdots \\ 0 & \cdots & \cdots & a_{n_a} & \cdots & a_1 & 1 \end{pmatrix} \text{ and } \mathbf{B} = \begin{pmatrix} b_1 & 0 & \cdots & \cdots & \cdots & \cdots & 0 \\ \vdots & \ddots & & & & & \vdots \\ b_{n_b} & & \ddots & & & & \vdots \\ 0 & & & \ddots & & & \vdots \\ \vdots & & & & b_1 & & \vdots \\ \vdots & & & & \vdots & \ddots & \vdots \\ 0 & \cdots & \cdots & \cdots & b_{n_b} & \cdots & b_1 \end{pmatrix}$$

with $\mathbf{y} = [y_1\, y_2 \cdots y_n]^T$ and $\mathbf{e} = [e_1\, e_2 \cdots e_n]^T$.

The parameters a_i and b_i have to be estimated . We need two different input/output sets. The first set is called the calibration set and the second one the validation set. The calibration couple is used to estimate a_i and b_i as they are the unknowns of the least-squares estimation problem of system (23). These are estimated with a QR factorization in Matlab®. After this calibration, the estimated parameters and the corresponding $(n_a, n_b, 0)$ ARX structure are validated with the validation set. The model expressed in equation (24), can be written in the following form

$$\mathbf{y} = \mathbf{Ce} = \mathbf{A}^{-1}\mathbf{Be} \tag{25}$$

We can notice that if n_a is null and $n_b = n$ then the ARX model is similar to a convolutive model presented in section 2.4.

Once the ARX model and the analytic physical model defined, we are now going to see the behaviour of both of them and their ability to fit each other.

2.6. Comparison of the double convolutive analytical model with the ARX model

Using the results obtained in part 2.4, equations (21a) and (21b) can be combined to eliminate $P'(t)$. They can be written in terms of lower triangular Toeplitz matrices:

$$\mathbf{M}(Z(t))\boldsymbol{\theta_m}(z,t) = \mathbf{M}(Z_m(z,t))\boldsymbol{\theta}(t) \tag{26}$$

where for example $\mathbf{M}(Z(t))$ is equal to :

$$\mathbf{M}(Z(t)) = \begin{pmatrix} Z_1 & 0 & \cdots & 0 \\ Z_2 & \ddots & & \vdots \\ \vdots & & \ddots & 0 \\ Z_n & \cdots & \cdots & Z_1 \end{pmatrix} \tag{27}$$

From equation (26), the expression of the load temperature for each time t_i (for $i = 2$ to n) can be written as a function of the previous outputs and the previous input values including the current one:

$$\theta_m(z, i) = -\sum_{j=1}^{i-1} a_j \theta_m(z, i-j) + \sum_{j=0}^{i-1} b_j \theta(i-j) \tag{28}$$

In fact, we can define a_j and b_j coefficients as

$$a_j = \frac{Z_{j+1}}{Z_1} \text{ for } (j = 1 \text{ to } (i-1)) \text{ and } b_j = \frac{Z_{m,j+1}}{Z_1} \ (j = 0 \text{ to } (i-1))$$

The analogy with the ARX definition (23) is therefore clear, if b_0 is null. Let us notice that if n observations of θ_m and θ are available, a higher number $(2n-1)$ of coefficients a or b are present in equation (23). So it is impossible to estimate all of them. The maximum number of parameters that can be identified is $n_a + n_b = n$.

3. Results for one heating power input

3.1. Outputs from the analytic model

Here, the simulation starts from an equilibrium state at 1173 K for the susceptor and the load has a non uniform temperature. The temperature distribution in the load is presented in figure 3. The gradient is due to the cooled hearth. The gradient is high in the first millimeters because of the low conductivity of the load.

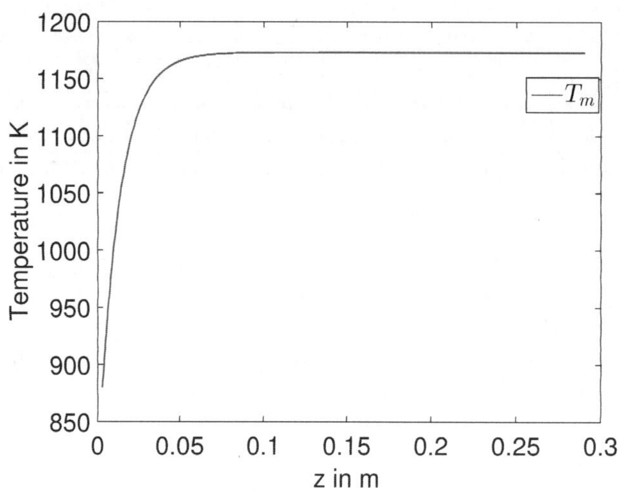

Figure 3. Temperature of the load function of z for a susceptor in a steady state at 1173 K

From this initial equilibrium, we are going to apply two profiles of relative electrical power. They are given in figure 4. The first one is the calibration input, that consists of different steps to get as much information as possible. The second input profile is the validation input. The relative electrical power can be negative because we built the model for $P'(t) = P(t) - P_{init}$. So, if the electrical power is stopped, we get $P'(t) = -P_{init} < 0$.

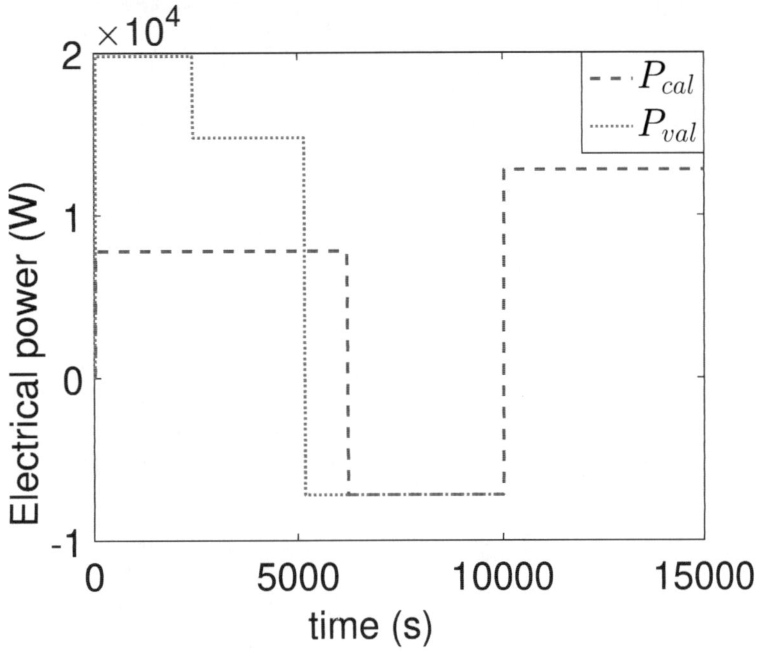

Figure 4. Profiles of electrical power as inputs for model calibration and validation

The analytical temperature responses for these inputs are presented in figure 5. We can clearly see the gradient in the first millimeter of the load, $T_m(0.0029, t) \ll T_m(0.0145, t)$. The load temperature is stable for higher elevations, and $T_m(0.087, t)$ and $T_m(0.276, t)$ almost overlap. We can also give the time profiles of the transmissivity $W_{0.087}(t)\Delta t$ at $z = 0.087m$ and the impedances $Z(t)\Delta t$, $Z_m(0.087, t)\Delta t$, for $\Delta t = 10s$, see figure 6. We can notice that $Z(t)\Delta t$ starts at a non-zero value because the susceptor is a lumped body and it reacts instantly to the heating power. On the contrary $Z_m(0.087, t)\Delta t$ starts at a zero value because the load has a non-null time response.

From the definition of a_j and b_j coefficients in part 2.6 and the value of the impedances, the a_j and b_j coefficients are estimated and presented in figure 7. Their values result from the ordinary least square estimate of the following model:

$$\boldsymbol{y} = \boldsymbol{D}(\boldsymbol{y})\boldsymbol{\beta} \quad \text{where} \quad \boldsymbol{\beta} = \begin{bmatrix} a_1 \\ \vdots \\ a_{n_a} \\ b_1 \\ \vdots \\ b_{n_b} \end{bmatrix} \quad \text{and} \quad \boldsymbol{y} = \begin{bmatrix} y_1 \\ \vdots \\ \vdots \\ \vdots \\ y_n \end{bmatrix} \tag{29}$$

where n observations y_i have been made for the output and where sensitivity matrix $\boldsymbol{D}$ can be derived from (23) and depends on $\boldsymbol{y}$.

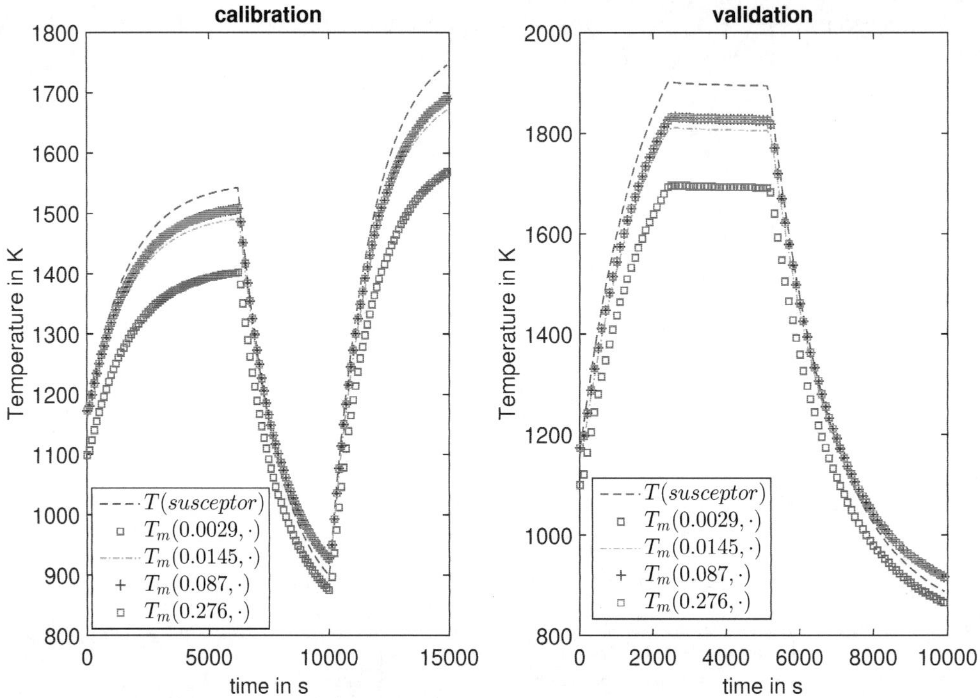

Figure 5. Time profiles of susceptor temperature and load temperatures at 4 levels of the load ($z = 0.0029m$, $z = 0.00145m$, $z = 0.087m$ and $z = 0.276m$) for calibration and validation phases

3.2. ARX results

3.2.1. Study without noise Different orders of ARX are studied in this section. We focus here on autoregressive models with $n_a = n_b$ and convolutive models with $n_a = 0$. The identification results, corresponding to inversion of equation (29) are presented in table 1 for the calibration and validation phases in terms of the square mean root estimation error (E_{RMS}) with

$$E_{RMS} = \|\boldsymbol{\theta_m}(z,t) - \boldsymbol{y}\| = \sqrt{\frac{1}{n}\sum_{j=1}^{n}(\theta_m(z,j) - y(j))^2} \quad (30)$$

where $\boldsymbol{y}$ is the output from the ARX model. This ARX model provides excellent fits even for small orders like (1,1,0). Table 1 shows clearly the interest of ARX models since the number of parameters to estimate is much smaller for an autoregressive model than for a convolutive model in order to get an equivalent result in terms of E_{RMS}.

The parameters a_j and b_j values are presented, as a function of the their order j, for different ARX orders $(n_a, n_b, 0)$, in figure 8. For the convolutive problem, the parameters b_j seem to converge to a stable solution when the ARX order increase. In the case of an autoregressive model, the a_j's and b_j's coefficients oscillate and do not seem to vary in a continuous way with their order, in contrast with their values derived from the analytical model shown in figure 7.

In fact, the b_j coefficients of the convolutive model and the matrix coefficients c_j of the $\mathbf{C}$ in equation (25) converge to the value of $W_{0.087}(t)\Delta t$ when their maximum order is increased, see figure 9.

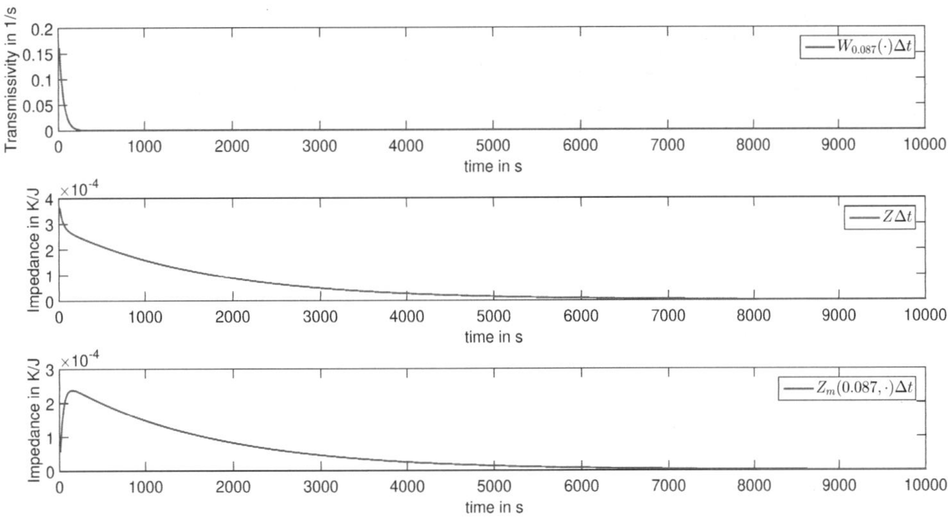

Figure 6. Time profiles of the different transfer functions multiplied by the time step $W_{0.087}(t)\Delta t$, $Z(t)\Delta t$ and $Z_m(0.087, t)\Delta t$

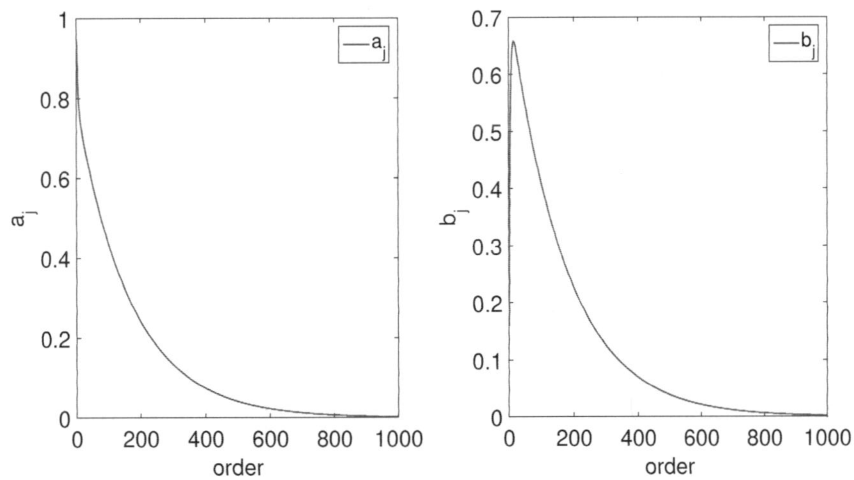

Figure 7. Profiles of a_j and b_j coefficients estimated from the analytical model, equation (28)

3.2.2. Study with noise The addition of a white noise with a 5 K standard deviation, which is the worst case in practice, does not impact the results. In this case we use Truncated Singular Values Decomposition (TSVD) to estimate the $n_a + n_b$ coefficients, if this number is large with, as a consequence, an ill-posed inversion of (29). This is made using Morozov's discrepancy principle. The results for the calibration and the validation phases are presented in figure 10. The E_{RMS} for the calibration phase is 6.48 K and 6.39 K for the validation phase which is equivalent to the standard deviation.

Table 1. Ability of the ARX model to fit the calibration and validation sequences at $z = 0.087m$ in the load

ARX order	Calibration (E_{RMS} in K)	Validation (E_{RMS} in K)
Autoregressive models		
(1,1,0)	$1.73\,10^{-6}$	$2.52\,10^{-6}$
(4,4,0)	$2.38\,10^{-7}$	$3.35\,10^{-7}$
(50,50,0)	$9.14\,10^{-10}$	$1.32\,10^{-9}$
Convolutive models		
(0,2,0)	1.20	1.71
(0,8,0)	0.36	0.52
(0,70,0)	$2.03\,10^{-6}$	$2.88\,10^{-6}$
(0,100,0)	$4.65\,10^{-9}$	$6.64\,10^{-9}$

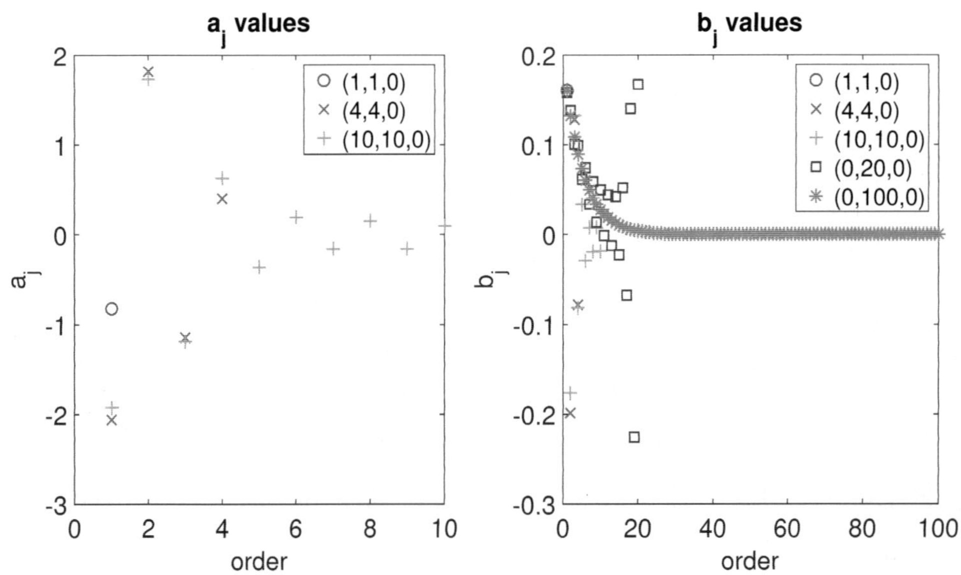

Figure 8. Estimated values of the parameters a_j and b_j function of the order j for different ARX order $(n_a, n_b, 0)$

4. Conclusion and perspectives

The ARX model has shown its capacity to fit the analytical model even with a significant white noise and it has been validated in a synthetic experiment different from the calibration one (with noise). The transfer functions of the analytic model based on a double convolution product (a "white box" model) can be recovered from the ARX coefficients a_j and b_j. Furthermore, the autoregressive models are efficient to represent the analytic model with a fewer number of parameters than the convolutive model.

The next step will be to explain experimental data for a more complex system with several susceptors at different temperatures with the help of ARX models.

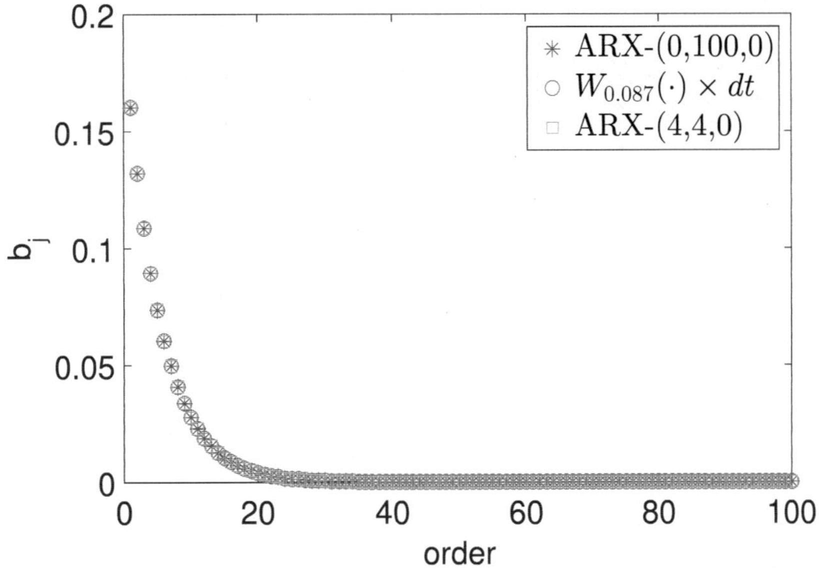

Figure 9. Estimated values of the b_j coefficients for ARX order (0,100,0), estimated values of the c_j coefficients for ARX order (4,4,0) and the transmissivity $W(0.087, j) \times dt$ value function of the occurrence j

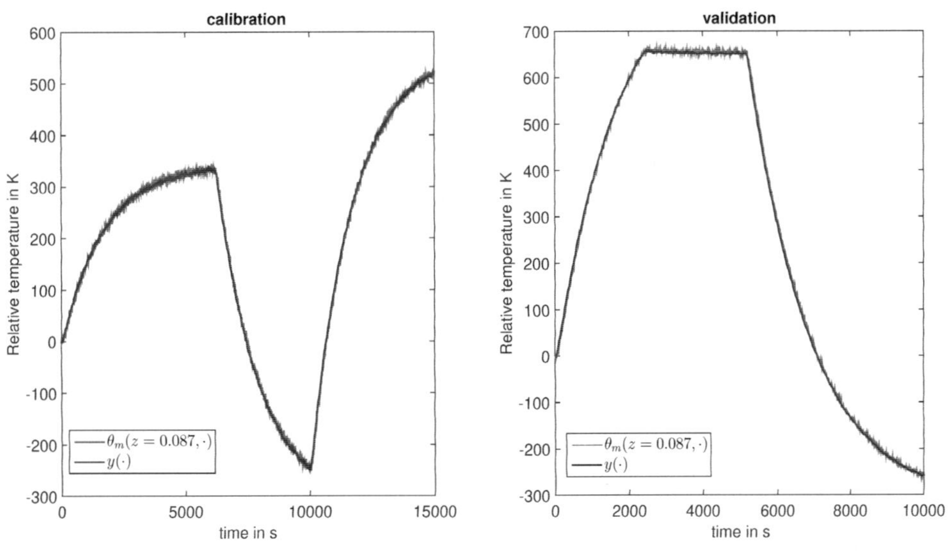

Figure 10. Time Profiles of $\theta_m(z = 0.087, t)$ and corresponding ARX outputs for the calibration and validation phases

References

[1] Loussouarn T, Maillet D, Remy B and Dan D 2017 Model reduction for experimental thermal characterization of a holding furnace Heat and Mass Transfer DOI 10.1007/s00231-017-2156-7.

[2] Bhat A A, Sujish D, Agarwal S, Muralidharan B, Padmankumar G and Rajan KK 2013 Electromagnetic and thermal modeling of vacuum distillation furnace, Proc. COMSOL conference proceedings (Bangalore: Comsol).

[3] Steinboeck A, Wild D, Kiefer T and Kugi A, 2010 A mathematical model of a slab reheating furnace with radiative heat transfer and non-participating gaseous media International Journal of Heat and Mass Transfer 53 pp 5933-46.

[4] Samyudia Y and Sibarani H 2006 Identification of reheat furnace temperature models from closed-loop data - an industrial case study, Asia-Pacific Journal of Chemical Engineering 1 pp 70-81.

[5] van Ditzhuijzen G, Staalman D and Koorn A 2002 Identification and model predictive control of a slab reheating furnacel Proc. IEEE Int. Conf. on Control Applications (Glasgow) vol 1 (New York: IEEE Xplore) pp 361-366.

[6] Uriz Jauregui F, Remy B, Degiovanni A and Verseux O 2013 Model identificaton for temperature extrapolation in aircraft powerplant systems International Journal of Thermal Sciences 64 pp 162-177.

[7] Ros-Moreno G, Trejo-Perea M, Castaeda-Miranda R, Hernndez-Guzmn V and Herrera-Ruiz G 2007 Modelling temperature in intelligent buildings by means of autoregressive models, Automation in Construction 16 pp 713-722.

[8] Al Hadad W, Maillet D and Jannot Y 2017 Modeling unsteady diffusive and advective heat transfer for lineardyna mical systems: a transfer function approach International Journal of Heat and Mass Transfer 115 pp 304-13.

[9] Chata F, Belut E, Maillet D, Keller F X and Taniere A 2017 Estimation of an aerosol source in forced ventilation through prior identification of a convolutive model International Journal of Heat and Mass Transfer 108 pp 623-33.

[10] Maillet D, Andr A, Batsale JC, Degiovanni A and Moyne M, 2000 Thermal quadrupoles: solving the heat equation through integral transforms (Chichester: Wiley)

[11] Hollenbeck K 1998 INVLAP.m: A matlab function for numerical inversion of Laplace trnasforms by the De Hoog algorithm URL http://www.isva.dtu.dk/staff/karl/invlap.htm

[12] Ljung L 1996 System identication: theory for the user, 2nd ed. (Upper Saddle River NJ: Prentice-Hall PTR).

IOP Conf. Series: Journal of Physics: Conf. Series **1047** (2018) 012007 doi:10.1088/1742-6596/1047/1/012007

Shale structure reconstruction by X-ray computerized tomography based on total variation regularization[1]

S S Luo[§] and Y F Wang[†,‡]

[§] School of Mathematics and Statistics, Henan University, Kaifeng 475001, P. R. China
[†] Key Laboratory of Petroleum Resources Research, Institute of Geology and Geophysics, Chinese Academy of Sciences, Beijing 100029, P. R. China
[‡] University of Chinese Academy of Sciences, Beijing 100049, P. R. China

E-mail: `yfwang@mail.iggcas.ac.cn`

Abstract. The microstructure knowledge of shale is the basis of shale gas exploration and production. In this paper, we applied parallel X-ray computerized tomography technology rather than conventional surface-based observation methods to recover the shale structure non-destructively. In order to avoid the X-ray hardening, we used the synchrotron radiation source to acquire the projection data. Total variation regularization method was applied to suppress the noise effect and improve the image contrast, and primary-dual algorithm was used to solve it. Experimental tests show that the proposed approach is a feasible and efficient method to reconstruct the shale microstructure precisely and accurately.

1. Introduction

Shale gas is trapped within the non-connected small shale fractures and is difficult to extract. With the advent of energy crisis and development of hydraulic fracturing technology, it attracted more and more attentions in the past decade. Shale gas is considered as one of the unconventional sources of natural gas. The exploration of shale gas becomes one of the most hot research fields gradually.

In order to successfully recover distribution of shale gas, we should get the microstructure of shale, which plays an important role in the gas exploration and production [1]. Traditional surface observations methods, such as optical and scanning electron microscopy, are inadequate to obtain three-dimensional information of the shale sample, and the shale samples are destructed [2]. Therefore, these methods are not ideal because the shale samples are hard to exploit and precious [3].

X-ray computerized tomography (XCT) technique is a well-known noninvasive method to visualize the interior structure, which has been applied in medicine, biology, industry and science successfully from its invention in the 1970's [4, 5]. Therefore, we attempt to use XCT to reconstruct the shale microstructure nondestructively.

In this paper, the parallel X-ray beams generated by synchrotron radiation (SR) source are used to acquire the projection data. The synchrotron radiation source, which can provide

[1] This paper is in honour of the distinguished Professor Graham Gladwell.

energy-tunable, monochromatized and naturally collimated X-ray beams, was generated at beamline BL14W1 XAFS of the Shanghai Synchrotron Radiation Facility (SSRF), by using NI (National Instrument) LabVIEW's Data logging and Supervisory Control (DSC) module communication with experimental physics and industrial control system (see [6] for details). One of the important reasons for using monochromatized beams is that it can avoid the X-ray hardening caused by polychromatic beams.

Because the 3-dimensional (3D) information can be obtained slice by slice, we only consider the 2-dimensional (2D) XCT image reconstruction problem. From the analytical point of view, the 2D XCT problem is to invert the Radon transform of the X-ray attenuation map $f(\boldsymbol{x})$ of the detected material [5, 7]

$$Rf(s,\theta(\alpha)) = \int_{\mathbb{R}} f(s\theta(\alpha) + t\theta^{\perp}(\alpha))dt, \tag{1}$$

where $\theta(\alpha) = (\cos\alpha, \sin\alpha)^t, \theta^{\perp}(\alpha) = (-\sin\alpha, \cos\alpha)^t \in \mathbb{S}^1$ and $s \in \mathbb{R}$. The f can be reconstructed by filtered back-projection (FBP) formula from 180° projection data

$$f(\boldsymbol{x}) = \frac{1}{2\pi^2}\int_0^{\pi}\int_{\mathbb{R}} \frac{Rf'(s,\theta(\alpha))}{\boldsymbol{x}\cdot\theta(\alpha) - s} ds d\alpha, \tag{2}$$

where the derivative of $Rf(s,\theta(\alpha))$ is about the first variable s.

Another reconstruction technique is algebraic method. Directly approximating the unknown attenuation map f as an $m \times n$ digital image (also denoted by f for simplicity), and vectorizing f by lexicographical order, we can formulate the XCT image reconstruction problem as linear system [5, 7, 8]

$$b = Af + \eta, \tag{3}$$

where $A \in \mathbb{R}^{M\times N}$ is called imaging matrix, $f \in \mathbb{R}^N$ ($N = mn$) and $b \in \mathbb{R}^M$ polluted by noise η are the unknown image and the observed projection data [5, 8]. For the unknown image f, the vector form $f \in \mathbb{R}^N$ and matrix form $f \in \mathbb{R}^{m\times n}$ are not distinguished hereafter, and its meaning is determined in the context.

However, neither FBP nor algebraic method could reconstruct ideal images from polluted or partial data since the XCT reconstruction problem is ill-posed [8, 10]. Regularization methods are the commonly used technique for handling ill-posed problems, and have gradually become the research hotspot in the field of XCT [9, 10, 11, 12]. The regularization methods are usually modeled as the following minimization problem

$$\arg\min_{f\in C} \left\{\lambda\phi(f) + D(Af, b)\right\}, \tag{4}$$

where ϕ is a convex function (called regularization term) and C is a convex set, representing the prior knowledge of f, and $D(\cdot,\cdot)$ is a distance function(called fidelity term), measuring how Af is fit to the projection data b. Here $\lambda > 0$ is called the regularization parameter.

The regularization term and fidelity term are at the core of regularization methods and studied extensively. There are a lot of choices for regularization term, such as total variation (TV)- based [10, 13, 14], sparse representations methods based on wavelet, tight-frame and dictionary basis for XCT image reconstruction [15, 16, 17]. In this paper, we use the total variation (TV) function as the regularization term. On one hand, shales are typically consisted of incompatible components, such as clay mineral and quartz grain. Therefore, we can assume the attenuation map f is a piecewise constant function. On the other hand, it is well known that total variation minimization method [18] forces the image (or function) to be a piecewise constant approximation [19, 20]. Therefore, we use the total variation function as the regularization term for the reconstruction of shale information. As for the fidelity term, we use l^2 norm as the distance function D, i.e. $D(Ax, b) = \frac{1}{2}\|Ax - b\|_2^2$, because the noise in the XCT projection data obeys Gaussian distribution. Furthermore, we impose a nonnegativity constraint on the solution, i.e. $C = \{f \in \mathbb{R}^N |, f_i \geq 0\}$, because the attenuation coefficients are always nonnegative physically. As for the selection of regularization parameter, we determined it by experiments on the simulated projection data with the same noise level as the real projection data.

Efficient algorithm is another key problem for the regularization XCT image reconstruction problem (4). Although several algorithms were developed for total variation minimization problem in the past decade [21, 22, 23, 24] , we use the primary-dual algorithm [23] to solve the proposed model in this paper.

We test the proposed method on stimulated data of Shepp-Logan phantom [26] and the experimental data (see [6] for details). The results show that the TV regularization method can suppress the noise effect and preserve the weak edges, which help us to distinguish different components in the shale sample. Therefore, the XCT technology is an efficient method to obtain the interior information of shale, and the TV-based regularization method is more suitable to the reconstruction of shale microstructure than filtered back-projection method.

The contributions of this paper are as follows: (1) We investigated a XCT-based technique to obtain the structure of shale nondestructively. (2) In the implementation, monocharmatic X-ray are used to avoid the hardening effecting caused by ploymathic X-ray. (3) TV minimization is adopted as the regularization term, because the attenuation map shale can be approximated by piecewise constant function accurately. And we used a strategy to select the proper regularization parameter.

The rest of this paper is organized as follow. In section 2, we illustrate the TV model for XCT image reconstruction and the primal-dual algorithm. Experiments on simulated and practical projection data are presented in section 3. In the last section, some conclusions and discussions are demonstrated.

2. Model and Algorithm

The TV-based model used in this paper is of the form

$$\min_{f \geq 0} \left\{ \lambda \mathrm{TV}(f) + \frac{1}{2}\|Af - b\|_2^2 \right\}, \tag{5}$$

where the user-defined regularization parameter $\lambda > 0$ is related to the noise level of data b. The total variation of a 2D image $f \in \mathbb{R}^{m \times n}$ is computed by

$$\mathrm{TV}(f) = \|(|\nabla f|)\|_1 = \sum_{i,j=1}^{m,n} \sqrt{(\nabla_1 f)_{ij}^2 + (\nabla_2 f)_{ij}^2}, \tag{6}$$

where the gradient ∇f of f is defined as $(\nabla f)_{i,j} = ((\nabla_1 f)_{i,j}, (\nabla_2 f)_{i,j})$ with $(\nabla_1 f)_{i,j} = \begin{cases} f_{i+1,j} - f_{i,j} & i < m \\ 0 & i = m \end{cases}$, $(\nabla_2 f)_{i,j} = \begin{cases} f_{i,j+1} - f_{i,j} & j < n \\ 0 & j = n \end{cases}$.

The selection of regularization parameter is very important for the performance of regularization methods. It is well known that the proper regularization parameter is related to the noise level of projection. We firstly estimate the noise level in the projection data in the domain in which the X-ray does not go through the shale (The ideal value should be zero). Then we determined a proper parameter by experiments of simulated data of Shepp-Logan phantom with the noise level.

Let $G(f) = \begin{cases} 0 & f \geq 0 \\ +\infty & \text{otherwise} \end{cases}$, $F_1(v) = \lambda \sum_{i=1,=1}^{m,n} |v_{ij}| = \lambda \sqrt{(v_{ij}^1)^2 + (v_{ij}^2)^2}$ for $v \in Z = \mathbb{R}^{m\times n} \times \mathbb{R}^{m\times n}$ and $F_2(w) = \frac{1}{2}\|w-b\|_2^2$ for $w \in \mathbb{R}^M$. Then the minimization problem (5) can be rewritten as an unconstrained minimization problem

$$\min_f F_1(\nabla u) + F_2(Af) + G(f). \tag{7}$$

The problem (7) is equivalent to the following saddle point problem [9, 23]

$$\min_f \max_{v,w} \{\langle \nabla f, v\rangle - F_1^*(v) + \langle Af, w\rangle - F_2^*(w) + G(f)\}. \tag{8}$$

Here the conjugate functions F_1^* and F_2^* are defined as

$$F_1^*(v) = \max_{p\in Z}\{\langle v, p\rangle_Z - F_1(p)\}, \tag{9}$$

$$F_2^*(w) = \max_{q\in \mathbb{R}^M}\{\langle w, q\rangle_{\mathbb{R}^M} - F_2(q)\}, \tag{10}$$

where the inner products $\langle p, q\rangle_Z = \sum_{i=1,j=1}^{m,n}(p_{ij}^1 q_{ij}^1 + p_{ij}^2 q_{ij}^2)$ for all $p, q \in Z$, and $\langle x, y\rangle_{\mathbb{R}^M} = \sum_{i=1}^M x_i y_i$ for all $x, y \in \mathbb{R}^M$.

The general frame of the primal-dual iteration procedure for the saddle point problem (8) can be formulated as Algorithm 1 (see [9, 23] for details). In Algorithm 1, A^t is the conjugate transpose operator of A and ∇^t is the conjugate operator of ∇ defined by

$$\langle v, \nabla x\rangle_Z = \langle \nabla^t v, x\rangle_{\mathbb{R}^N}. \tag{11}$$

The proximal mapping $\text{prox}_\sigma[H]$ of a given convex function H is defined as

$$\text{prox}_\sigma[H](v) = \min_p\{H(p) + \frac{1}{2\sigma}\|p - v\|_2^2\}. \tag{12}$$

Algorithm 1 Primal-dual algorithm for (8)

Initialization: $k = 0$, $\theta = 1$, $f^k = \bar{f}^k = \mathbf{0} \in \mathbb{R}^N$, $v^k = \mathbf{0} \in Z$, $y^k = \mathbf{0} \in \mathbb{R}^M$, τ, $\sigma > 0$ and $\tau\sigma \left\| \begin{pmatrix} A \\ \nabla \end{pmatrix} \right\|_2^2 \leq 1.$

for $k = 0, 1, \cdots$, do

$v^{k+1} = \text{prox}_\sigma[F_1^*](v^k + \sigma\nabla \bar{f}^k);$

$y^{k+1} = \text{prox}_\sigma[F_2^*](y^k + \sigma A\bar{f}^k);$

$f^{k+1} = \text{prox}_\tau[G]\left(f^k - \tau(\nabla^t v^{k+1} + A^t y^{k+1})\right);$

$\bar{f}^{k+1} = f^k + \theta(f^{k+1} - f^k);$

In order to use the primal-dual algorithm, one only needs to compute the conjugate functions F_1^*, F_2^* and the proximal mappings $\text{prox}_\sigma[F_1^*], \text{prox}_\sigma[F_2^*]$ and $\text{prox}_\tau[G]$. It has been proved that [9, 23]

$$F_1^*(z) = \begin{cases} 0, & |z_j| \leq \lambda, \ \forall j \\ +\infty, & \text{otherwise} \end{cases}, \tag{13}$$

$$F_2^*(y) = \frac{1}{2}\|y\|_2 + \langle y, b\rangle_{\mathbb{R}^M}. \tag{14}$$

IOP Conf. Series: Journal of Physics: Conf. Series **1047** (2018) 012007 doi:10.1088/1742-6596/1047/1/012007

The proximity operators of F_1^*, F_2^* and G are

$$\text{prox}_\sigma[F_1^*](z) = \frac{\lambda z}{\max\{\lambda, |z|\}}, \tag{15}$$

$$\text{prox}_\sigma[F_2^*](y) = \frac{y - \sigma b}{1+\sigma}, \tag{16}$$

$$\text{prox}_\tau[G](f) = \max\{f, 0\}. \tag{17}$$

Here the divisions in (15) and (16) are componentwise. Therefore, the primal-dual algorithm for the original problem (5) can be illustrated as Algorithm 2.

Algorithm 2 Primary dual algorithm for TV model

Step 1 Initialization: $k = 0,\ \theta = 1,\ f^k = \bar{f}^k = \mathbf{0} \in \mathbb{R}^N$, and $v^k = \mathbf{0} \in Z,\ y^k = \mathbf{0} \in \mathbb{R}^M,\ \tau,\ \sigma >$ 0 and $\tau\sigma \left\| \begin{pmatrix} A \\ \nabla \end{pmatrix} \right\|_2^2 < 1$.

Step 2 Perform the following iterations until convergence:

$$y^{k+1} = \frac{y^k + \sigma(A\bar{f}^k - b)}{1+\sigma}, \tag{18}$$

$$v^{k+1} = \frac{\lambda(v^k + \sigma\nabla\bar{f}^k)}{\max\{\lambda, |v^k + \sigma\nabla\bar{f}^k|\}}, \tag{19}$$

$$f^{k+1} = \max\left\{0, f^k - \tau(\nabla^t v^{k+1} + A^t y^{k+1})\right\}. \tag{20}$$

Step 3 Update: $\bar{f}^{k+1} = f^{k+1} + \theta(f^{k+1} - f^k)$, $k = k+1$, and go to Step 2.

By the definition (11), one can derive the computing formula of ∇^t as follows [25]

$$(\nabla^t v)i,j = -\begin{cases} v^1_{i+1,j} - v^1_{i,j} & 1 < i < m \\ v_{i,j} & i = 1 \\ -v^1_{i-1,j} & i = m \end{cases} - \begin{cases} v^1_{i,j+1} - v^1_{i,j} & 1 < j < n \\ v_{i,j} & j = 1 \\ -v^1_{i,j-1} & j = n \end{cases}. \tag{21}$$

3. Experimental results

In this section, we applied our mathematical model and algorithm on simulated and practical data, and compared the results with that of the traditional FBP method. Beside the visual comparison of the reconstructed image, we also compared the mean square errors (MSEs) for the simulated data. The MSE is defined by

$$\text{MSE}(f) = \sqrt{\frac{1}{m \times n} \sum_{i=1,j=1}^{m,n} \left(f_{ij} - f^0_{ij}\right)^2}, \tag{22}$$

where f^0 and f are the original and estimated images, respectively. In our experiments, for Algorithm 2, we choose the two parameters $\sigma = \tau = 0.5$.

3.1. Experiments on simulated data

The simulated data were generated on the Shepp-Logan phantom [26] at 180 directions with 201 equal distance lines. In order to illustrate the noise suppressing effect, we compared the FBP method and the proposed TV method for contaminated projections, which were obtained

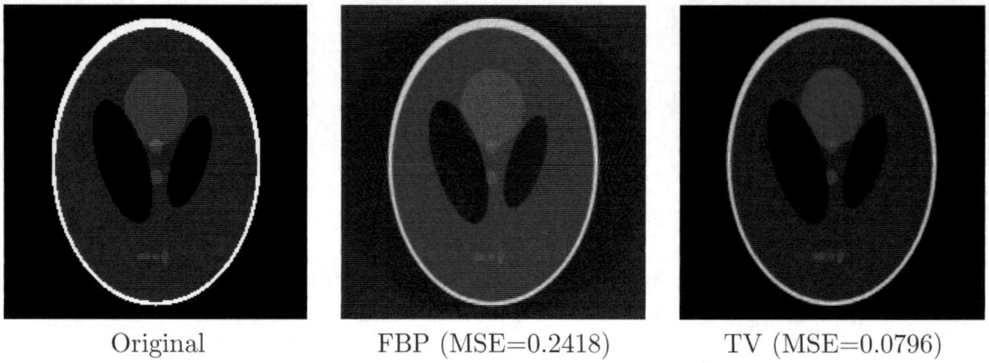

Figure 1. Reconstructed images with simulated data of Shepp-Logan phantom using different methods.

by adding zero mean Gaussian noise (with standard variance 0.005) to the noiseless projection data. The regularization parameter is chose as $\lambda = 0.0005$.

The results by FBP method (2) and the TV method are shown in Figure 1. By visual comparison, we can observe that the TV method is superior to FBP with respect to fine structure reconstruction and noise suppression. The MSE comparison also shows that the TV method is much better than the traditional FBP method.

3.2. Experiments on real data

The experiment was carried out at BL13W1 beamline at Shanghai Synchrotron Radiation (SR) Facility (see reference [6] for details on the setup and shale sample information). The estimated noise level is about 0.005. The regularization parameter $\lambda = 0.0005$. In Figure 2, we present the reconstructed images of one slice of the tested shale sample, and the parts in the rectangle were enlarged for comparison.

From Figure 2, we can observe that TV method is superior to the FBP method visually. Firstly, the image contrast by FBP method is much lower than that by the TV method. Secondly, the TV method suppresses the noise effect and preserves the edge between different components successfully. Therefore, we may conclude that the combination of the SR sources X-ray and TV regularization is a feasible and efficient tool to recover the shale interior information non-destructively.

4. Conclusions and Future works

In this paper, we applied the X-ray computerized tomography technique to obtain the shale interior structure. In order to suppress the noise in the projection data, the total variation regularization method was used to reconstruct shale structures from XCT projection data. The experiments on stimulated and real data showed that the TV method was superior to the filtered back projection method in noise suppression and edge preservation. Based on the observation of the shale, in the future, we will study the segmentation-based method to reconstruct a piecewise constant attenuation function directly, not by the aid of total variation minimization.

Acknowledgements

We appreciate the constructive, detailed and helpful comments by the reviewers for our manuscript. The research is supported by National Natural Science Foundation of China under grant numbers 91630202, 41325016 and 11401171, Strategic Priority Research Program of the Chinese Academy of Science (Grant No.XDB10020100) and the China Scholarship Council.

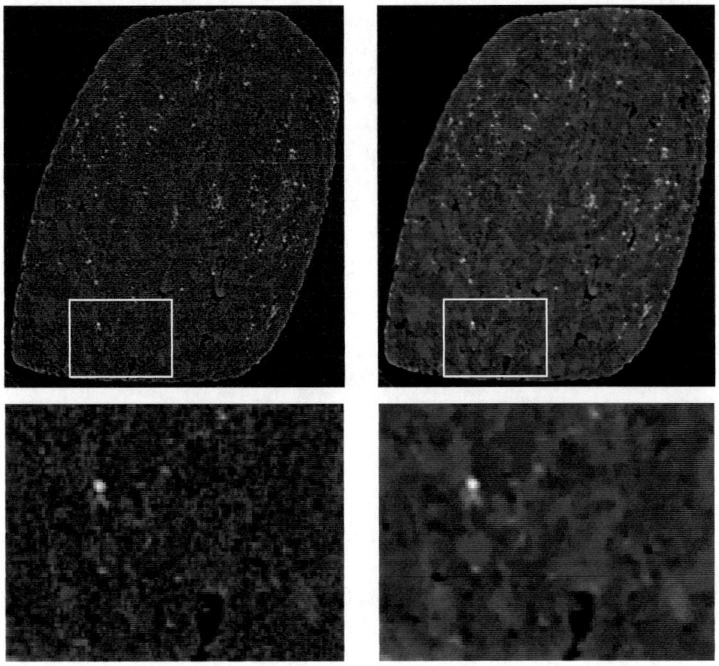

Figure 2. Reconstructed images from experimental data. First row: reconstructed images from projection data; Second row: zoom-in images in the rectangle of the corresponding images in the first row.

References

[1] Riepe L, Suhaimi M H, Kumar M, Knackstedt M A, *et al.* 2011 Application of high resolution micro-CT-imaging and pore network modeling (pnm) for the petrophysical characterization of tight gas reservoirs-a case history from a deep clastic tight gas reservoir in Oman *Proc. SPE Middle East Unconventional Gas Conference and Exhibition* Society of Petroleum Engineers (Muscat, Oman).

[2] Ketcham R A and Carlson W D 2001 Acquisition, optimization and interpretation of x-ray computed tomographic imagery: applications to the geosciences *Computers & Geosciences*, **27** 4 pp. 381–400.

[3] Yang Y S, Gureyev T, Tulloh A, Clennell B and Pervukhina M 2009 Feasibility of a data-constrained prediction of shale microstructures *Proc. 18th World IMACS and MODSIM Congress*, (Cairns, Australia), pp. 342–347.

[4] Kak A C and Slaney M 2001 *Principles of Computerized Tomographic Imaging*. Philadelphia:SIAM.

[5] Herman G T 2009 *Fundamentals of Computerized Tomography: Image Reconstruction from Projections* 2 ed. Springer.

[6] Wang Y, Luo S, Wang L, Wang J and Jin C 2017 Synchrotron radiation-based L1-norm regularization on micro-CT imaging in shale structure analysis, *Journal of Inverse and Ill-posed Problems*, **25** 4 pp. 483–498.

[7] Natterer F and Wubbeling F 2001 *Mathematical Methods in Image Reconstruction* Philadelphia: SIAM.

[8] Jiang M and Wang G 2002 Development of iterative algorithms for image reconstruction *Journal of X-Ray Science and Technology*, **10** 10 pp. 77–86.

[9] Sidky E Y, Jorgensen J H and Pan X C 2012 Convex optimization problem prototyping for image reconstruction in computed tomography with the Chambolle-Pock algorithm, *Physics in Medicine and Biology*, **57** 10, pp. 3065–3091.

[10] LaRoque S J, Sidky E Y and Pan X C 2008 Accurate image reconstruction from few-view and limited-angle data in diffraction tomography *Journal of the Optical Society of America* **25** 7 pp. 1772–82.

[11] Herman G T and Davidi R 2008 Image reconstruction from a small number of projections, *Inverse Problems*, **24** 4 045011.

[12] Davidi R, Herman G T and Censor Y 2009 Perturbation-resilient block-iterative projection methods with application to image reconstruction from projections, *International Transactions in Operational Research*, **16** 4 pp. 505–24.

[13] Butnariu D. Davidi R. Herman G. T. and KazantsevI. G. 2007 Stable convergence behavior under perturbations of a class of projection methods for convex feasibility and optimization problems, *IEEE Journal of Selected Topic Signal Process*, **1** 4 pp.540-547.

[14] Zhao Z. Yang J. Jiang M. 2015 A fast algorithm for high order total variation minimization based interior tomography, *Journal of X-ray science and technology*, **23** 3 pp. 349–364.

[15] Zhou W. Cai J. Gao H. 2013 Adaptive tight frame based medical image reconstruction: a proof-of-concept study for computed tomography, *Inverse problems*, **29** 12 125006.

[16] Dong B. Li J. Shen, Z. 2013 X-ray CT image reconstruction via wavelet frame based regularization and Radon domain inpainting, *Journal of Scientific Computing*, **54** 2-3 pp.333–349.

[17] Li S. Cao Q. Chen Y. Hu, Y. Luo L. Toumoulin C. 2014 Dictionary learning based sinogram inpainting for CT sparse reconstruction, *Optik-International Journal for Light and Electron Optics*, **125** 12, pp.2862–2867.

[18] Rudin L, Osher S and Fatemi E 1992 Nonlinear total variation based noise removal algorithms *Physica D-nonlinear Phenomena*, **60** 1 pp. 259–268.

[19] A. Buades, B. Coll and J. M. Morel 2006 The staircasing effect in neighborhood filters and its solution, IEEE Trans. Image Process., **15** 1499-1505.

[20] A. Chambolle and P. L. Lions 1997 Image recovery via total variation minimization and related problems, Numer. Math. **76** 167-188.

[21] Goldstein T and Osher S 2009 The split Bregman method for ℓ_1-regularized problems *SIAM Journal on Imaging Sciences* **2** 2 pp. 323–343.

[22] Yin W, Osher S, Goldfarb D and Darbon J 2008 Bregman iterative algorithms for ℓ_1-minimization with applications to compressed sensing *SIAM Journal on Imaging Sciences* **1** 1 pp. 143-168.

[23] Chambolle A and Pock T 2011 A first-order primal-dual algorithm for convex problem with apllications to imaging *Journal of Mathematical Imaging and Vision* **40** 1 pp. 120–145.

[24] Micchelli C A, Shen L Xand Xu Y S 2011 Proximity algorithms for image models: denoising, *Inverse Problems* **27** 4 pp. 45009–45038.

[25] Chambolle A 2004 An algorithm for total variation minimization and applications *Journal of Mathematical Imaging and Vision* **20**, 1 pp. 89–97.

[26] Shepp L A and Logan B F 1974 The Fourier reconstruction of a head section *IEEE Trans. Nuclear Science* **NS-21** 3 pp. 21–43

IOP Conf. Series: Journal of Physics: Conf. Series **1047** (2018) 012008 doi:10.1088/1742-6596/1047/1/012008

Bayesian approach to single-tree detection in airborne laser scanning – use of training data for prior and likelihood modeling

Teemu Luostari[1], Timo Lähivaara[1], Petteri Packalen[2] and Aku Seppänen[1]

[1] Department of Applied Physics, University of Eastern Finland, P.O. Box 1627, FIN-70211 Kuopio, Finland

[2] School of Forest Sciences, University of Eastern Finland, P.O. Box 111, FIN-80101 Joensuu, Finland

E-mail: teemu.luostari@uef.fi, timo.lahivaara@uef.fi, petteri.packalen@uef.fi, aku.seppanen@uef.fi

Abstract. This paper introduces a novel computational approach to handling remote sensing data from forests. More specifically, we consider the problem of detecting an unknown number of trees based on airborne laser scanning (ALS) data. In addition to detecting the locations of individual trees, their heights and crown shapes are estimated. This detection-estimation problem is treated in the Bayesian inversion framework. We use simplified, rotationally symmetric models for the tree canopies to model the echoes of laser pulses from the canopies. To account for the associated modeling errors, we use training data consisting of ALS data and field measurements to build a likelihood function which models statistically the propagation of a laser beam in the presence of a canopy. The training data is utilized also for constructing empirical prior models for the crown height/shape parameters. As a Bayesian point estimate, we consider the *maximum a posteriori* estimate. The proposed approach is tested with ALS measurement data from boreal forest, and validated with field measurements.

1. Introduction

Airborne laser scanning (ALS) is a widely used tool for remote sensing of forest [7, 9, 11]. The applications include, e.g., the forest inventory and ecology. ALS is usually performed using an aeroplane, but it is also possible to use a helicopter or unmanned aerial vehicles, such as drones. ALS is based on light detection and ranging (LiDAR) technology, where laser beams are directed towards ground within a device-specific scan angle; these beams reflect from surfaces of objects they encounter, such as tree crowns and ground (see Figure 1). The times of the pulses reflected back from the tree crowns/ground are recorded and transformed into distance information, which together with constantly determined position and orientation of the ALS device allow for calculating coordinates on the Earth's surface. The ALS point clouds thus include accurate (yet discrete) information of the ground level, and simultaneously versatile information on the structure of the forest.

Because canopy surfaces are not solid but formed by separate leaves, a reflected laser wave can contain more than one intensity peaks, which are registered as separate echoes from the

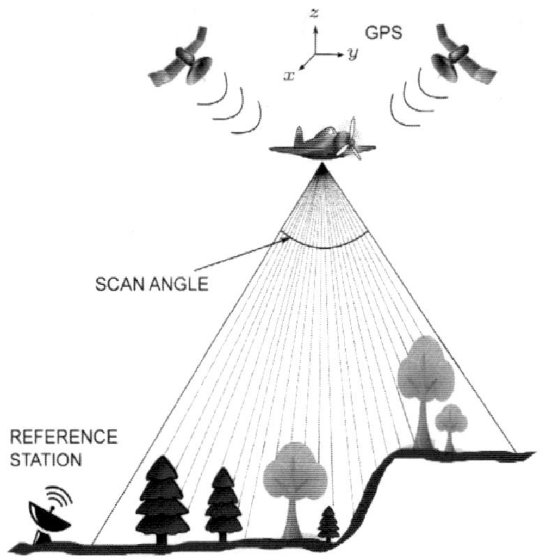

Figure 1. A schematic figure of ALS, cf. [10]. ALS device is sending laser pulses and records multiple echoes from the trees and ground surface. While scanning the forest, the correct position of the aeroplane is tracked using GPS.

same scanning direction. The collected ALS points are thus identified as *only*, *first of many*, *intermediate* and *last echoes*. Figure 2 shows an example of ALS-based point cloud data from a boreal forest. Point clouds corresponding to only echoes and first of many echoes are shown separately, and approximate (field measured) locations of trees are drawn in the same figures. The figure clearly indicates a qualitative difference between the spatial distributions of the two point clouds: Most of the only echoes are reflected from the ground or from tops (centers) of tree crowns, while the first of many echoes mostly result from reflections from the crowns but not from their centers.

The interpretation of ALS data can be divided into two categories: the area-based approaches [7, 9, 11] and single-tree detection [7, 4, 9, 11, 1, 5, 10]. While in the widely used area-based approach, the plot-level statistics of forest attributes are estimated directly from the statistics of ALS data using regression-type methods, the single-tree detection aims at deriving the plot-level statistics from tree -level information. Potentially, the latter approach – which accounts for the spatial distribution of the ALS point clouds – could improve the accuracy of the ALS based forest inventory, and enable deriving more versatile information (such as distinguishing tree species) from the data. This, however, requires development of the computational methods used in the analysis of the ALS data.

In this paper, the problem of individual tree detection is cast in the framework of Bayesian inverse problems. As in a recent work [4], simplified, rotationally symmetric models for the tree canopies are used for modeling the echoes of laser beams from the canopies. However, while in [4], an *ad hoc*, additive noise model was used for constructing a Gaussian likelihood model for the ALS observations, the aim of the present work is to study whether a feasible likelihood model could be constructed systematically by analyzing ALS data from a set of training plots. The training data is utilized also for constructing empirical prior models for the crown height/shape parameters. The feasibility of the approach is tested with ALS measurement data from boreal forest, and validated with field measurements.

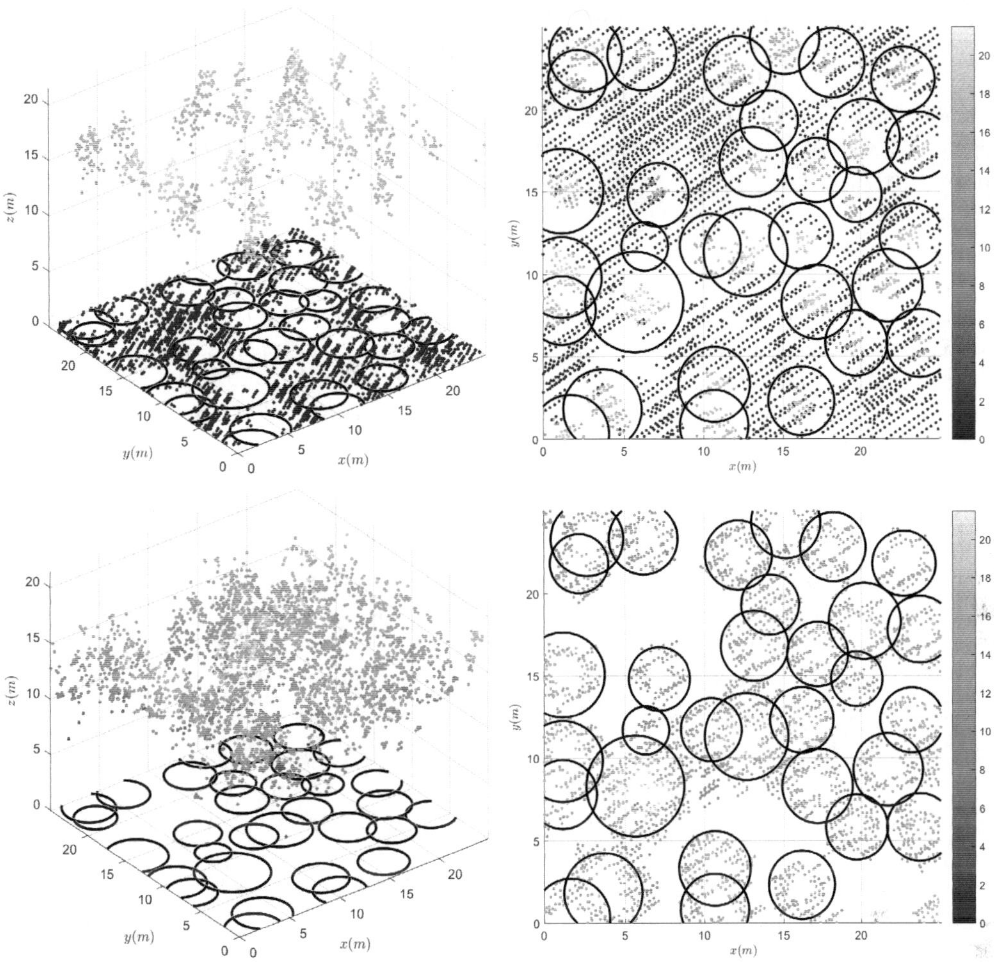

Figure 2. An illustration of real ALS data. The ALS observations consist of coordinates of points where the laser beams intersect with the surfaces of trees or ground. The *only echoes* are illustrated in the top row and *first of many echoes* in the second row. The left column shows side views of the two point clouds and in the right column, the observations are shown from top. The black circles illustrate the field-measured locations of the trees – the centers of the circles correspond to trunk locations on the ground and the radii of the circles approximate the radii of the canopies.

2. Materials

We consider managed boreal forest plots as were also investigated in [4, 8, 10], i.e. Eastern Finland latitude 62°31'N, longitude 30°10'E. The ALS data were collected by using an Optech Gemini laser scanning system in 2009. The scanning height was approximately 600 m above ground level and the scan angle was 26°. Multiple echoes were recorded and pulse repetition frequency was 125 kHz. In this paper, we consider first of many echoes and only echoes, as illustrated in Figure 2. The densities of data points were approximately $8/\text{m}^2$ for each type of echoes. The sizes of forest plots vary from 20 m × 20 m up to 30 m × 30 m consisting of Scots pines, Norway spruces and deciduous trees.

For validation purposes, the locations and sizes of the trees were also measured manually from field. More specifically, only trees higher than 4 m or having diameter at breast height larger than 4 cm were accounted for. The field measurement data includes tree locations, tree heights, diameters at breast height and tree species and it was collected during 2010. In the present work, the field measurement data is used also as a training data, for constructing the likelihood and prior models for Bayesian inference. Note however, that different plots were selected for training the model and validating the solutions of the inverse problems. In future studies the models and estimates should also be cross-validated with data from different forest types.

3. Computational methods

In this section, a novel approach to detecting individual trees and estimating their heights and crown shapes is proposed. First, in Section 3.1, we describe the parametrization of trees and a computational model that approximates the formation of ALS point clouds. In Section 3.2, statistics of the uncertainty of the ALS observations are estimated based on a set of training data, and an approximative likelihood model is written. In Section 3.3, the likelihood function is combined with a field measurement -based prior model for the model unknowns (crown shape parameters) to form the posterior density for the unknowns. As an estimate for the individual tree parameters, the *maximum a posteriori* estimate is considered.

3.1. Parametrization and observation model

The parametrization of individual trees is illustrated in Figure 3. We denote the horizontal location of a tree trunk on the ground by (x, y) and the height of the tree by h. The crown of the tree is approximated as a rotationally symmetric object and the vertical profile of the crown radius is modeled as

$$R(h_s) = c_r \sin(h_s)^{a_t}, \tag{1}$$

where c_r denotes the radius of the tree crown at the lower limit of the living crown s_h, $h_s = (\pi h_v)/(2c_h)$, $h_v \in [0, c_h]$, $c_h = h - s_h$, and a_t is a species-specific shape parameter; for pines, birch and spruces, a_t is 0.3755, 0.2463 and 0.3825, respectively, see [4].

In this paper, we aim at estimating the parameters x, y, h and c_r based on ALS data. In principle, it would be possible to consider also the lower limit of the living crown s_h (or alternatively, the crown height c_h) as an additional unknown in the model. However, the sensitivity of the ALS measurements to the ratio of s_h and c_h is very low, and hence we write a species-specific, deterministic model for the dependence of c_h and the tree height h:

$$c_h = \alpha_t h + \beta_t. \tag{2}$$

Here, α_t and β_t are fitting parameters based on analysis of field data. This approximation can be used for rewriting variable h_s as $h_s = (\pi h_v)/(2(\alpha_t h + \beta_t))$, which allows for expressing the vertical profile of the tree crown (Equation (1)) as the function of the tree height h.

In addition to tree location, size and shape parameters x, y, h and c_r, the tree species t is considered as an unknown in the model. Here, t is a discrete variable which has four possible realizations: $t \in \{\text{pine}, \text{spruce}, \text{deciduous}, \text{none}\}$. Here, the last value ("none") signifies a case where the corresponding tree model is to be deleted – this option is needed, because the number of trees is unknown. We denote the vector consisting of all model unknowns by θ:

$$\theta = [\theta_1, \ldots, \theta_M]^T, \tag{3}$$

where $\theta_n = (x_n, y_n, h_n, c_{r_n}, t_n)$, $n \in \{1, \ldots, M\}$ is the index of a tree, and M is an estimate for the number of trees.

Denote the index of a laser beam in a data set corresponding to one forest plot by j. To construct a computational model for the formation of an ALS observation, we approximate for

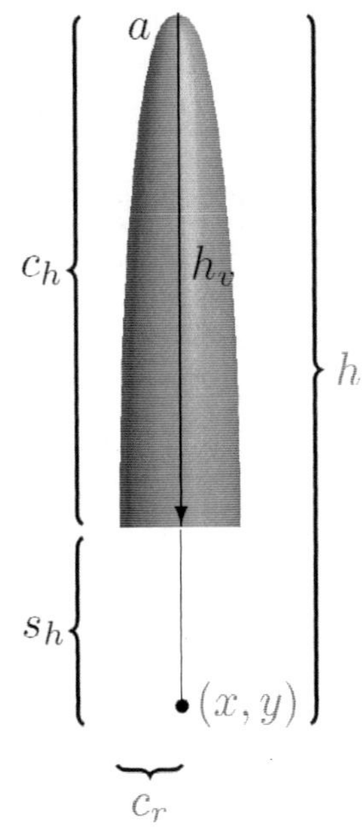

Figure 3. Illustration of a rotationally symmetric approximation of the tree crown and parametrization. The primary unknowns in the inverse problem are marked with red font: the horizontal location of a tree trunk on the ground (x, y), tree height h and c_r, the radius of the tree crown at the lower limit of the crown s_h.

simplicity that the ALS beams are vertically oriented. If the horizontal location (x^j, y^j) of a beam reflection point intersects with the area under the crown of a single tree n, then the height of the ALS reflection point z^j is modeled as the height of the nth tree crown surface at (x^j, y^j): i.e.,

$$z^j = g_j(\theta) = g_{n(x^j,y^j,\theta)}(x^j, y^j, \theta). \tag{4}$$

Here we have highlighted that the index of the tree, n, that a beam coincides with depends on both the horizontal location (x^j, y^j) and the parameter vector θ. Further, if the beam reflection point intersects with more than one tree, then the highest canopy surface point at the location (x^j, y^j) is considered as the reflection point, i.e.,

$$z^j = g_j(\theta) = \max_{m \in n(x^j,y^j,\theta)} \{g_m(x^j, y^j, \theta)\}, \tag{5}$$

and if the beam does not intersect with any tree crown, ALS measurement modeled as a ground echo, i.e., $z^j = g_j(\theta) = 0$.

Combining all vertical coordinates z^j into a vector $z = [z^1, \ldots, z^N]$, we can write the approximative computational model for the observations as

$$z = g(\theta), \tag{6}$$

where $g(\theta) = [g_1(\theta), \ldots, g_N(\theta)]$. In principle, the problem of estimating the tree parameters θ could be considered as an ordinary non-linear least-squares (LS) problem induced by the model (6). However, this problem would be severely ill-posed, as its solutions are non-unique and unstable. Moreover, as it turns out in the next section, the modeling errors associated with the computational model (6) have non-trivial statistics (non-homogeneous, multimodal and non-zero mean), which would yield additional difficulty in the LS fitting of tree crown surfaces to the observed ALS data. In the following sections, tree crown fitting problem is considered in the framework of Bayesian inverse problems.

3.2. Construction of prior and likelihood models using training data
In this section, we write the prior probability density for the tree location/shape parameters θ and the likelihood function of the ALS observations z on the basis of training data. That is, we model both θ and z as random variables, and approximate their statistics using models derived from ALS and field measurements.

3.2.1. Prior model The field measurements consist of tree locations, tree species, tree heights and trunk diameters at breast height. As in article [4], we approximate the crown radii c_r from these measurements using models presented in [6]. Figure 4 (left) shows a scatter plot of the observed/approximated tree heights h and crown radii c_r. In this plot, different colors represent different tree species (pine, spruce and deciduous). The figure indicates a strong correlation between parameters h and c_r. Moreover, the differences between the scatter plots corresponding to three tree species are relatively small. Based on this training data set, we model the parameters h and c_r as mutually correlated species-independent Gaussian random variables, and approximate their expectations and covariance matrix by sample means and covariance. The Gaussian joint prior probability density for the model parameters h and c_r is illustrated in Figure 4 (right). Further, the tree location parameters x and y are modeled as mutually independed Gaussian random variables; the expectation of each location parameter is chosen to be equal to the initial estimate corresponding to each tree (see end of this section), and the variances of all location parameters are chosen to be $(5\text{m})^2$. The tree species parameter t is modeled as a uniformly distributed discrete random variable, i.e., $P(t) = \frac{1}{4}, \forall t \in \{\text{pine}, \text{spruce}, \text{deciduous}, \text{none}\}$, where $P(\cdot)$ denotes the probability. Using the above approximations, the probability density (prior density) of θ can be written in the form

$$\pi(\theta) \propto \exp\left(-\frac{1}{2}(\tilde{\theta} - \tilde{\theta}_*)^T \Gamma_{\tilde{\theta}}^{-1} (\tilde{\theta} - \tilde{\theta}_*)\right), \tag{7}$$

where $\tilde{\theta}$ is a vector consisting of those model unknowns that are modeled as continuous random variables, that is, $\tilde{\theta} = [\tilde{\theta}_1, \ldots, \tilde{\theta}_N]^T$, $\tilde{\theta}_n = (x_n, y_n, h_n, c_{r_n})$. Further, $\tilde{\theta}_*$ and $\Gamma_{\tilde{\theta}}$ are the expectation and covariance matrix of $\tilde{\theta}$, respectively. Note here that since the prior probabilities of tree species are assumed to be equal and independent from x, y, h and c_r, the prior density of θ is simply $\pi(\theta) = \frac{1}{4}\pi(\tilde{\theta})$, i.e., $P(t)$ only acts as a scaling factor in the prior density of θ.

3.2.2. Likelihood model Figure 2 shows the spatial distribution of first of many echoes and only echoes in one example plot. The figure also illustrates the field measured locations of trees in the plot: the black circles approximate the extents of the tree crowns. The figure reveals that the above approximation of the tree crowns being rotationally symmetric objects, the surfaces of which reflect the ALS pulses, is erroneous. Indeed, Figure 2 (top right) shows that for the only echoes, a large number of ground reflections ($z \approx 0$) are obtained from under the tree canopies, and Figure 2 (bottom right) shows that some of the first of many echoes at high altitude ($z = 5, \ldots, 20$ m) have reflected from tree branches outside the approximate rotationally symmetric tree crown models. In addition, the ALS data outside areas under the canopy are contributed by local variations of the ground height and the understory vegetation. In this section, we analyze the ALS observations in the training set, and combine their statistics with the

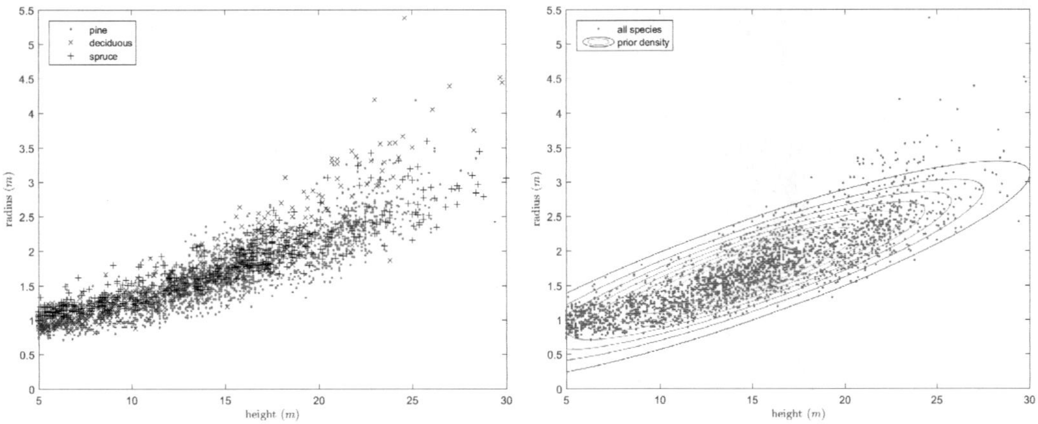

Figure 4. An illustration of mutually correlated parameters h and c_r for different tree species (left), and a contour plot of the fitted Gaussian prior density (right).

computational model of ALS observations written in Section 3.1 to formulate an approximative likelihood model for ALS.

From the field measured tree locations and sizes corresponding to each training plot, we a get realization of the parameter vector θ. Based on these observations, we also divide the horizontal planes of the training plots into following three subsets:

- I^1: areas under rotationally symmetric tree crown models
- I^2: "buffer zones" $I^{2,1}$ and $I^{2,2}$ within $0, \dots, 25$ cm and $25, \dots, 50$ cm from I^1, respectively.
- I^3: The remaining areas, i.e., those areas which are not in the proximity of any tree.

These zones are illustrated in Figure 5. The figure also shows the subdivisioning of the tree crown area I^1 into areas of inner circles $I^{1,1}$ and outer rings $I^{1,2}$; these subsets are used in the analysis of only echoes. Here, we set the inner circle $I^{1,1}$ radius to be $c_r/4$. We analyze the statistics of the following variable r^j calculated from the set of ALS observations (x^j, y^j, z^j) in the set of training plots:

$$r^j = \begin{cases} \frac{z^j - g_j(\theta)}{h_{j,\theta}}, & \forall\ (x^j, y^j) \in I^1(\theta) \\ \frac{z^j}{h_{j,\theta}}, & \forall\ (x^j, y^j) \in I^2(\theta) \\ z^j, & \forall\ (x^j, y^j) \in I^3(\theta) \end{cases} \tag{8}$$

where $h_{j,\theta}$ denotes the height of the tree, the crown model surface of which is intersected by the jth laser beam (for $(x^j, y^j) \in I^1(\theta)$) or the buffer zone of which is intersected by the jth laser beam (for $(x^j, y^j) \in I^2(\theta)$). Thus, for an ALS data point (x^j, y^j, z^j) corresponding to an observation on the area of a tree crown, we consider a scaled residual between the observed height of reflection z^j and the height given by the computational model: $g_j(\theta)$. In the buffer zone $I^2(\theta)$, where $g_j(\theta) = 0$, the scaled residual is obtained as $\frac{z^j}{h_{j,\theta}}$, and in the ground area $I^3(\theta)$, residuals $r^j = z^j - g_j(\theta) = z^j - 0$ are considered.

Figure 6 shows the histograms of r^j corresponding to zone $I^1(\theta)$ for only echoes (top row) and first of only echoes (bottom). The histograms show that the residuals are heavily concentrated to negative values; i.e., most of the ALS beams are reflected *below* the modeled tree crown surfaces. Moreover, as anticipated, the statistics of the only echoes and first of only echoes differ from

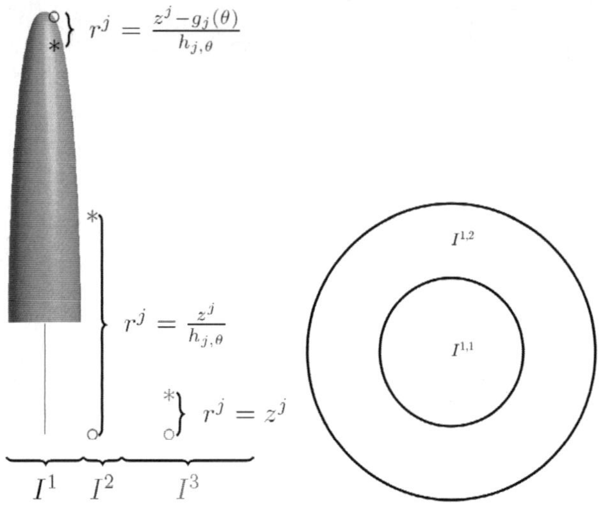

Figure 5. Left: An illustration of zones I^1, I^2 and I^3 and (scaled) residuals r^j. ALS data is illustrated using symbol '*' and the modeled observations with symbol '◦'. Right: For the only echoes, the tree crown zone I^1 is further divided to two sub-zones $I^{1,1}$ and $I^{1,2}$.

each other significantly. In particular, the only echoes feature peaks around $r^j = -1$, which corresponds to a ground observation in the location of a tree crown.

We model the (scaled) residuals r^j as random variables and approximate their probability densities at each zone I^i by sums of exponential functions

$$\pi_{r^j}^{I^i}(r^j|\theta) = f_i(r^j;\theta) = \beta_i \sum_{k=1}^{k_i} a_{i,k} \exp\left(-\left(\frac{r^j - b_{i,k}}{c_{i,k}}\right)^2\right), \tag{9}$$

where $a_{i,k}$, $b_{i,k}$ and $c_{i,k}$ are defined by the fitting $f_i(r^j;\theta)$ to the histogram corresponding to r^j at zone I^i, k_i is the number of exponential functions, and β_i is a scaling coefficient. Here, however, we note that for the only echoes in I^1, we manually decreased the histogram peaks corresponding to ground observations $r^j = -1$; this was made to diminish the errors caused by inaccuracies in the field measurements of tree locations. The fitted probability density functions corresponding to zone I^1 are illustrated in Figure 6.

Next, we write the probability density of an observation z^j corresponding to a given set of parameters as $\pi(z^j|\theta) = \pi_{r^j}^{I^i}(r^j|\theta) = f_i(r^j;\theta)$, where $f_i(r;\theta)$ is the approximate sample-based probability density of r^j, given by Equation (9). Assuming that observations z^j, z^ℓ are mutually independent for all $j \neq \ell$, the likelihood $\pi(z|\theta)$ of the observation vector $z = [z^1, \ldots, z^N]^{\mathrm{T}}$ is

$$\pi(z|\theta) = \prod_{j=1}^{N} \pi(z^j|\theta) = \prod_{j=1}^{N} f_{i(x^j,y^j,\theta)}(r^j;\theta). \tag{10}$$

Here, we have highlighted that the zone index i depends on the horizontal coordinates (x^j, y^j) of the ALS observation point and the model parameters θ.

3.3. *Bayesian inversion*

Combining the prior probability density in Equation (7) and the likelihood in Equation (10), the conditional density of θ given z, or the posterior density, can be written using Bayes' formula

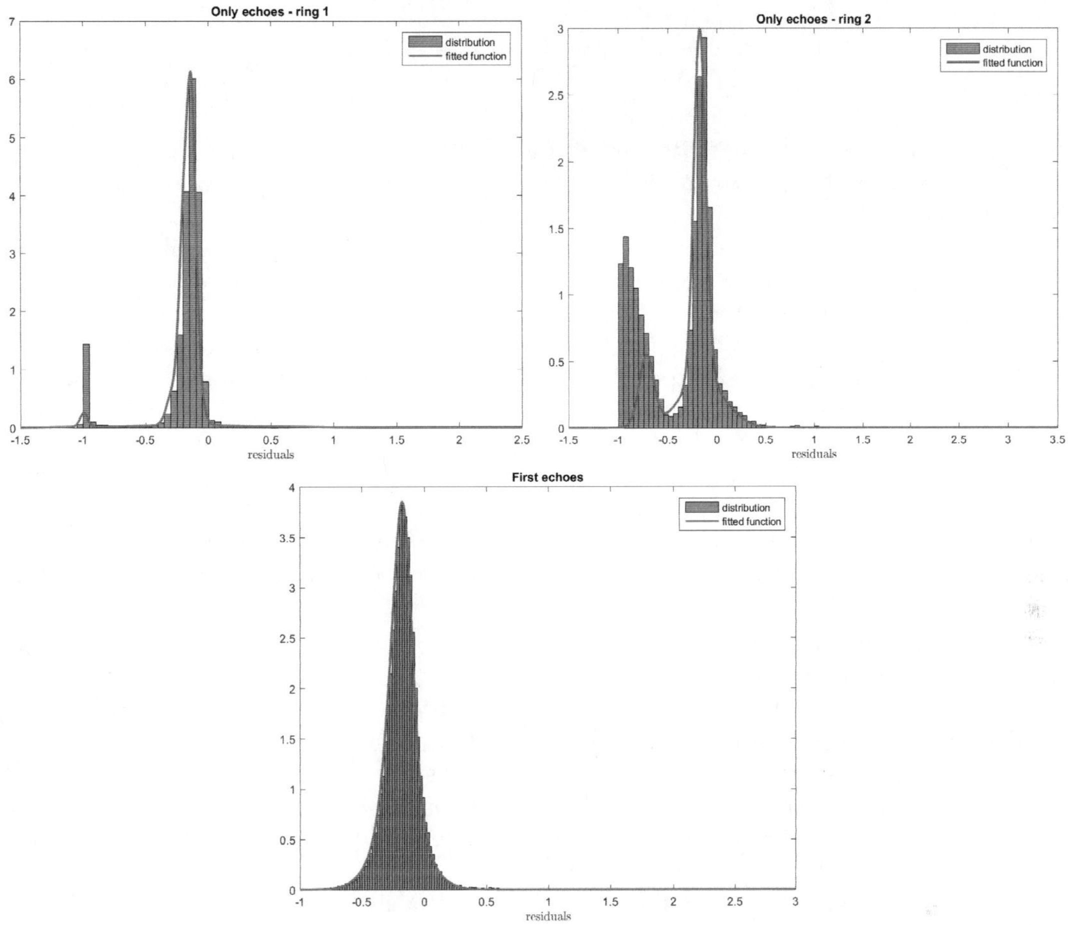

Figure 6. Example of histograms of the residuals and fitted likelihood functions in the zone I^1. Top: only echoes in $I^{1,1}$ (ring 1), only echoes in $I^{1,2}$ (ring 2) and bottom: first echoes in I^1.

as [2]

$$\begin{aligned}
\pi(\theta|z) &= \frac{\pi(z|\theta)\pi(\theta)}{\pi(z)} \\
&\propto \pi(z|\theta)\pi(\theta) \\
&\propto \prod_{j=1}^{N} f_{i(x^j,y^j,\theta)}(r^j;\theta)\exp\left(-\frac{1}{2}(\tilde{\theta}-\tilde{\theta}_*)^T\Gamma_{\tilde{\theta}}^{-1}(\tilde{\theta}-\tilde{\theta}_*)\right) \\
&\propto \exp\left(\sum_{j=1}^{N}\log(f_{i(x^j,y^j,\theta)}(r^j;\theta)) - \frac{1}{2}(\tilde{\theta}-\tilde{\theta}_*)^T\Gamma_{\tilde{\theta}}^{-1}(\tilde{\theta}-\tilde{\theta}_*)\right) \qquad (11)
\end{aligned}$$

and the *maximum a posteriori* (MAP) estimate gets the form

$$\theta_{\mathrm{MAP}} = \arg\max_{\theta} \pi(\theta|z)$$

$$= \arg\min_{\theta} \left\{ -\sum_{j=1}^{N} A_j(z_j, \theta) + \frac{1}{2} ||L_{\tilde{\theta}}(\tilde{\theta} - \tilde{\theta}_*)||_{L^2}^2 \right\}, \tag{12}$$

where $A_j(z_j, \theta) = \log(f_{i(x^j, y^j, \theta)}(r^j; \theta))$ and $L_{\tilde{\theta}}^T L_{\tilde{\theta}} = \Gamma_{\tilde{\theta}}^{-1}$.

In this paper, the MAP estimate is computed by using a simple random search algorithm, where at each iteration step: First, a new value for the discrete tree species variable, t_n^{new}, corresponding to nth tree model is drawn randomly from the uniform distribution $P(t_n) = \frac{1}{4}, \forall t_n \in \{\text{pine}, \text{spruce}, \text{deciduous}, \text{none}\}$. If t_n^{new} gets the value 'none', the nth tree is removed from the model, and in other cases, the tree species dependent parameters a_t, α_t and β_t are chosen to correspond to $t = t_n^{\text{new}}$. Next, a Gaussian distributed random vector $\xi_n \in \mathbb{R}^4$ is added to the parameters $\tilde{\theta}_n$ corresponding to nth tree model. If the functional in Equation (12) decreases, the parameters are changed to t_n^{new} and $\tilde{\theta}_n + \xi_n$, otherwise t_n and $\tilde{\theta}_n$ remain unchanged. The initial estimate for the iteration is computed by setting an excessive number of tree models into the areas of point clouds, by applying a tree detection algorithm introduced in [9] two times: first to the original ALS data z, and subsequently to a reduced data set where the ALS data points within the areas of canopy models fitted in the first round are removed. Hence, the (initial) number of trees M varies between plots, depending of the initial estimate.

4. Results and discussion

The proposed computational method was tested with data from seventeen field plots. The results from Plot 1 are shown in Figure 7. The figure compares the field measured trees (top row) with the MAP estimates (second row). In bottom left of the figure, both the field measured and estimated tree locations are marked. The size of each circle is proportional to the measured/estimated height of the tree. This illustration reveals that the MAP estimate fails to detect only one of the trees (marked with a red circle) and gives three falsely detected trees (three green circles without gray counterpart). Also seven other trees without field measured counterpart are detected (black circle), but these tree models are on the perimeter of the plot and hence the field measurements have clearly neglected the trees (cf. perimeters of the image in Figure 7 (top left): large number of ALS observations above ground without field measured counterpart); these tree estimates are thus not accounted for in the analysis of the success of the method. To assess the feasibility of detection, we calculated a success rate (SR) as in [4]:

$$\text{SR} = 100 \cdot \frac{N_{\text{correct}} - N_{\text{false}}}{N_{\text{field}}} \%, \tag{13}$$

where N_{correct} refers to correctly detected trees, N_{false} refers to false positives (estimation gives tree but field measurement does not) and N_{field} is the number of field measured trees. For Plot 1, SR $\approx$ 90 %. Figure 7 (bottom right) shows the estimated tree heights vs. field measured heights corresponding to trees in Plot 1. This figure illustrates the feasibility of the tree height estimation; for all detected trees, the height estimates are in good correspondence with the field measured heights.

The results corresponding to Plot 2 are shown in Figure 8. In this case, SR $\approx$ 73 %, i.e., the success rate of tree detection is somewhat smaller than in the case of Plot 1. The illustration in Figure 8 (bottom left) indicates that the undetected trees are rather small (small gray patches labelled with red circle) and/or very close to another (detected) tree. This is an expected result: In cases where two or more trees are very close to each other, the distinguishability of the trees becomes difficult. This is especially the case if one of the trees is significantly smaller than the others – ultimately, if the crown of a small tree is entirely covered by the crown of a taller tree, the ALS data does not carry any information on the presence of the smaller tree.

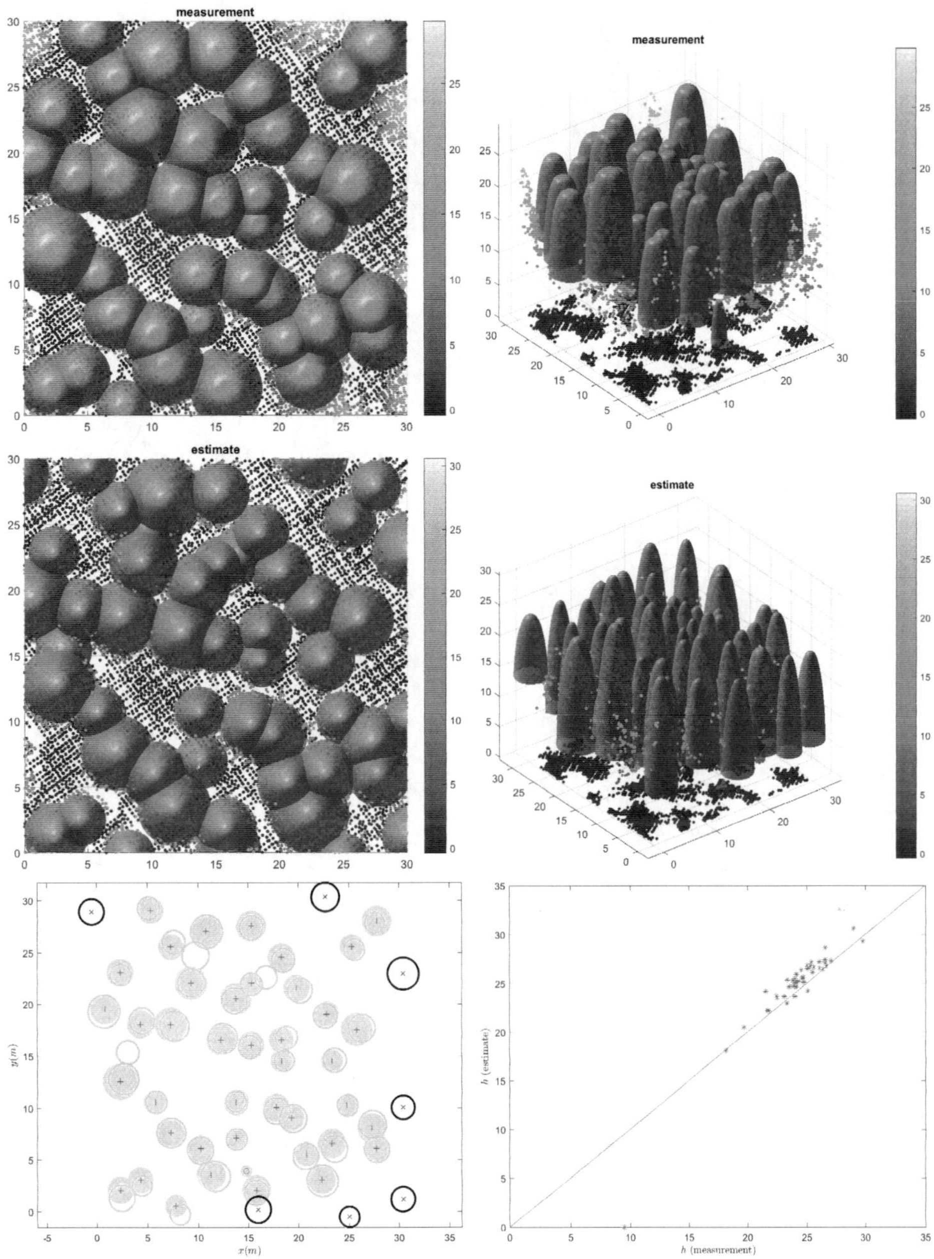

Figure 7. Results for Plot 1. Top row: field measured trees (illustrated as rotationally symmetric tree crown shapes) and ALS data (colored dots). Middle row: MAP estimate for the trees and ALS data. Bottom left: field measured (gray circles) and estimated (green circles) tree locations; the blue plus signs '+' indicate correctly detected trees and red circles '∘' indicate missed trees. The black circle with a cross '×' is an estimated tree near the boundary that is not taken into account when computing the success rate. Bottom right: Estimated tree heights vs. the field measured heights. SR ≈ 90%. In all subfigures, the units of the coordinate axes and colorbars are meters.

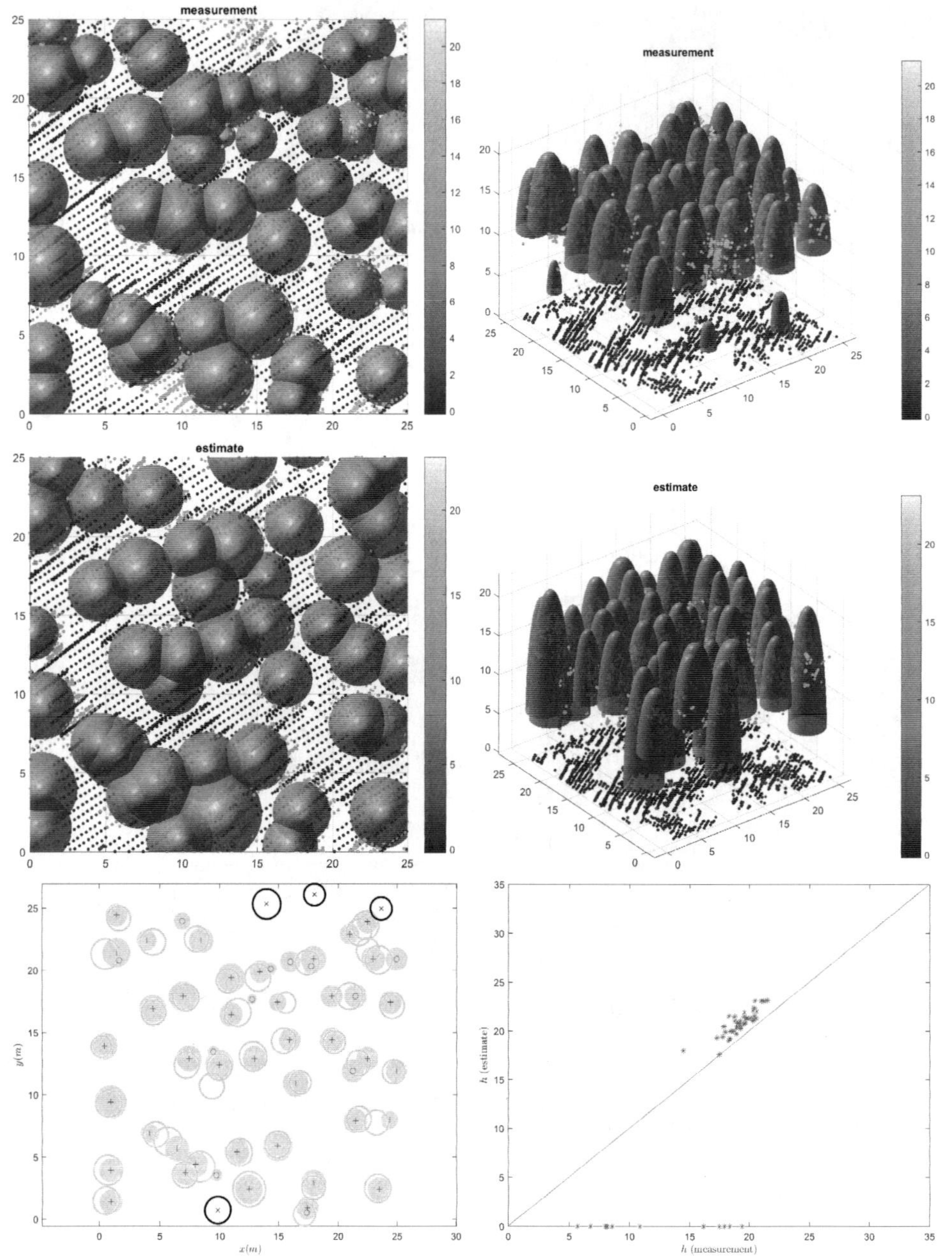

Figure 8. Results for Plot 2. Top row: field measured trees (illustrated as rotationally symmetric tree crown shapes) and ALS data (colored dots). Middle row: MAP estimate for the trees and ALS data. Bottom left: field measured (gray circles) and estimated (green circles) tree locations; the blue plus signs '+' indicate correctly detected trees and red circles 'o' indicate missed trees. The black circles with a cross '×' are estimated trees near the boundary which are not taken into account when computing the success rate. Bottom right: Estimated tree heights vs. the field measured heights. SR ≈ 73%. In all subfigures, the units of the coordinate axes and colorbars are meters.

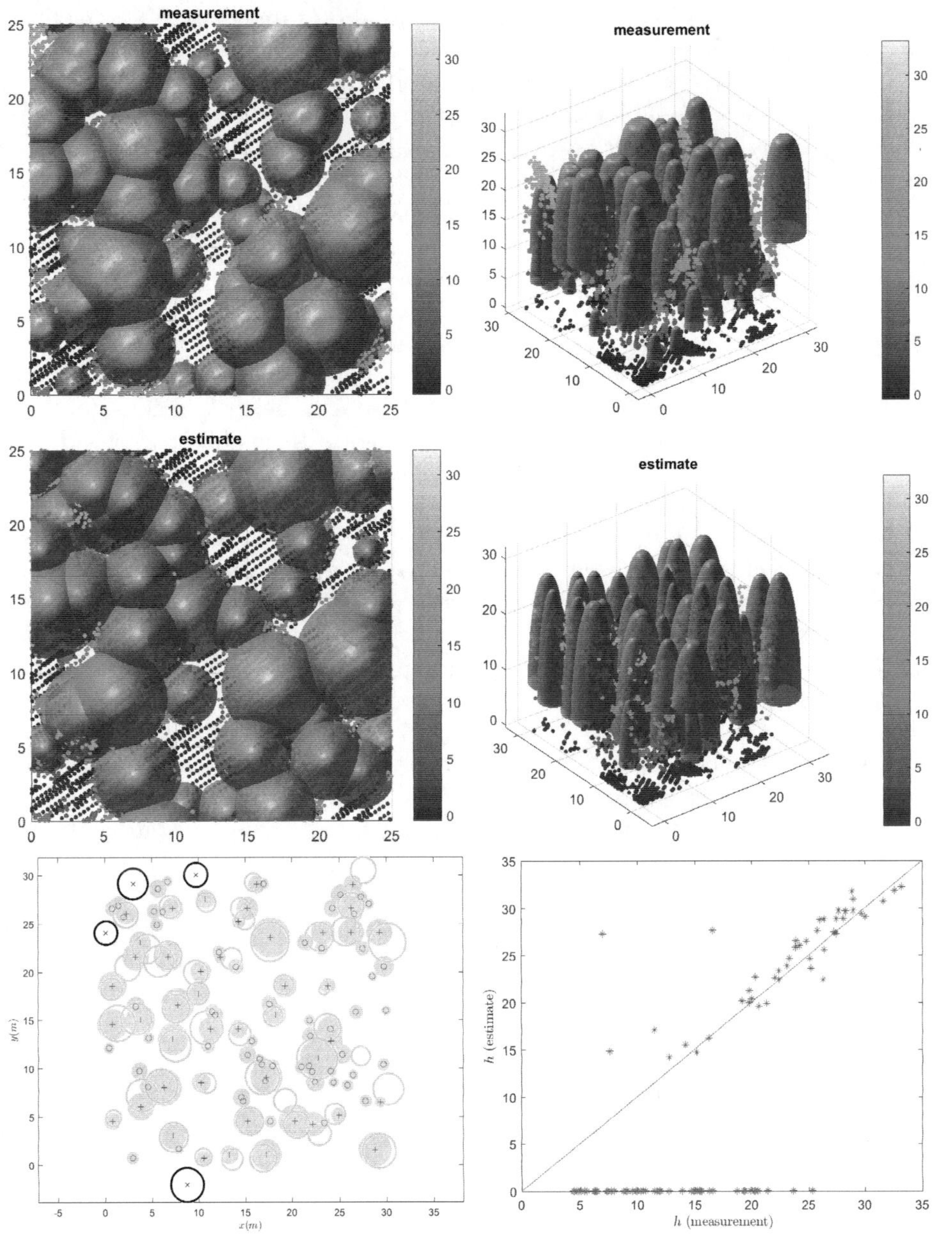

Figure 9. Results for Plot 3. Top row: field measured trees (illustrated as rotationally symmetric tree crown shapes) and ALS data (colored dots). Middle row: MAP estimate for the trees and ALS data. Bottom left: field measured (gray circles) and estimated (green circles) tree locations; the blue plus signs '+' indicate correctly detected trees and red circles '○' indicate missed trees. The black circle with a cross '×' is an estimated tree near the boundary that is not taken into account when computing the success rate. Bottom right: Estimated tree heights vs. the field measured heights. SR ≈ 41%. In all subfigures, the units of the coordinate axes and colorbars are meters.

Figure 9 illustrates the results corresponding to the third example plot. This case represents one of the lowest tree detection successes in the set of seventeen plots evaluated in this study, SR being approximately 41 %. The analysis of the bottom left chart in Figure 9 reveals the reason for the low SR: Plot 3 contains a large number of small trees, many of them being partly or entirely covered by crowns of larger trees. This observation is confirmed by comparison of the top views of the field measured and estimated trees (Figure 9, left column, first and second row, respectively): Indeed, from top (the view of LiDAR), the patterns of field measured and detected tree crowns look very similar to each other despite the low SR – because most of the undetected trees are hidden by the highest layer of the canopy.

In the seventeen test plots, the average SR of tree detection was 63 %. Figure 10 shows histograms of field measured and estimated tree heights in the seventeen test plots. The histogram corresponding to field measurements (Figure 10, top) indicates a layered structure of the forest, featuring highest peaks around 4-7 m and 16-21 m heights. The latter peak is well traced by the ALS-based estimates described above (Figure 10, middle): especially at heights over 20 m, the histogram corresponding to ALS-estimates resembles the shape of the histogram corresponding to field measured tree heights very well. In smaller tree heights ($h \approx$ 13-19 m), the numbers of trees are somewhat underestimated, and peak corresponding to the layer of smallest trees ($h \approx$ 4-7 m) is completely missed by the ALS-based estimates. This result confirms that the undetected trees were majorly the smallest trees of the plots, as in Plots 2 and 3.

The undectability of smaller trees is a feature that causes uncertainty to the tree detection, if addional data is not available. However, it can be possible to account for the undectability of the smaller trees, by post processing the estimated tree statistics in plot level: For a novel stochastic approach to handling this uncertainty, we refer to article [3].

Finally, we also computed the ALS-based estimates corresponding to a Gaussian likelihood [4] for comparison. For these estimates, the training data -based likelihood model described in Section 3.2.2 was replaced by a Gaussian model. Table 1 shows the success rates of tree detection corresponding to both likelihood models for seventeen test plots. On avarage, the success rate of tree detection corresponding to the Gaussian likelihood is slightly lower than for the training data -based likelihood model; with the Gaussian likelihood, the mean SR is approximately 60 %. In four plots, the Gaussian likelihood yields better SR performance than the new, training data -based likelihood model. It is not fully clear for the authors why this occurs. However, trees that are not detected by the estimates based on the new likelihood model are practically always small trees which are (at least partly) covered by taller trees (cf. Figs 8–9). It is possible that in some cases the misfit of the Gaussian likelihood model produces extra trees in the estimates, and these extra trees are coincidentally near small trees covered by taller trees. Indeed, estimates corresponding to both likelihood models can include such coincidentally located trees, especially in cases of very densely populated plots.

More importantly, in the case of the Gaussian likelihood, tree heights were overall severerly underestimated, as indicated also by the histogram shown in the bottom row of Figure 10: In this histogram, the peak of the tall trees is shifted to negative direction by a few meters. This is caused by the fact that the Gaussian model represents poorly the errors in the observation model – cf. especially the top row in Figure 6 illustrating the training data based residuals. This result further supports the use of the proposed training data -based likelihood model in

Table 1. Success rates (SR, Eq. (13)) of tree detection in seventeen test plots corresponding to two likelihood models: the new, training data-based likelihood and Gaussian likelihood.

Plot	1	2	3	4	5	6	7	8	9	10	11	12	13	14	15	16	17	mean
New	90	73	41	46	67	69	47	62	90	65	61	85	75	87	53	33	30	63
Gaussian	86	79	47	46	40	64	49	60	80	48	59	73	75	87	64	32	29	60

IOP Conf. Series: Journal of Physics: Conf. Series **1047** (2018) 012008 doi:10.1088/1742-6596/1047/1/012008

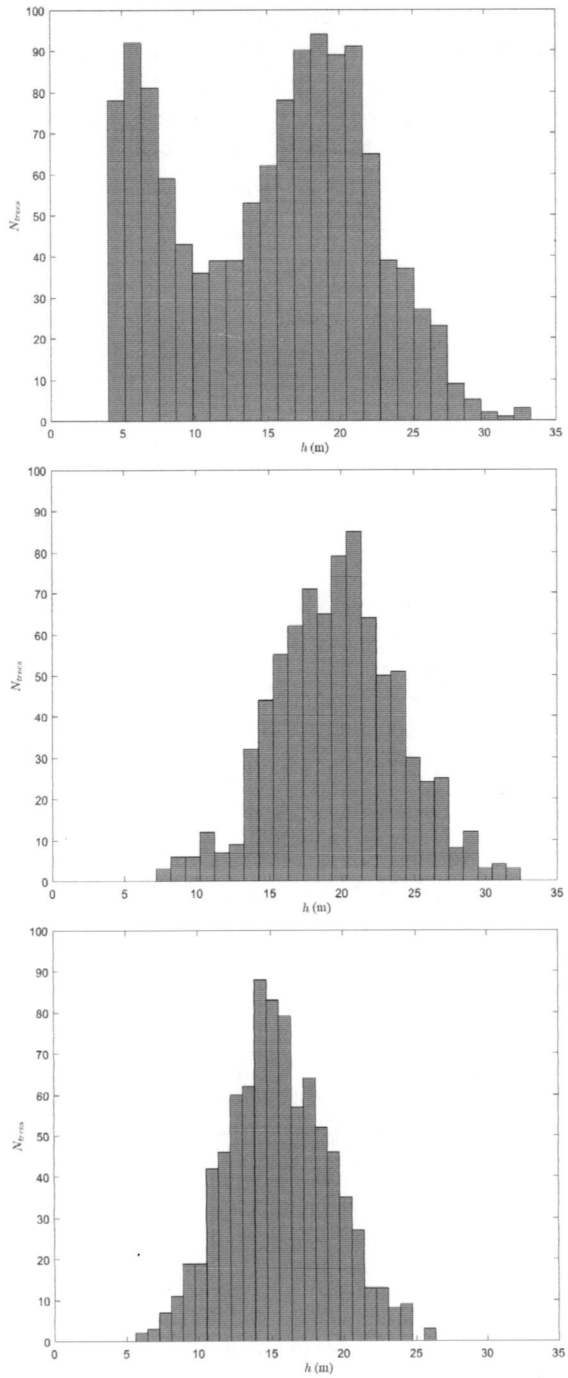

Figure 10. Histograms of the tree height. Top: Field measured tree heights. Middle: Estimated tree heights corresponding to training data -based likelihood model. Bottom: Estimated tree heights corresponding to Gaussian likelihood model.

individual tree detection, in cases where such training data is available.

In the iterative solution of the MAP estimates, the number of iterations needed for convergence and, consequently, the computation times varied between plots. On average, about 2500 iterations were computed, and the computation times corresponding to each plot were about 12 h in Matlab environment. However, we emphasize that the computational methods were not optimized in terms of efficiency, and we expect that the computation times can be reduced by orders of magnitude, e.g, by improving the implemented random search algorithm, parallelizing part of the computations and using an alternative programming language.

5. Conclusions

In this paper, we studied the problem of single tree detection using ALS data. We proposed a Bayesian approach, where the prior and likelihood models were constructed on the basis of training data consisting of ALS and field measurements. The approach was tested using real ALS data and verified with field measured tree locations and sizes. The results demonstrate the feasibility of the approach: In the selected test cases, the large trees were detected at high rate, and the tree height estimates were feasible. The results also demonstrate how the small trees, especially when fully of partly covered by the crowns of taller trees, remain often undetected. Handling the uncertainty related to undetected small trees needs to be handled separately, e.g. by taking the approach proposed in [3]. Nevertheless, the Bayesian inversion accompanied with training data -based statistical models holds potential for becoming a computational tool for the analysis of ALS data in single tree level, and allowing for more reliable and versatile information on the forest structure.

Acknowledgments

The project is funded by Academy of Finland (projects: 270174, 295341, 295489, 303801) and Finnish Centre of Excellence in Inverse Problems Research (250215).

References

[1] Andersen, H.-E., Reutebuch, S. E., and Schreuder, G. F. Bayesian object recognition for the analysis of complex forest scenes in airborne laser scanner data.

[2] Kaipio, J., and Somersalo, E. *Statistical and computational inverse problems*, vol. 160. Springer Science & Business Media, 2006.

[3] Kansanen, K., Vauhkonen, J., Lähivaara, T., and Mehtätalo, L. Stand density estimators based on individual tree detection and stochastic geometry. *Canadian Journal of Forest Research 46*, 11 (2016), 1359–1366.

[4] Lähivaara, T., Seppänen, A., Kaipio, J. P., Vauhkonen, J., Korhonen, L., Tokola, T., and Maltamo, M. Bayesian approach to tree detection based on airborne laser scanning data. *IEEE Transactions on Geoscience and Remote Sensing 52*, 5 (May 2014), 2690–2699.

[5] Micheas, A. C., Wikle, C. K., and Larsen, D. R. Random set modelling of three-dimensional objects in a hierarchical bayesian context. *Journal of Statistical Computation and Simulation 84*, 1 (2014), 107–123.

[6] Muinonen, E. Metsikön heijastussuhteen ennustaminen geometrisella latvustomallilla. *Licenciate of Science thesis (in Finnish), University of Joensuu, Faculty of Forest Sciences* (1995).

[7] Næsset, E. Estimating timber volume of forest stands using airborne laser scanner data. *Remote Sensing of Environment 61*, 2 (1997), 246 – 253.

[8] Packalen, P., Vauhkonen, J., Kallio, E., Peuhkurinen, J., Pitkänen, J., Pippuri, I., Strunk, J., and Maltamo, M. Predicting the spatial pattern of trees by airborne laser scanning. *International Journal of Remote Sensing 34*, 14 (2013), 5154–5165.

[9] Pitkänen, J., Maltamo, M., Hyyppä, J., and Yu, X. Adaptive methods for individual tree detection on airborne laser based canopy height model. *International Archives of Photogrammetry, Remote Sensing and Spatial Information Sciences 36*, 8 (2004), 187–191.

[10] Vauhkonen, M., Tarvainen, T., and Lähivaara, T. Inverse problems. In *Mathematical Modelling*, S. Pohjolainen, Ed. Springer International Publishing, 2016, pp. 207–227.

[11] Yu, X., Hyyppä, J., Holopainen, M., and Vastaranta, M. Comparison of area-based and individual tree-based methods for predicting plot-level forest attributes. *Remote Sensing 2*, 6 (2010), 1481–1495.

IOP Conf. Series: Journal of Physics: Conf. Series **1047** (2018) 012009 doi:10.1088/1742-6596/1047/1/012009

An inverse problem approach to computational active depth from defocus

A Ma and A Wong

Vision and Image Processing Lab, Systems Design Engineering, University of Waterloo, 200 University Ave W, Waterloo, ON N2L 3G1

E-mail: avery.ma@uwaterloo.ca

Abstract. A novel methodology for inferring depth measurements using active quasi-random colored point projection patterns is presented by recourse to computational inverse modelling within the context of depth from defocus. The forward problem is considered that when capturing objects using a camera, light reflected off of objects at different depths will result in blurry observations in the acquired image, whereas the inverse problem seeks to estimate the depth from the knowledge of the blurry observations. In the proposed approach, a colored quasi-random point projection pattern is projected onto the scene of interest, and each projection point in the image captured by a cellphone camera is analyzed using an inverse model to estimate the depth at that point. The proposed method has a relatively simple setup, consisting of a camera and a projector, and enables depth inference from a single capture. We conclude with qualitative and quantitative results that highlight the strong potential for enabling active depth sensing in a simple, efficient manner.

1. Introduction

The problem of robust and fast 3D capture and reconstruction of objects and scenes has always been one of the core research tasks for image processing and robotic applications. Depth information is crucial for many applications including 3D reconstruction, action recognition, human-computer interfacing, etc. As such, depth cameras have received much attention both academically and in industry with constant advancements to depth camera technology.

Traditionally, active depth sensing techniques include methods such as projector-camera structured light systems, laser scanners, ToF cameras, and IR projected point patterns such as the Microsoft Kinect [1]. While recent efforts have lowered the cost of depth camera sensors, depth sensing still requires specialized complex hardware that are not compact. Motivated by these challenges, in this paper we present a system for inferring sparse depth measurements from a single camera capture by leveraging active quasi-random point projections and camera defocus.

We consider a model of forward problem: when capturing objects using a camera, light reflected off of objects at different depths will result in blurry observations in the acquired image. The corresponding inverse problem can be established where an estimation of the depth is sought from the knowledge of the blurry observations. Specifically, the forward and inverse model can be formulated mathematically as:

$$m = f(d) \tag{1}$$

$$d = f^{-1}(m) \tag{2}$$

where m is the camera measurement data of the projected pattern and d is the depth of the surface on which the pattern is projected. The proposed method of inferring sparse depth has a relatively simple setup, thus can potentially lead to very compact and low cost active depth sensing systems.

The proposed method involves projecting a colored quasi-random point pattern onto the scene of interest. The colored quasi-random point pattern can for instance be generated by using a light source with a point pattern mask, resulting in a more compact and self-contained depth sensing device. The detected projected points on the imaged scene are each analyzed to determine the minimum eigenvalue of the related covariance matrix, and the set of these minimum eigenvalues is used to estimate depth via an inverse model.

2. Related Work

Depth from defocus (DfD) methods generally estimate depth by analyzing the difference in blurriness of two images captured at different focal lengths [2, 3, 4], with different methods using different filters for determining the measure of blur. A major drawback to such DfD methods is the unreliable detection of blur, especially in untextured areas of the image. This problem is mitigated in active depth sensing, where an optical projection is used to find correspondences for triangulation. A review of structured light patterns for depth measurement is provided by Salvi *et al.* [5]. Therefore, using active projection patterns does not depend on the objects in the scene and is also effective in untextured regions. On the other hand, active depth sensing systems suffer from occlusion and require complex hardware. As such, we are motivated in the proposed system to take advantage of the benefits from both DfD and active depth sensing methods to design a system that has a simple setup yet is reliable in the depth measurements.

The concept of DfD using active projections has been explored in the literature. Pentland *et al.* [6] used evenly spaced line projections to determine depth from line spread. This simple method is able to create low resolution depth maps. Nayar *et al.* [7] used a dual sensor plane with optimized projection and camera setup to produce a dense depth map and reduce front/back focal ambiguity. Ghita *et al.* [8] used a dense projected pattern with a tuned local operator for finding the relationship between blur and depth. Moreno *et al.* [9] used an evenly spaced point pattern with defocus to obtain an approximate depth map used for automatic image refocusing. These methods use a high density projection pattern which require either a projector or more specialized calibrated hardware.

Promising results were shown in the preliminary approach by Ma *et al.* [10]. In the previous work, blurry projected points as captured by the camera were modelled using the standard deviation of a circularly-symmetric 2D Gaussian. However, in traditional projectors, it occurs that one-pixel point source from the projector does not converge into an ideal point after transmission through the projector-lens system. This causes skewness of the projected point, which leads to erroneous depth inference results. As such, an enhanced active DfD technique that addresses this issue is highly desired.

3. Proposed Method

The proposed system can be described as follows. A colored quasi-random point pattern is projected onto the scene, which is then captured by a camera. The camera's focus is fixed such that the degree of focus of each point in the quasi-random point pattern as it appears in the captured image is dependent on the depth of the surface. The inverse model relating the depth and the blurriness of the projected point pattern is characterized in a calibration step, and used to infer the depth at each point to produce sparse depth measurements. A one-time calibration step is required to learn the inverse model.

IOP Conf. Series: Journal of Physics: Conf. Series **1047** (2018) 012009 doi:10.1088/1742-6596/1047/1/012009

3.1. Calibration

The purpose of the calibration procedure is to characterize the relationship between point projection measurement and depth, and establish the mathematical formulation for the inverse model $f^{-1}(m)$. When out of focus, a projected point will appear blurred, with the degree of blurriness correlated with the depth of the scene at that point. The blurry projected points as captured by the camera are modelled using an elliptical Gaussian intensity map, and the minimum eigenvalue of the elliptical Gaussian covariance is used the characterize the depth. It was noticed that the maximum eigenvalue of the covariance matrix corresponds to the magnitude of the skew caused by projector distortion, whereas the minimum eigenvalue is significantly less affected under the distortion. As such, the minimum eigenvalue can better preserve the geometric information of the actual projected points. The blur effect of a projected point is visualized in Figure 1.

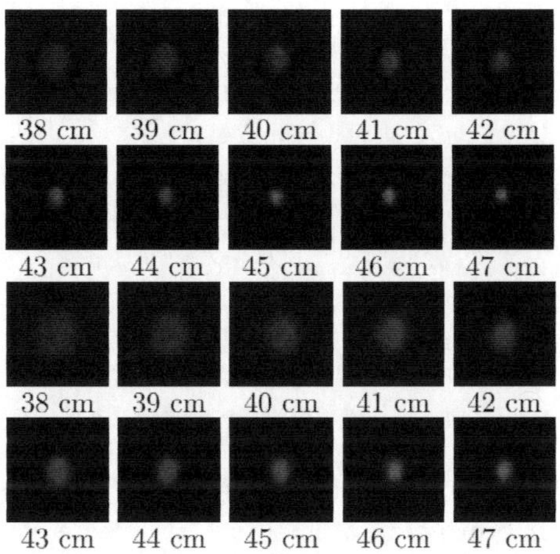

Figure 1: Projected blue and red points on an object at various distances (38cm to 47cm) away from the setup, as captured by the camera

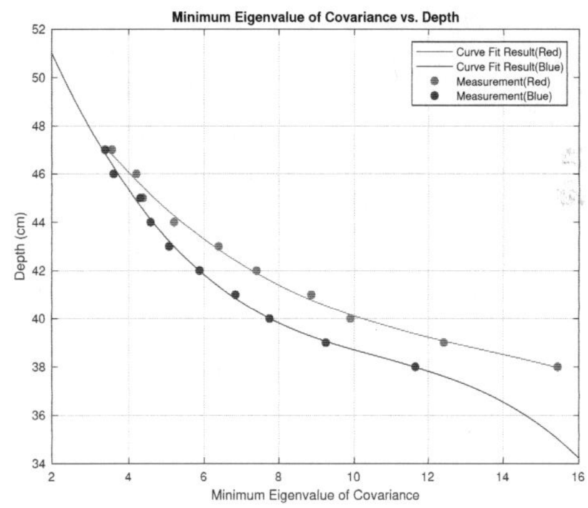

Figure 2: Inverse model characterizing minimum eigenvalue of covariance matrix vs. distance

This eigenvalue-depth model can be used as an inversion operator mapping the range of minimum eigenvalues to a unique depth estimation corresponding to each projected point and the sparse depth map can be obtained. To learn the inverse model, a point pattern is projected onto various planes with known distances away from the projector-camera setup. The measured data points corresponding to minimum eigenvalue of the covariance matrix vs. depth can then be used to construct the inverse model characterizing the relationship between point projection measurement and the distance away from setup. To obtain a continuous curve for the inverse model, regression with a third order polynomial function is used to fit the data points:

$$f^{-1}(m) = a_3 * \lambda(m)^3 + a_2 * \lambda(m)^2 + a_1 * \lambda(m) + a_0 \tag{3}$$

where $\lambda(m)$ is the minimum eigenvalue of the covariance matrix given the camera measurement at the corresponding location. The third order polynomial function is used to fit the data points because it is the lowest order of a polynomial function to provide best fit to the series of data points. It can be observed from Figure 2 that the eigenvalue-depth curve is dependent on the color of the projection pattern. This is due to the discrepancy in the spectral response between channels in the camera sensor. The third order polynomial regression is sought by fitting the data to the model:

$$argmin(\sum(d_i - f^{-1}(m_i))^2) \tag{4}$$

3.2. Sparse Depth Estimation Pipeline

With the calibrated model, the proposed system can then be used to estimate sparse depth of the scene. To this end, the proposed depth recovery method can be divided into 4 main stages outlined in Figure 3 and described as follows.

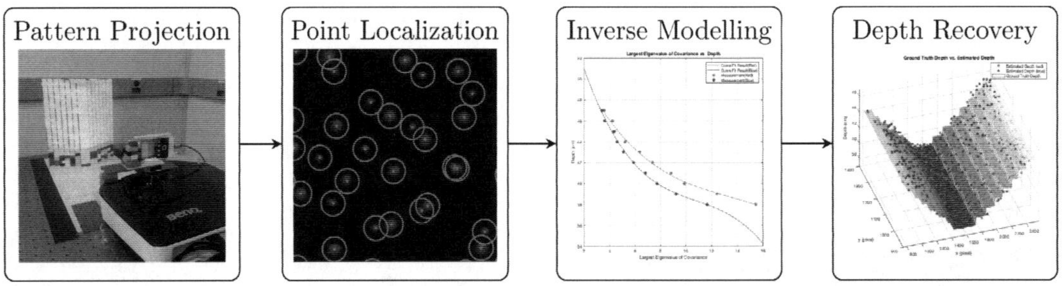

Figure 3: Flowchart of the depth recovery pipeline

Active Quasi-random Point Projection: A colored quasi-random point pattern is projected onto the scene. The use of projection pattern with different wavelengths can provide increased spatial resolution beyond single-wavelength approaches. The camera sensor captures images in three unique ranges of the visible spectrum: red, green and blue. The red and blue channels lie the furthest apart in the spectrum among the three channels, which makes them easily separable from each other. As such, the red and blue projection patterns are selected to achieve improved reconstruction results with an accurate point localization. In the proposed approach, Poisson-disc sampling method was utilized to generate the quasi-random point pattern. Poisson-disc sampling method produces random points that are tightly packed together, but no closer than a specified minimum distance [11]. Compared to other random sampling methods, Poisson-disc method significantly reduces the chance of having overlaps between blurred projected points of the same wavelength, which would result in erroneous depth recovery. Moreover, the size of the projection point is chosen to minimize overlapping of the blurred projected points at the closest depth level.

Point Localization: After the projected point pattern has been captured by the camera, colored patterns are separated by their corresponding wavelength. Otsu's method is used to obtain a binary map consisting of regions corresponding to the projected points [12].

Estimation of Minimum Eigenvalue of the Covariance Matrix: Based on the detected regions of interest of the projected points, the minimum eigenvalue of the covariance matrix of each projected point is estimated by using an intensity-weighted approach, where the number of samples at each location is approximated by multiplying the pixel intensity by a factor of 10,000. An elliptical Gaussian can be formulated to model the generated sample populations:

$$p(\underline{x}) = \frac{1}{(2\pi)^{\frac{n}{2}}\,|\Sigma|^{\frac{1}{2}}} exp\left\{-\tfrac{1}{2}((\underline{x}-\underline{\mu})^T\Sigma^{-1}(\underline{x}-\underline{\mu}))\right\} \tag{5}$$

where μ is the sample mean and Σ is the sample covariance. The minimum eigenvalue of the covariance matrix can be computed as follows:

$$\lambda = min\left\{\lambda_{1,2} : det(\Sigma - \lambda_{1,2}I) = 0\right\} \tag{6}$$

Recovery of Sparse Depth Map: Based on the minimum eigenvalue of the covariance matrix of a projected point, the inverse model can then be used to infer the depth corresponding to that projected point. By performing this on all projected points in the quasi-random point projection pattern, the sparse depth map can be obtained.

4. Experimental Setup

The main goal of this current realization of the proposed technique is to build a compact and portable system to obtain depth information of the scene. For this purpose, the scene is imaged using an iPhone 5S camera (resolution: 3264 × 2448) and the quasi-random point pattern is projected using a BENQ MH630 Digital Projector (resolution: 1920 × 1080).

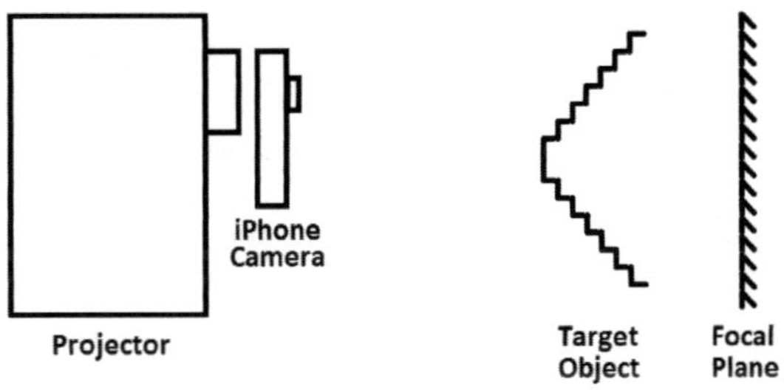

Figure 4: The diagram of the experimental setup

The experimental setup is described in Figure 4. The iPhone camera is mounted 3 cm before the projector. The camera is in-line with the projection lens. The focal plane of both the projector and the camera is 50 cm away from the projector.

5. Results

The test scene was a two-way staircase with 1 cm step-size away from the setup, and the quasi-random point density is approximately 0.20% of the projector resolution. Also, the possible depth range of the inference system is limited to the calibrated depth values.

Figure 5 shows the depth recovery of the object compared to the ground truth surfaces. To evaluate the performance of the proposed method, the root mean square error (RMSE) and correlation coefficient were computed between the estimated depth and the ground truth depth at corresponding locations.

To evaluate the reconstruction errors, the RMSE value is calculated as:

$$RMSE = \sqrt{\frac{1}{n}\sum_{i=1}^{n}(d_i - \hat{d}_i)^2} \tag{7}$$

over the n captured point pattern. The RMSE value of the system is $\pm 0.51cm$, which demonstrates that the proposed depth recovery pipeline is capable of achieving 3D reconstruction with high accuracy. Figure 6 shows the top view of the sparse depth recovery of the test scene.

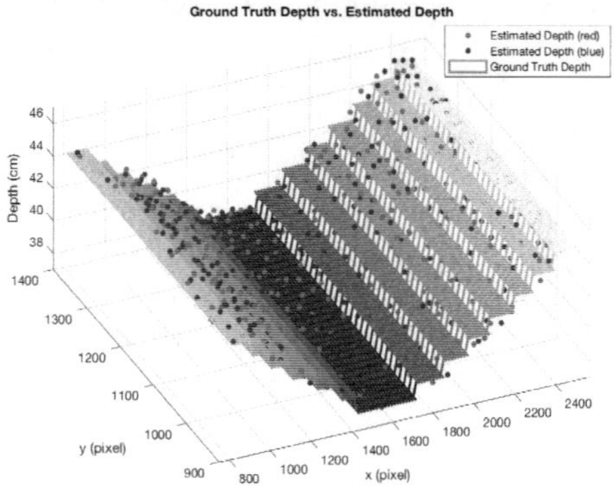

Figure 5: Sparse depth estimation of the staircase with ground truth surfaces

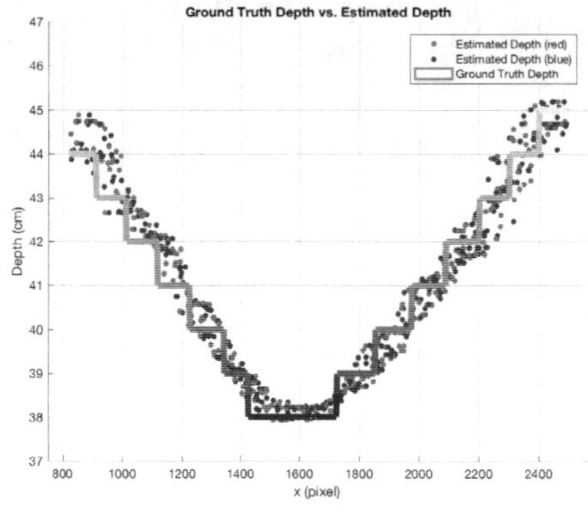

Figure 6: Top view of the reconstructed sparse depth estimation of the staircase

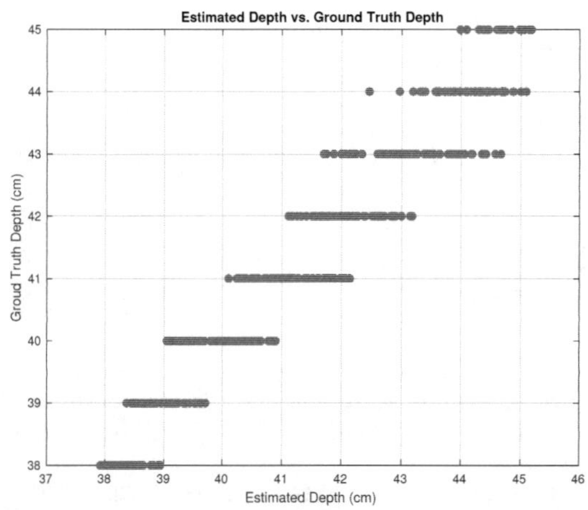

Figure 7: Estimated depth vs. ground truth depth

The correlation coefficient was computed to be **0.9726**, which demonstrates that a strong positive correlation between the ground truth depth and the estimated depth. This strong correlation between the ground truth depth map and the estimated depth can also be observed in the sample point visualization shown in Figure 7.

Figure 8 illustrates the depth estimation of a smiley face made of LEGO. The RMSE value between the ground truth and the estimated depth of the LEGO smiley face is $\pm 0.14cm$, with a high correlation coefficient of **0.9803**. Both visual appearance and quantitative results demonstrate that the proposed method can also achieve great performance when imaging complex geometric shapes and objects. Figure 9 shows the actual scene of the smiley face imaged by the camera. It is worth mentioning that the above results are limited to surfaces that are parallel to the image plane, and the results are expected to degrade for angled surfaces since they are not considered in the calibration stage.

6. Conclusion

This work presented a sparse depth measurement system based on active quasi-random colored point projection and depth from defocus. The new enhanced method extends upon previous

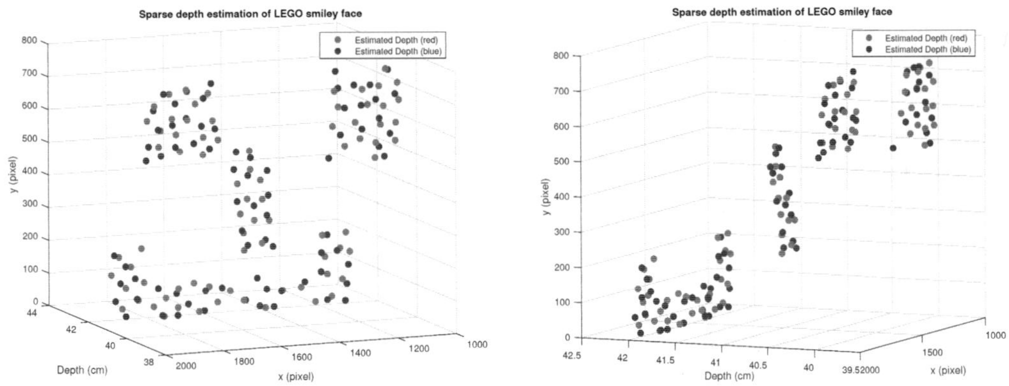

Figure 8: Sparse depth estimation of the LEGO smiley face

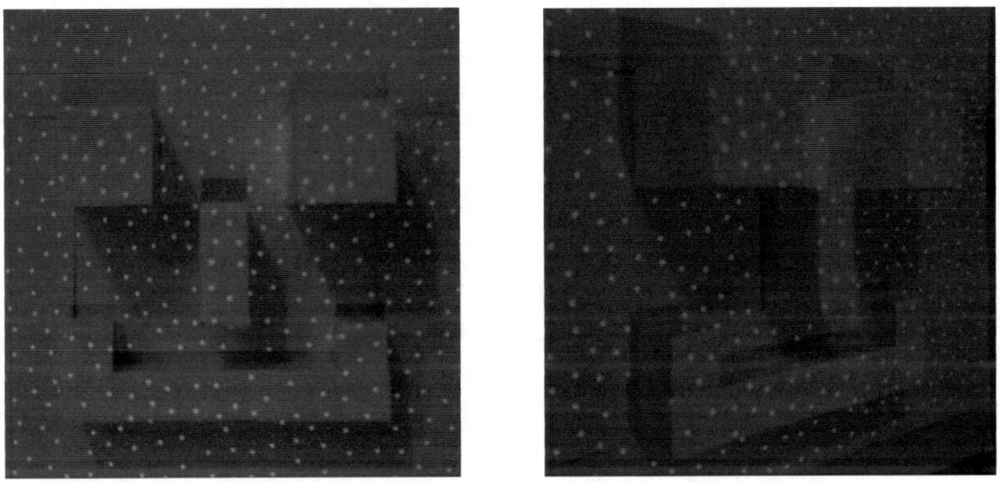

Figure 9: Actual scene of the LEGO smiley face

approach by leveraging an extended inverse model that models the projected pattern as an elliptical Gaussian function as opposed to a symmetric Gaussian, to better account for the distortions of the pattern after projection. Colored point projection drastically increases the density of the reconstructed sparse map. The main advantage of the proposed method is its simplicity in hardware and computation, requiring merely a camera and projected point pattern. There are many additional future directions that can be explored. Deep learning approach can be investigated to better model the relationship between the point projection measurement and its corresponding depth. Different surface orientations can be considered during the calibration stage to improve the robustness of the method when imaging angled planes. In addition, different spatial-temporal, data-guided quasi-random point projection patterns at various wavelengths can be examined to optimize reconstruction accuracy while facilitating for fast acquisitions.

Acknowledgments
This work was supported by the Natural Sciences and Engineering Research Council of Canada, Canada Research Chairs Program, and the Ontario Ministry of Research and Innovation.

IOP Conf. Series: Journal of Physics: Conf. Series **1047** (2018) 012009 doi:10.1088/1742-6596/1047/1/012009

References

[1] Microsoft 2016 Kinect Hardware (online) https://developer.microsoft.com/en-us/windows/kinect/hardware
[2] Xiong Y and Shafer S A 1993 Depth from focusing and defocusing *Proc. CVPR'93* pp 68-73
[3] Pentland A, Darrell T, Turk M and Huang W 1989 A simple, real-time range camera *Proc. CVPR'89* pp 256-261
[4] Watanabe M and Nayar S K 1998 Rational filters for passive depth from defocus *International Journal of Computer Vision* **27** pp 203-225
[5] Salvi J, Pages J and Batlle J 2004 Pattern codification strategies in structured light systems *Pattern recognition* **37** pp 827-849
[6] Pentland A, Scherock S, Darrell T and Girod B 1994 Simple range cameras based on focal error *JOSA A* **11** pp 2925-2934
[7] Nayar S K, Watanabe M and Noguchi M 1996 Real-time focus range sensor *IEEE Transactions on Pattern Analysis and Machine Intelligence* **18** pp 1186-1198
[8] Ghita O, Whelan P F and Mallon J 2005 Computational approach for depth from defocus *Journal of Electronic Imaging* **14** 023021
[9] Moreno-Noguer F, Belhumeur P N and Nayar S K 2007 Active refocusing of images and videos *ACM Transactions On Graphics* **26** pp 67
[10] Ma A, Li F and Wong A 2016 Depth from Defocus via Active Quasi-random Point Projections *Journal of Computational Vision and Imaging Systems* **2** 1
[11] Bridson R 2007 Fast Poisson disk sampling in arbitrary dimensions *SIGGRAPH sketches* pp 22
[12] Liu J, Li W and Tian Y 1991 Automatic thresholding of gray-level pictures using two-dimension Otsu method *Proc. International Conference on Circuits and Systems* pp 325-327

IOP Conf. Series: Journal of Physics: Conf. Series **1047** (2018) 012010 doi:10.1088/1742-6596/1047/1/012010

A machine-learning approach for noise reduction in parameter estimation inverse problems, applied to characterization of oil reservoirs

C. Minutti, S. Gomez, G. Ramos

National University of Mexico, Institute of Applied Mathematics, Mexico City, Mexico

E-mail: carlos.minutti@iimas.unam.mx

Abstract. Reservoir characterization is an inverse problem where parameter values associated to the properties of the porous media, are estimated. In this problem, the pressure data and its log-derivative curves are used in the fitting process. However, the numerical differentiation needed to generate the log-derivative curve, is an ill-posed problem where the noise in the data can be largely propagated. This noise can produce several spurious local minima in the objective function and prevents to get a precise approximation to the parameters. Therefore, an efficient noise reduction method is required to achieve the desired accuracy in the parameter values.

In this work, we explore three noise reduction methods by applying them to well-test data. These methods are based on multi-step finite differences and splines, but using a machine-learning approach.

In addition, a new method is proposed which produce an optimal curve which is a linear combination of multiple data curves, that are approximations to the solution of the inverse problem.

We show how these methods which use information from the forward model are more efficient for noise reduction. The proposed methods can be used in many inverse problems where there is a mathematical model meant to describe the measured data.

1. Introduction

In many inverse problems, where the parameters of a forward model are identified through measured data, the noise in the data might be a major problem, especially in complex models (i.e. with a large number of parameters).

The characterization of oil reservoirs through well tests analysis is one of these problems, where the log-derivative curve (the derivative of the measured pressure in the well with respect to the logarithm of time), has also to be fitted as it indicates several characteristics of the reservoir. However, since the numerical derivation is an ill-posed problem, the noise of the data is highly propagated, making it difficult to achieve a good estimation of the parameters of the model.

For reservoir characterization, the forward model that we are using is a Triple Porosity-Double Permeability (TP-DP) model [2], in which it is necessary to identify up to 10 parameters for a real case (see [5], [6] and [8] where reservoir characterization using the TP-DP model, is described). The solution of the model for a given set of parameters, produces the curve of the pressure in the well under study. Due to the number of parameters to be identified, this model can give more information of the porous media than the conventional double-porosity model of Warren and Root [11]. Thus, it is necessary a noise reduction method to be able to produce the so called type-curves of the pressure and the log-derivative, that have to be fitted to produce accurate estimate of the parameters of the model.

The Bourdet algorithm [1] based in a finite difference approach, has been the standard method to calculate the log-derivative curve of well-test data, but it has shown to be inefficient when the noise level is high.

Lane *et al.* [7] address the problem of noise reduction in well-test data for a simpler forward model, using splines in which the smoothing parameters are optimized. Later, Escobar *et al.* [3] also address the problem and compare different noise reduction methods, which are based on finite difference schemes (mainly the Bourdet algorithm with different values of L and the derivative of interpolated polynomials) and splines, concluding that splines is the most effective method.

The approach of using multi-step finite differences and/or splines, to reduce the noise in the data, is well extended in many different areas. However, the finite differences based on the interpolation of polynomials at several points, considered by Escobar *et al.* actually increase the noise in the data.

In this study, we compare three methods of noise reduction: Multi-step finite differences, Splines and a Mixture of sub-optimal curves. In both, splines and multi-step finite differences, we propose a machine-learning approach, that uses information from the forward model to determine the optimal parameters of each of the methods. Being splines a method widely used in different areas to reduce noise in the data, we also study the effect of another type of splines: regression splines.

In the first part of the paper we present the methodology of how the classic algorithms of multi-step finite differences and splines are adapted to a machine-learning approach and the proposed method of mixture of sub-optimal curves is described. In the second part, we present the results obtained when applying these noise reduction methods to a large data set, measuring the error of the method and also its effect in the parameter estimation error of the inverse problem.

2. Methodology

Considering a machine-learning approach, we use the forward model (the TP-DP model) to generate a first group of 1,000 random synthetic data sets, for which Gaussian noise was subsequently added. In each of the methods to be studied, this data set is used to *train* the algorithms, that is, to determine the optimal value of the parameters that minimize the error between the noisy and noise-free data. Later a second group of 1,000 random synthetic data sets is generated, for which noise is also added. This data set is the *cross-validation* data set and we use this data to compare the performance of the proposed algorithms, with the classic approach and also with the optimized parameters. The purpose of determining the optimal parameters of each algorithm with a training data set, is to help the noise reduction method to reproduce better the kind of data created by the forward model in the inverse problem.

The error of the noise reduction algorithm at each data point is measured by:

$$Error = \left| \frac{d'_{algorithm} - d'_{noise-free}}{d'_{noise-free}} \right| \tag{1}$$

where $d'_{noise-free}$ is the log-derivative of the data without noise (in the synthetic examples) and $d'_{algorithm}$ is the log-derivative of the obtained data, after applying the noise reduction algorithm to the noisy data.

To compare the performance of each method we use the mean and median value of $\log_{10}(Error)$ because, as we are measuring a relative error, we are more interested in the order of magnitude of the error. Also because the mean is more sensitive to the existence of outliers than the median, when relative errors are used, $\log_{10}(Error)$ has a more symmetric distribution, so it is less affected by outliers.

2.1. *Multi-step finite differences*

In the case of noisy data, it does not make sense to use finite difference schemes that are based on interpolation of polynomials (as it is the case of the schemes presented by Fornber [4]), since they increase the noise of the numerical derivative (this happens in Escobar *et al.*). Instead, we use the schemes presented by Savitzky and Golay [10], where the base polynomial, fits the data in the least squares sense, following an idea similar to the Lanczos derivative. Based on that, we choose a 9-point scheme because this is the maximum number of points showing a relevant improvement in the noise reduction for a set of data samples. The finite difference 9-point formula of Savitzky and Golay [10] is:

$$f_0' \approx \frac{(f_1 - f_{-1}) + 2(f_2 - f_{-2}) + 3(f_3 - f_{-3}) + 4(f_4 - f_{-4})}{60h} \tag{2}$$

We propose here another 9-point finite difference method:

$$f_0' \approx \frac{1}{4}\left(\omega_1 \frac{f_1 - f_{-1}}{2h} + \omega_2 \frac{f_2 - f_{-2}}{4h} + \omega_3 \frac{f_3 - f_{-3}}{6h} + \omega_4 \frac{f_4 - f_{-4}}{8h}\right) \tag{3}$$

where $\sum_{i=1}^{4} \omega_i = 1$. The weights values ω_i are determined by minimizing the error (equation 1) obtained with the training data set. Subsequently, the errors of both methods are compared using the cross-validation data set.

2.2. *Splines*

For splines, we consider three cases. First, a classic approach where a cubic spline is used, and the smoothing parameter is determined by cross-validation, that is, for each one of the 1,000 data sets, a part of the data is used to determine the spline coefficients, and the other part to determine the smoothing parameter, selecting in each case the value that minimizes the mean squared error.

In the second approach, we use optimization to determine a single smoothing parameter value, which minimizes the error of the entire training data set and the same value will be used in the cross-validation data set.

In the third approach we use regression splines (see [9]) where, unlike the cubic splines, it is possible to use different types of base functions to approximate the data. In order to have a good result, the base functions must be similar or related to the functions that are generated by the forward model which are used to fit the measured data. In the case of the TP-DP model, the asymptotic approximations of the model at early and late times (which appear in Camacho *et al.* [2]), are used as base functions. The smoothing parameter is determined in the same way as in the previous case, using optimization to get the value that minimizes the error of the entire training data set.

The performance of each algorithm is determined by measuring the error of the equation (1) in the cross-validation data set.

2.3. *Mixture of sub-optimal curves*

We propose here a method that we call *mixture of sub-optimal curves*, which consists of using the noisy (pressure) data to perform a series of quick optimizations to solve the inverse problem from different starting points, in order to obtain approximations to local optima, that is, to obtain different values of the parameters of the forward model, which can approximate the data in a rough way. These approximated data are the sub-optimal curves.

These sub-optimal curves are combined into one curve using a linear combination, where the weight of each curve is determined by minimizing the quadratic error between the combined curve and the noisy data. Based on the multiple linear regression model, we can get the vector of the weights (b) using the regression normal equation:

$$b = (X'X)^{-1}X'y \tag{4}$$

where, in this model, X is a matrix whose columns are the different sub-optimal curves values and y is a vector which contains the noisy data.

A regularization parameter (λ) is added to the model in order to avoid problems of singular matrices (due to similar sub-optimal curves) and also to avoid overfitting, which can lead to reproduce the noise in the data. Thus, applying Tikhonov regularization, the final model to estimate the weights of each curve is:

$$b = (X'X + I\lambda)^{-1}X'y \tag{5}$$

where I is the identity matrix.

The noise-reduced curve and its log-derivative are obtained with $b'X$ and $b'X_d$, where X_d is the matrix whose columns are the log-derivative of the columns of X.

Both, the best number of sub-optimal curves to be used and the value of the regularization parameter, are obtained by optimizing (minimizing) the error of the training data set. For this case, 40 sub-optimal curves were obtained from the TP-DP model for each of the 1,000 cases in the training data set.

Once these parameters are determined, the process is applied to the cross-validation data set in order to estimate the error of the proposed method.

2.4. *Training process and noise reduction comparative*

The following is a summary of the training process for each algorithm and how the noise reduction comparison is performed.

(i) A training and a cross-validation data sets are created. Each set has 1,000 synthetics data sets which are generated by solving the forward model using random parameter values and adding random noise to the data.

(ii) For the cubic splines method, the regularization parameter is estimated dynamically for each data set by cross-validation, where part of the data is used to find the optimal value of the regularization parameter by minimizing the quadratic error.

(iii) For the trained algorithms, the training data set is used to obtain the (static) parameter values of each method by minimizing the error between the noisy and the noise-free data, using optimization.

- In the case of multi-step finite differences, the optimization is used to estimate the optimal weights of each finite difference.
- In the case of the spline based algorithms, the optimization is used to estimate the optimal value of the regularization parameters.
- In the case of the mixture of sub-optimal curves, the optimization is used to estimate the number of sub-optimal curves to be used as well as the regularization parameter value.

(iv) The cross-validation data set is used to estimate the error of each method applying equation (1), where the weights and regularization parameters (previously estimated with the training data set) are used. This data set is also used to estimate the error of the classic methods.

2.5. *Error of the forward model parameters*

In order to measure the effect of these noise reduction methods in the estimation of the parameters of the forward model, we solve the inverse problem using a optimization following the procedure described in Minutti *et al.* [8] to obtain an estimation of the parameters of the forward model for the synthetic data in the cross-validation data set.

Additionally, we provide as a starting point to the optimization algorithm, the true value of the parameters with which the noise-free data was generated. Because the optimization algorithm has as starting point the true value of the parameters, any error in the parameter estimates is due to the noise in the data.

To measure the error of each parameter we use the following formula:

$$err(\hat{\theta}) = \frac{|\theta - \hat{\theta}|}{\max_\theta - \min_\theta} \tag{6}$$

where θ is the true value of the parameter, $\hat{\theta}$ is the estimated value given by the optimization algorithm, and $\max_\theta - \min_\theta$ is the length of the interval of possible values that the parameter can take. Thus, equation (6) measure the absolute error expressed between 0 to 1.

3. Results

Using the cross-validation data set, the mean and median of $\log_{10}(Error)$ are determined for each of the methods described above, obtaining the values of Table 1, being the classic methods the 9-point scheme Savitzky and Golay [10] (FD-9_CLS) and the cubic splines (SPL_CLS).

Table 1. Mean and median value of $\log_{10}(Error)$ for the tested algorithms

	Multi-step finite differences		**Smoothing Splines**			**Mixture of sub-optimal curves**
	FD-9_CLS	FD-9	SPL_CLS	SPL	REG_SPL	MIX_OPT_C
Mean	-1.6257	-1.7406	-2.1575	-2.1846	-2.2604	-2.5310
Median	-1.5021	-1.6342	-2.1716	-2.1875	-2.2846	-2.5974

It can be observed that, as in multi-step finite differences and also in splines, the methods where the parameters of the algorithms were determined using a training data set (FD-9, SPL, REG_SPL), are more efficient than using a classic method (FD-9_CLS, SPL_CLS). It is also observed that the method of regression splines (REG_SPL) is more efficient than the other two methods of splines, and the method of mixture of sub-optimal curves is better than any other of the analyzed algorithms. Specifically, the mixture of sub-optimal curves (MIX_OPT_C) has only 53.6% of the noise of regression splines, the second best method ($\frac{10^{-2.5310}}{10^{-2.2604}} \approx 0.5363$), showing that the use of information of the forward model, produces a more efficient method for noise reduction.

Figures 1 and 2 show the boxplots and histograms for the methods where the algorithms were trained: Multi-step finite differences (FD-9), cubic splines with optimized smoothing parameters

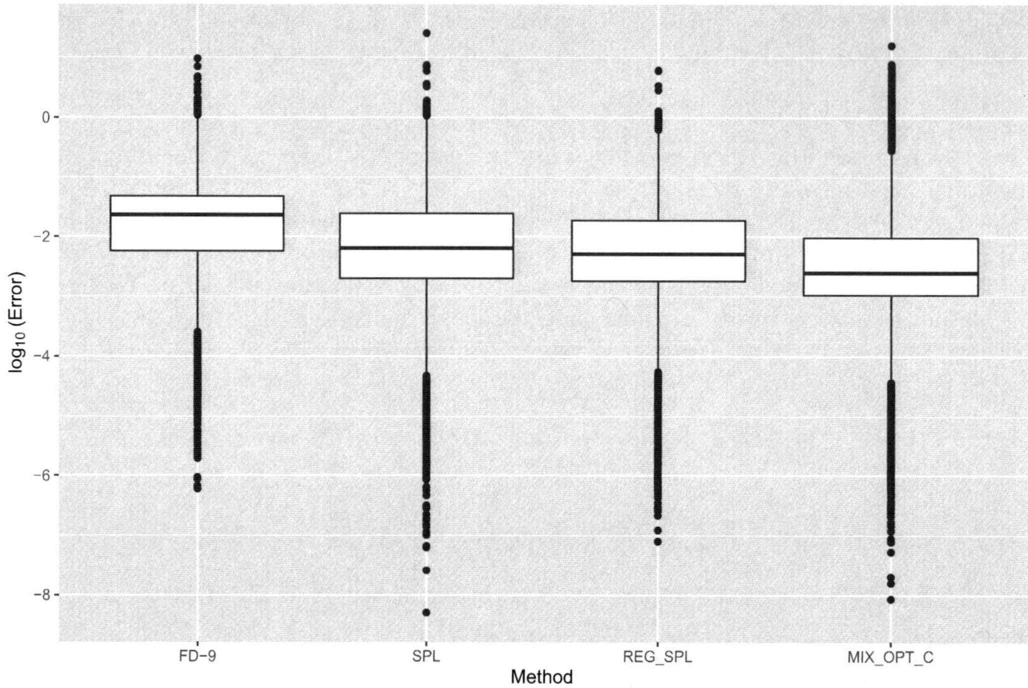

Figure 1. Boxplot of $\log_{10}(Error)$ for the trained algorithms.

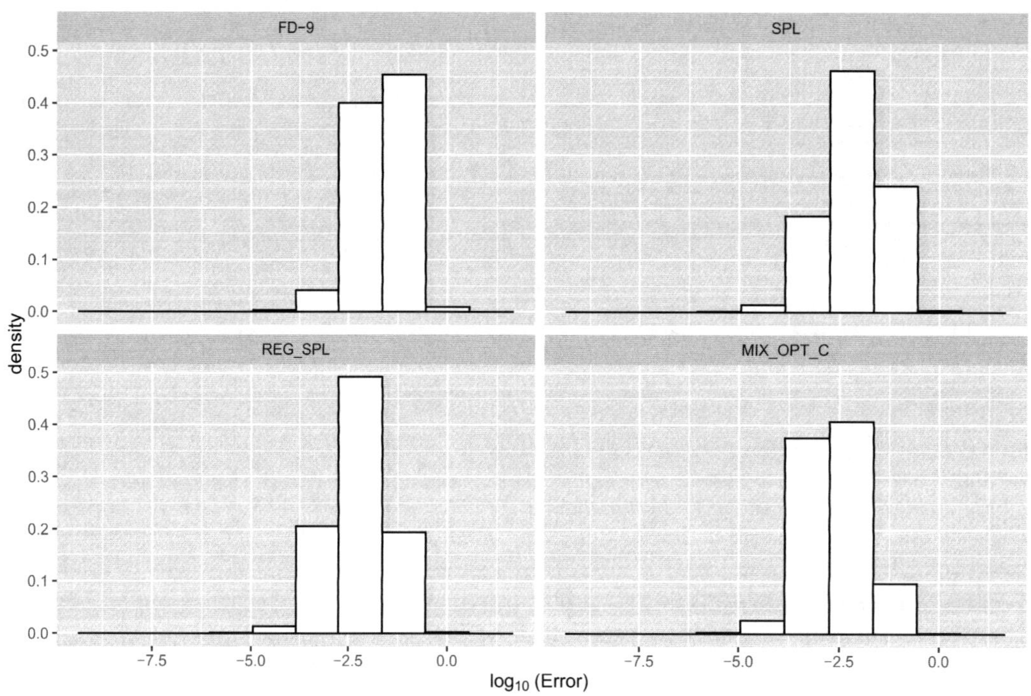

Figure 2. Histogram of $\log_{10}(Error)$ for the trained algorithms.

(SPL), regression splines with optimized smoothing parameters (REG_SPL) and the mixture of sub-optimal curves (MIX_OPT_C).

In Figures 3-5 we present three examples taken from the cross-validation data set, where the pressure and its derivative (log-derivative) are plotted on a log-log scale with the result of the three best noise reduction methods: trained splines (cubic and regression splines), and mixture of sub-optimal curves. The noisy data and its derivative, calculated with a finite difference method (Bourdet), is also presented.

The major noise reduction is obtained with the mixture of sub-optimal curves. Cubic splines and regression splines have problems of oscillations, especially in the valleys of the log-derivative curves.

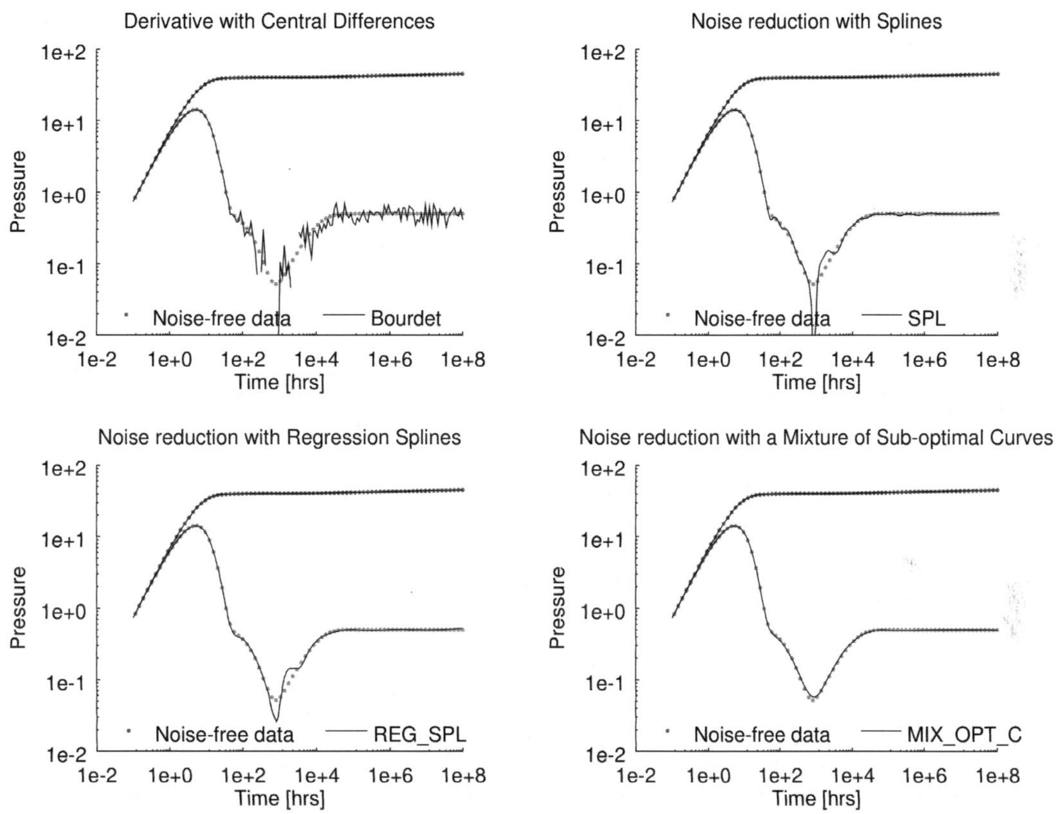

Figure 3. Example 1: Noise reduction with the three best methods.

In Table 2 we show the mean estimation error (using equation 6) of the parameters of the forward model, when we solve the inverse problem using the noisy data in the cross-validation data set and also when we use the noise-reduced data given by the trained algorithms. As we provide to the optimization algorithm with the true value of the parameters as a starting point, the estimation error is due to the noise in the data.

For each parameter a significant reduction in the estimation error can be observed, compared with the noisy data. For parameters like ω's, λ_{mf}, λ_{mv} and κ, the mean estimation error is one-third of the error of the noisy data when the mixture of sub-optimal curves was used.

Although regression splines was a better noise reduction method than cubic splines, there was not much difference in the mean estimation error of the parameters.

Results in Table 2 show that:

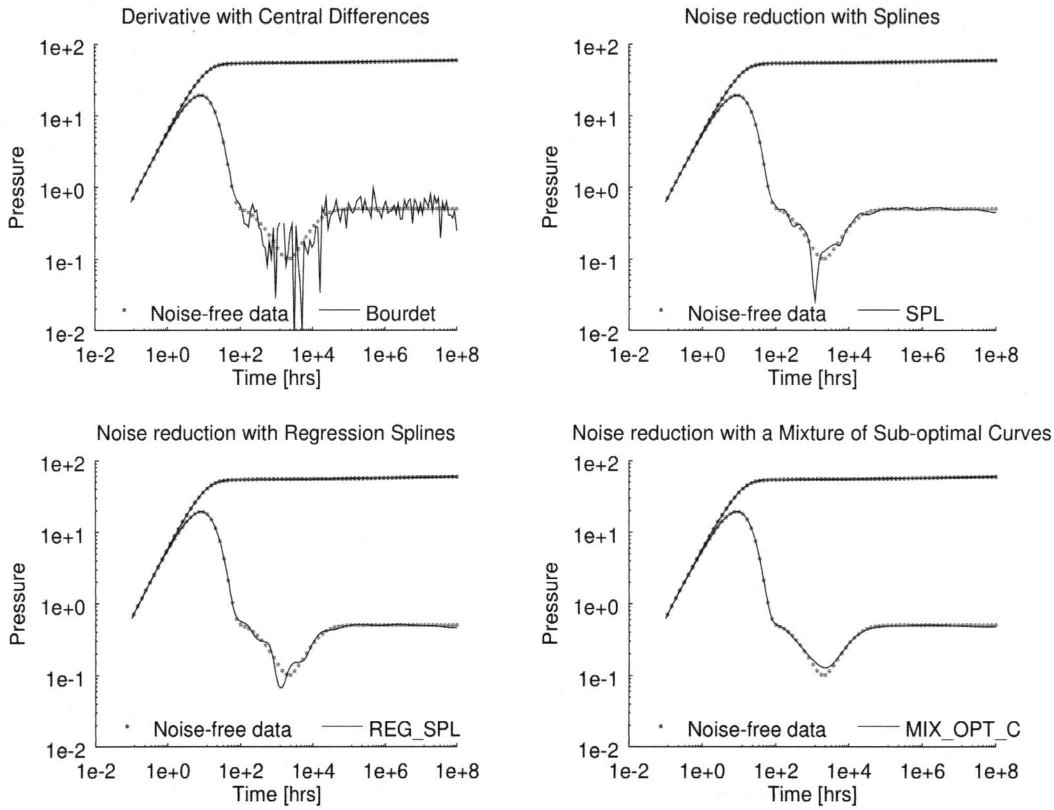

Figure 4. Example 2: Noise reduction with the three best methods.

Table 2. Mean estimation error of the model parameters with different noise reduction methods

Method	ω_f	ω_v	λ_{mf}	λ_{mv}	λ_{vf}	κ	s	C_D
NOISY DATA	0.1862	0.1844	0.1008	0.0966	0.1615	0.1567	0.0148	0.0011
SPL	0.0955	0.0962	0.0403	0.0355	0.0775	0.0670	0.0012	0.0003
REG_SPL	0.0923	0.0945	0.0395	0.0365	0.0800	0.0639	0.0011	0.0003
MIX_OPT_C	0.0666	0.0623	0.0360	0.0340	0.0719	0.0488	0.0005	0.0001

- If a suitable noise reduction method is not used, even with a good initial guess of the parameters (or a good optimization algorithm), due to the noisy data we can only expect a limited accuracy in the estimation of the parameters of the inverse problem.
- Even if a noise reduction method performs better (i.e. smaller error between the noisy data and the noise-free data), it may not have a significant effect on the inverse problem, this is the case for `SPL` and `REG_SPL`, where `REG_SPL` has a smaller error than `SPL` (Table 1), but the improvement in the mean estimation error of the parameters is almost the same. This may mean that part of the improvement done by the algorithm could be in an area (a time interval in our pressure curve) where the data may have little sensitivity to the parameters.
- The `MIX_OPT_C` method does not have the problem mentioned above, because we are performing quick optimizations, solving the inverse problem many times and it is because of this that the shape of the curve changes mainly in areas that are sensitive to the parameters.

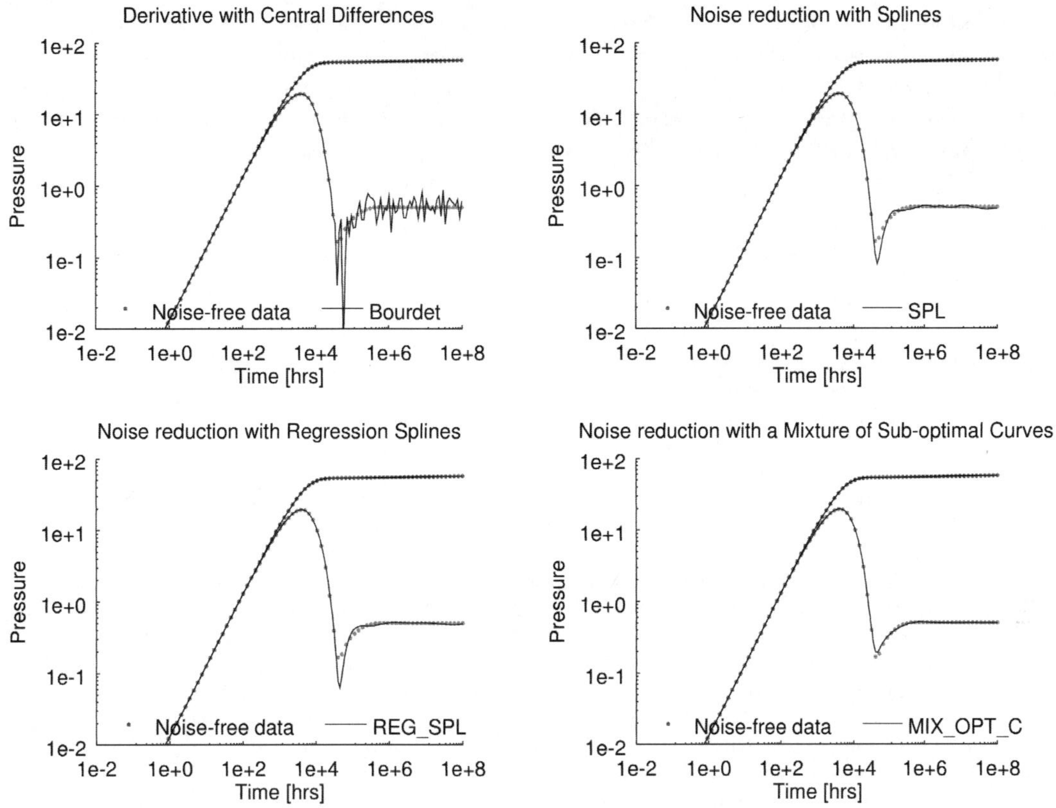

Figure 5. Example 3: Noise reduction with the three best methods.

4. Conclusions

By training the algorithms with a synthetic data set it can be seen a smaller error between the the noisy and the noise-free data compared with a classic approach. Especially when a regression spline (with base functions derived from the forward model) is used, we obtain better results than using a classic cubic spline.

The proposed method, it mixture of sub-optimal curves, combines the idea of using base functions from the forward model and the training of the algorithm in order to get a set of optimized parameters to the problem. This method was the most efficient for noise reduction, having only 53.6% of the noise that has the regression splines, being the former the second best method.

When the inverse problem was solved for the cross-validation data set, we observe the impact of the different noise reduction methods in the parameter estimation problem, being the mixture of sub-optimal curves also the best method, having for many parameters, up to one-third of the estimation error, as compared with the error of the noisy data.

Because in many inverse problems there is a mathematical model which is used to approximate or reproduce a real phenomena (forward model), the methods presented in this work can be applied to solve them, as the forward model can be used to train the algorithms to reduce the noise in the observed data. Especially mixing different sub-optimal solutions of the inverse problem is a good approach to get noise-reduced data which is more consistent with the real phenomena, and because of that it can be better exploited in an inverse problem.

These methods, based in the forward model, are especially important when there are many parameters to be identified, because it increases the likelihood of adjusting the solution of the inverse problem to noise in the data. So, a noise reduction method which is consistent with the forward model helps to improve the accuracy of the estimated parameters.

In our application (characterization of oil reservoirs), these parameter estimates can be used to take decisions to forecast production. It is because of this that a more accurate estimation of the parameters can have a large impact in the production of a well.

References

[1] Bourdet D, Ayoub J A and Pirard Y M 1989 Use of Pressure Derivative in Well Test Interpretation. em Society of Petroleum Engineers. doi:10.2118/12777-PA

[2] Camacho-Velazquez R, Vasquez-Cruz M A, Castrejon-Aivar R and Arana-Ortiz V 2005 Pressure Transient and Decline Curve Behaviors in Naturally Fractured Vuggy Carbonate Reservoirs. *Society of Petroleum Engineers.* doi:10.2118/77689-PA.

[3] Escobar F H, Navarrete J M and Losada H D 2004 Evaluation of Pressure Derivative Algorithms for Well-Test Analysis. *Society of Petroleum Engineers.* doi:10.2118/86936-MS

[4] Fornber B 1988 Generation of Finite Difference Formulas on Arbitrarily Spaced Grids. *Mathematics of Computation*, Vol. 51, No. 184. pp. 699 – 706, doi:10.2307/2008770

[5] Gomez S, Camacho R, Vasquez M, Ramos G, del Castillo N and Mesejo J A 2014 Well Test Characterization of Naturally Fractured Vuggy Reservoirs, with a Global Optimization Method. *Offshore Technology Conference.* doi:10.4043/24762-MS

[6] Gomez S, Ramos G, Mesejo A, Camacho R, Vasquez M and del Castillo N 2014 *Well test analysis of naturally fractured vuggy reservoirs with an analytical triple porosity - double permeability model and a global optimization method*, In Oil Gas Sci. Technol. - Rev. IFP Energies nouvelles. pp. 653 – 671

[7] Lane H S, Lee W J and Watson A T 1991 An Algorithm for Determining Smooth, Continuous Pressure Derivatives From Well-Test Data. *Society of Petroleum Engineers.* doi:10.2118/20112-PA

[8] Minutti C, Ramos G, Gomez S, Camacho R, Vzquez M and Castillo N del 2016 Robust Characterization of Naturally Fractured Carbonate Reservoirs Through Sensitivity Analysis and Noise Propagation Reduction. *Society of Petroleum Engineers.* doi:10.2118/181182-MS

[9] Monahan J F 2011 Numerical Methods of Statistics. Cambridge University Press. chapter 7 pp 159–163

[10] Savitzky A and Golay M J E 1964 Smoothing and Differentiation of Data by Simplified Least Squares Procedures. Anal. Chem. *In Analytical Chemistry*, Vol. 36, No. 8. pp. 1627-1639, doi:10.1021/ac60214a047

[11] Warren J E and Root P J 1963 The Behavior of Naturally Fractured Reservoirs. *Society of Petroleum Engineers.* doi:10.2118/426-PA

IOP Conf. Series: Journal of Physics: Conf. Series **1047** (2018) 012011 doi:10.1088/1742-6596/1047/1/012011

Study of defects of elastic thermal protection by inverse methods of nonlinear acoustics

A V Nenarokomov[1,3], O M Alifanov[1], K A Nenarokomov[1], D M Titov[1] and V S Finchenko[2]

[1] Moscow Aviation Institute, Russia
[2] Lavochkin Association, Khimki, Russia
[3] Author to whom any correspondence should be addressed: aleksey.nenarokomov@mai.ru

Abstract. The purpose of this research was to develop an approved experimental technique for remote (non-contact) diagnostics of structural defects in elastic materials of thermal protection of inflatable re-entry vehicles [1]. The performance of equipment is based on nonlinear interaction of acoustic beam of finite amplitude in analyzed materials with structural defects. Such problems are of great practical importance in the study of properties of composite materials used as non-destructive elastic surface coating in objects of space technology, power engineering etc. The most promising direction in non-destructive diagnostics (research methods) for the elastic composite materials is to use the procedure of inverse problems [2-8]. Corresponded inverse problem is considered as minimization of least-squares residual functional for calculated and measured acoustic pressure at external surface of specimen. The minimization of the residual functional is executed by conjugate gradient method. Determination of spatial distribution of structural acoustic nonlinearity in the analysed specimen allows spotting the defects. In the paper the implementation of experimental technique for remote diagnostics of subsurface flaws in elastic materials is considered. Results of analysis of computational effectiveness of inverse problems algorithms are included in this paper. Experimental validation of the method is based on experimental testing of specimens with a-priori installed artificial defects.

Nomenclature

Latin symbols

d	Thickness of specimen
c	Velocity of sound
$f(\tau)$	Acoustic pressure measurement
$\overline{g^s}$	Vector of gradient minimization method at current iteration
J	Least-square minimized functional
$\overline{J}'^{(s)}_p$	Gradient of the functional J at current iteration
M	Number of pressure sensors
N_ε	Dimension of approximated function

IOP Conf. Series: Journal of Physics: Conf. Series **1047** (2018) 012011 doi:10.1088/1742-6596/1047/1/012011

$p(x,\tau)$	Acoustic pressure
$p_0(x)$	Initial temperature
s	Number of iteration
x	Spatial variable

Greek symbols

γ^s	Step descent
δ_f	Integral error of temperature measurements
$\varepsilon(x)$	Nonlinearity coefficient of acoustic wave propagation
$\varphi_k(x)$	Basic functions of approximation
ρ	Current density of material
$\sigma_m(\tau)$	Measurement variance
τ	Time
$\tau_{\max}$	Final time
$\psi(x,\tau)$	Adjoint variable for acoustic pressure
$\frac{\partial^2 \xi_1}{\partial \tau^2}$	Function, which determines the value of the gradient of acoustics pressure

1. Introduction

Remote (non-contact) and non-destructive diagnostics of specimens of elastic composite thermal protective materials of inflatable re-entry vehicles (figure 1)were executed to estimate nonlinearity coefficient in domain of possible defects of materials based on inverse method of nonlinear acoustics [8]. The value of the nonlinearity coefficient can be used to estimate geometric characteristics of defects [9-10]. The experimental specimens of elastic material were submitted for acoustic tests as a slab about 60x60x6 mm (figure 2) with thickness-to-length ratio not less than 1:10, such ratio of the geometrical characteristics, provide in the course of testing a pressure distribution in the specimen close to one-dimensional. Based on these speculations, the mathematical model of wave propagation in the specimen can be covered the following equations [8]:

$$\frac{\partial^2 p}{\partial \tau^2} - c^2 \frac{\partial^2 p}{\partial x^2} = \frac{1}{c^2 \rho} \varepsilon(x) \frac{\partial^2 p}{\partial \tau^2}, \quad x \in (0,d), \tau \in (0,\tau_{\max}] \tag{1}$$

$$p(x,0) = p_0, \; x \in [0,d] \tag{2}$$

$$\frac{\partial p}{\partial \tau}(x,0) = 0, \; x \in [0,d] \tag{3}$$

$$-\frac{\partial p}{\partial x}(0,\tau) = \rho_0 \frac{\partial^2 \xi_1}{\partial \tau^2}, \; \tau \in (0,\tau_{\max}] \tag{4}$$

$$\frac{\partial^2 p}{\partial x^2}(d,\tau) = 0, \; \tau \in (0,\tau_{\max}] \tag{5}$$

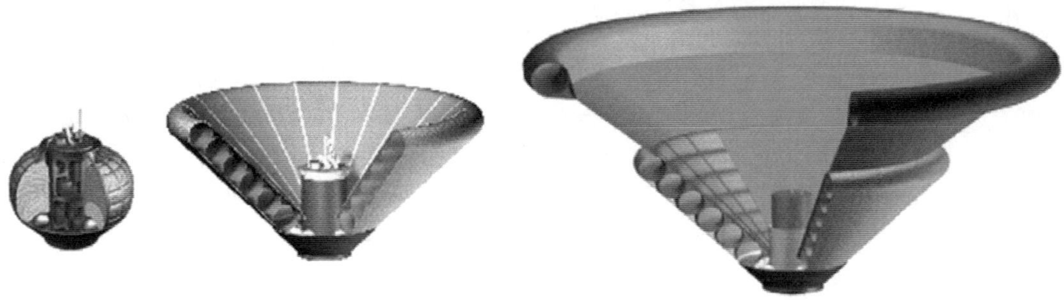

a b c

Figure 1. Re-entry vehicle with inflatable structure: a - initial position, b - 1st stage, c – 2nd stage

Figure 2. Specimen of material

In model (1)-(5) the nonlinearity coefficient $\varepsilon(x)$ is unknown. The physical meaning of this coefficient is spatial distribution of uniformity or non-uniformity of the material under analysis. $\varepsilon(x)$ is equal 0 for perfect structure of materials.

The results of pressure measurements at the left side (boundary) of the material's specimen are assigned as necessary additional information to solve an inverse problem

$$p^{\exp}(0,\tau) = f(\tau),\ \tau \in (0, \tau_{\max}) \tag{6}$$

Writing down a least-square discrepancy of the calculated and experimental pressure values in points of measuring we receive the residual functional

$$J(\varepsilon(x)) = \int_{0}^{\tau_{\max}} (p(0,\tau) - f(\tau))^2 d\tau \tag{7}$$

To construct an iterative algorithm of the inverse problem solving a conjugate gradient method has been used [9]. This inverse problems is conditionally well-posed, therefore special regularization technique is not necessary. The uniform mash with number of points N_ε is formed in

domain $x \in [0, d]$
$\omega_\varepsilon = \{x_k = (k-1)\Delta x, \ \ k = \overline{1, N_\varepsilon - 1}\}$, and the search function can be approximated as

$$\varepsilon(x) = \sum_{k=1}^{N_\varepsilon} \varepsilon_k \varphi_k(x) \tag{8}$$

where $\varphi_k(x)$, k=1,..., N_ε are the basic approximating functions (B-splines, piecewise functions, etc.), the coefficients of approximation ε_k, k=1,..., N_ε are unknown. In this case the functional (7) is reduced to the function of N_ε variables.

$$J(\bar{\varepsilon}) = \int_0^{\tau_{\max}} (p(0,\tau) - f(\tau))^2 d\tau \tag{9}$$

Following [9] the gradient of the residual functional can be presented as

$$\frac{\partial J}{\partial \varepsilon_k} = -\int_0^{\tau_{\max}} \int_0^d \psi(x,\tau) \frac{1}{c^2 \rho} \frac{\partial^2 p}{\partial t^2} dx d\tau \tag{10}$$
$$k = 1, \ldots N_\varepsilon$$

where $\psi(x,\tau)$ is the solution of adjoint problem to linear form of (1) – (5):

$$\frac{\partial^2 \psi}{\partial \tau^2} - c^2 \frac{\partial^2 \psi}{\partial x^2} = \frac{1}{c^2 \rho} \varepsilon(x) \frac{\partial^2 \psi}{\partial \tau^2}, \ x \in (0, d), \ t \in [0, \tau_{\max}) \tag{11}$$

$$\psi(x, \tau_{\max}) = 0, \ x \in [0, d] \tag{12}$$

$$\frac{\partial \psi}{\partial t}(x, \tau_{\max}) = 0, \ x \in [0, d] \tag{13}$$

$$c^2 \frac{\partial \psi}{\partial x}(0, \tau) + 2(p(0,\tau) - f(\tau)) = 0, \ \tau \in [0, \tau_{\max}) \tag{14}$$

$$\frac{\partial^2 \psi}{\partial x^2}(d, \tau) = 0, \ \tau \in [0, \tau_{\max}) \tag{15}$$

The boundary conditions in inverse problems of non-linear acoustics are based on results of measurements and can be Dirichlet or Von Neumann conditions. Type of boundary conditions should satisfy the uniqueness of considered inverse problem. The uniqueness usually demands the minimum number of measurement in one experiment. And so to estimate the coefficient $\varepsilon(x)$ in (1)-(7) it is necessary to measure the gradient of pressure at the left surface and to provide arbitrary conditions at the right surface, and to measure pressure at the left boundary as additional information to solve inverse problem.

The corresponded scheme of testing is presented in figure 3. In such case we can to estimate the spatial distribution of defect according estimations of $\varepsilon(x)$.

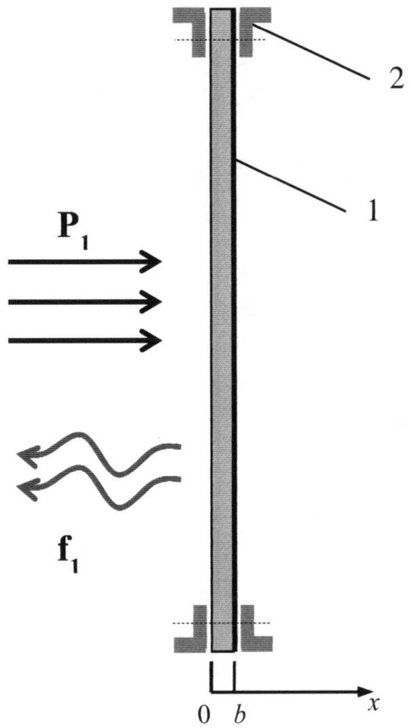

Figure 3. Scheme of testing: 1 – specimen of material, 2 – tensing frame of experimental module, p_1 – acoustics loading at specimen, f_1 - transient measurement of pressure at the front boundary.

2. Analysis of influence of experiment conditions to the accuracy of inverse problems solution

The successive application of remote (non-contact) diagnostics of structural defects in elastic materials demands correct development of computational algorithm of experimental data processing. At this stage of research computational experiments become the most effective. In this case all coefficients of mathematical model (1)-(5) are supposed to be known as well as size and position of defect and the direct problem of acoustics (1)-(5) was solving. After that the additional "experimental measured" pressure (6) in the point of sensor's installation are formed based on the pressure distribution from the solution of (1)-(5) and corresponded inverse problem of estimation of $\varepsilon(x)$ (which at this stage is supposed to be unknown). This inverse problem is formulated as minimization of least-squares residual functional for calculated and measured acoustic pressure at external surface of specimen (7). The minimization of the residual functional is executed by conjugate gradient method [9]. Such approach provides analysis of influence of errors in input data to the accuracy of inverse problem's solution, as well as search of optimal experimental conditions.

The stochastic errors in "experimental measurements" for simulation of inverse problem's solving are formed by

$$f_m(\tau) = \bar{f}_m(\tau)\left(1 + \varpi\,\delta_f(\tau)\right), \quad m = \overline{1, M}, \tag{16}$$

where $\bar{f}_m(\tau)$ is "exact" pressure measurement from the direct problem solution,; ϖ is produced by pseudorandom number generator with Gaussian distribution, expected value 1 and dispersion 0; $\delta(\tau)$ is maximum relative error.

The "exact" value of parameter $\tilde{\varepsilon}(x)$ (which was used for direct problem solving) is presented in figure 4. The initial value ε_0 at the iterative procedure of $\varepsilon(x)$ estimating was used 0.5. The thickness of specimens was supposed as 0.005 m. The relative error of inverse problem solution was calculated as

$$\delta\varepsilon = \|\varepsilon(x) - \tilde{\varepsilon}(x)\|_{L_2} / \|\tilde{\varepsilon}(x)\|_{L_2} \quad (17)$$

where ε is solution of simulated inverse problem, $\tilde{\varepsilon}$ is "exact" value of parameter.

The result of "reconstruction" of unknown parameter $\varepsilon(x)$ and comparing calculated pressure $p(0,\tau)$ with "measured" $f(\tau)$ in the case without errors of measurements ($\delta_f = 0$) are presented in figure 4. If $\varepsilon(x)$ is not equal to 0, therefore some defect is positioned at these points. There are the results of inverse problems solution for three different initial approximations. The results are practically identical and are equal the "exact" value. This fact confirms the uniqueness of this nonlinear inverse problem solution.

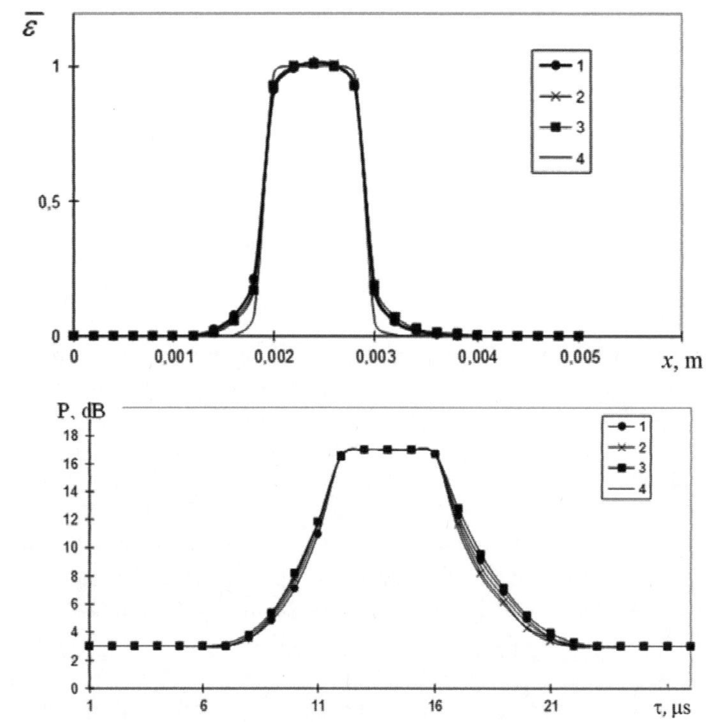

Figure 4. $\varepsilon(x), p(0,\tau)$ and $f(\tau)$ in the case without of errors of "measurements": 1 – $\varepsilon(x)$ and $p(0,\tau)$ if $\varepsilon_0 = 0.1$, 2 – $\varepsilon(x)$ and $p(0,\tau)$ if $\varepsilon_0 = 0.7$, 3 – $\varepsilon(x)$ and $p(0,\tau)$ if $\varepsilon_0 = 1.2$, 4 – "exact" value $\tilde{\varepsilon}(x)$ and "measured" pressure $f(\tau)$ respectively.

Let us consider the influence of the number of parameters of approximation N_ε (8) to the accuracy of inverse problem solution for various levels of errors in input data δ_f, as well as compare the different types of approximations. The results of using various basic functions of approximation (B-splines, piecewise functions (8)) are presented in figures 5 – 6. These results confirm the sufficient computational stability of developed algorithm to the random errors of input data.

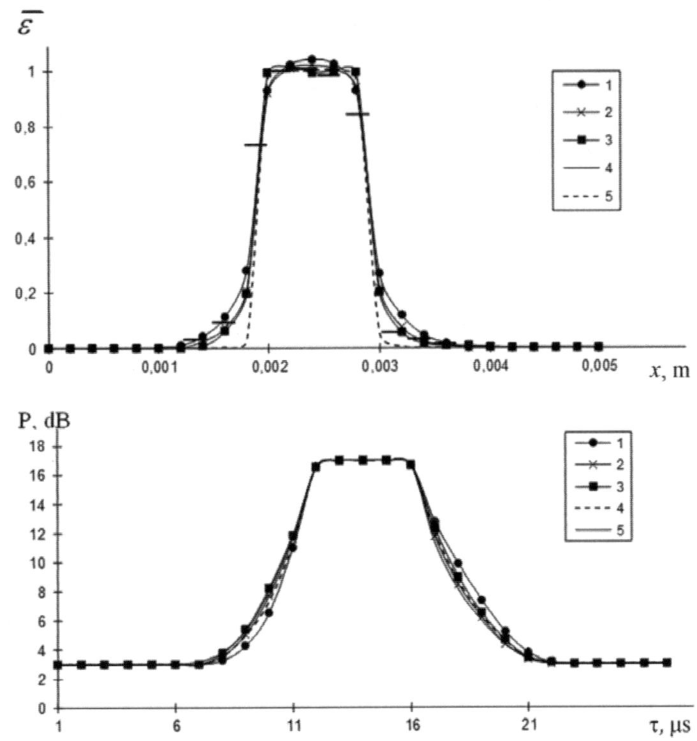

Figure 5. Estimated values of $\varepsilon(x)$, $p(0,\tau)$ and $f(\tau)$ for input data with relative error $\delta_f = 3\%$: 1, 2, 3 – approximation by cubic B-splines with numbers of approximating parameters N_ε 5, 9 and 13 respectively, 4 - approximation by piecewise functions ($N_\varepsilon = 20$), 5 – "exact" value $\tilde{\varepsilon}(x)$ and "measured" pressure $f(\tau)$ respectively.

Than the influence of errors in values of function $\frac{\partial^2 \xi_1}{\partial \tau^2}$ (4), which determines the value of the gradient of acoustics pressure at the external surface of specimen is analyzed. The bias errors with relative level $\delta = 5\%$ were simulated by

$$\frac{\partial^2 \xi_1}{\partial \tau^2}(\tau) = \frac{\partial^2 \bar{\xi}_1}{\partial \tau^2}(\tau)\left(1 + \delta_\xi\right) \tag{18}$$

The result of inverse problems solving in the case of errors in the left boundary condition (18) are presented in figure 7. The presented results demonstrated very small influence of these errors to the shape of the desired function, as well as the influence to the value of it is not small. Taking into account that the developed method is the remote diagnostics of defects, just shape of function $\varepsilon(x)$ is important to estimate the coordinates and sizes of defects.

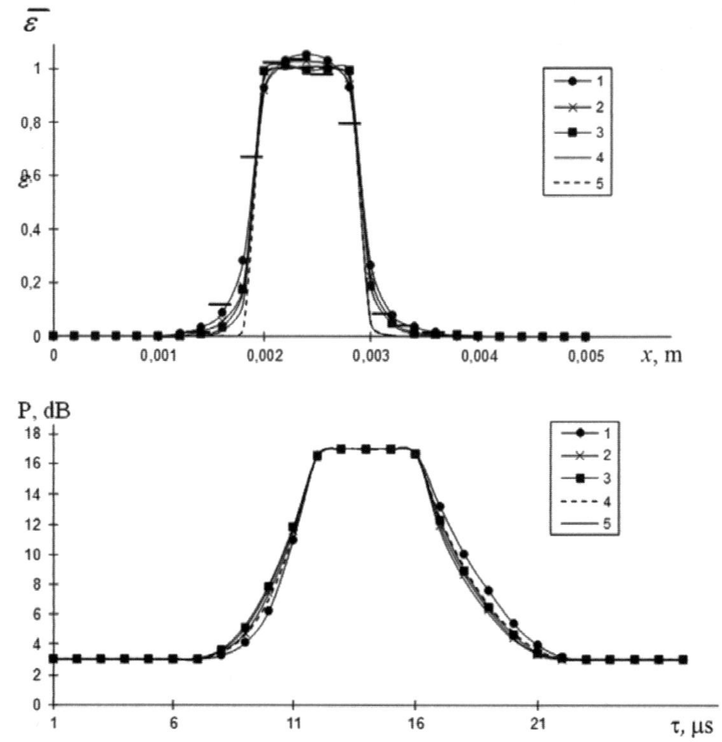

Figure 6. Estimated values of $\varepsilon(x)$, $p(0,\tau)$ and $f(\tau)$ for input data with relative error $\delta_f = 10\%$: 1, 2, 3 – approximation by cubic B-splines with numbers of approximating parameters N_ε 5, 9 and 13 respectively, 4 - approximation by piecewise functions (N_ε = 20), 5 – “exact” value $\tilde{\varepsilon}(x)$ and “measured” pressure $f(\tau)$ respectively.

The next very important question for iterative methods of inverse problems solving of conditionally well-posed problems [11] as considered problem (1)-(7) is the criterion of the stopping of iteration. From the pure mathematical point of view the minimization procedure can be executed till zero value of functional (7), but for ill-posed problems the discrepancy criterion [8] based on iterative regularization approach is usually used. In spite of the problem (1-7) is conditionally well-posed, the input data with experimental errors are used for its solving. This is the question of reasonability of using the discrepancy criterion for stopping the iterative procedure for the conditionally well-posed problem. In the case of implementation of this criterion the functional (7) is minimized till condition

$$J(\bar{\varepsilon}) \le \delta^2 \tag{19}$$

where $\delta^2 = \int_0^{\tau_{\max}} \sigma^2(\tau)\,d\tau$ is integral error of measurement $f(\tau)$, and σ is known dispersion of measurements. This approach is not regularization, it just improve the accuracy of calculation.

The values of minimized functional and correspondent values of estimated function for using of various approximations are presented in figures 8-10. These results also confirm the reasonability of choice the optimal number of approximation parameters based on (19): the number N_ε should be minimal, which provide the achievement of level δ^2.

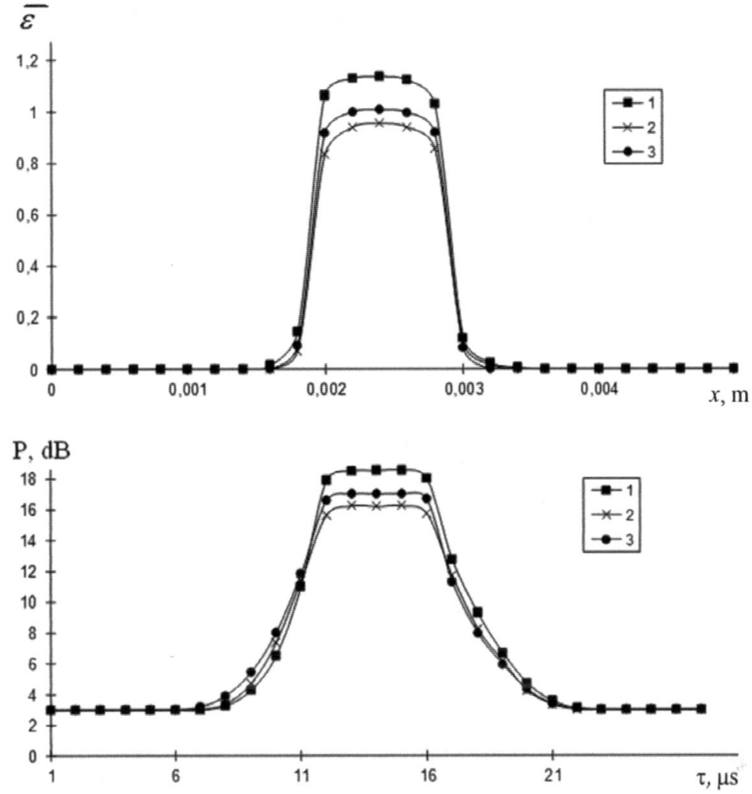

Figure 7. Estimated values of $\varepsilon(x)$, $p(0,\tau)$ and $f(\tau)$ for errors in the left boundary condition $\dfrac{\partial^2 \xi_1}{\partial \tau^2}$: 1, 2 – with errors $\delta_\xi = +10\%$ and $\delta_\xi = 0 - 10\%$ respectively, 3 – "exact" value $\tilde{\varepsilon}(x)$ and "measured" pressure $f(\tau)$ respectively.

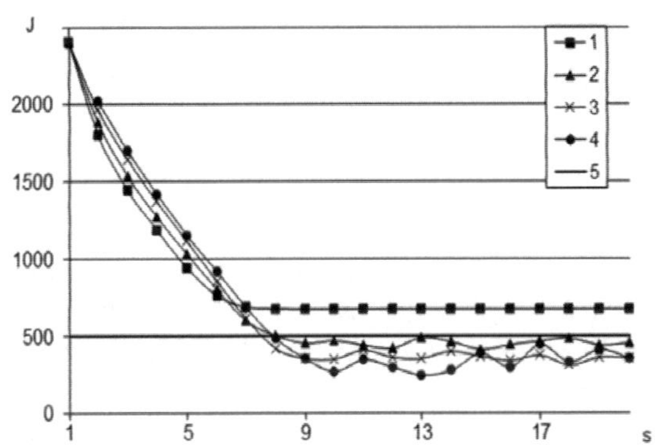

Figure 8. Influence of the number of approximation parameters N_ε to the value of minimized functional J as the function of the number of iteration s : 1, 2, 3 – approximation by cubic B-splines with numbers of approximating parameters N_ε 5, 9 and 13 respectively, 4 - approximation by piecewise functions (N_ε = 20), 5 – value of integral error of measurement δ^2.

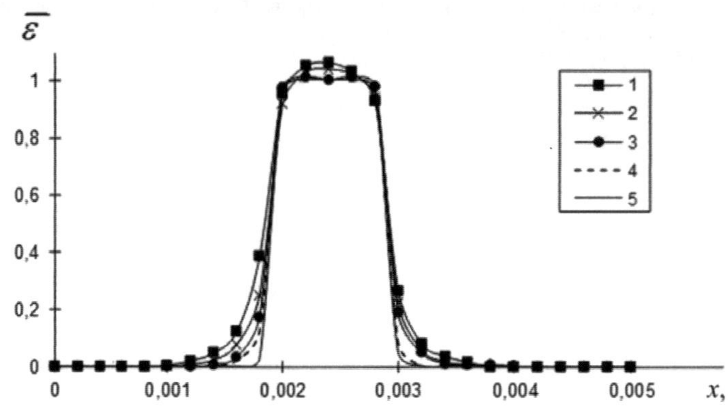

Figure 9. Estimated values of $\varepsilon(x)$: 1 – B-splines with N_ε = 5 (minimization without criterion (19)), 2 – B-splines with N_ε = 5 (stopping criterion (19)), 3 – B-splines with N_ε = 9 (minimization without criterion (19)),4 - B-splines with N_ε = 5 (stopping criterion (19)), 5 – "exact" value $\tilde{\varepsilon}(x)$.

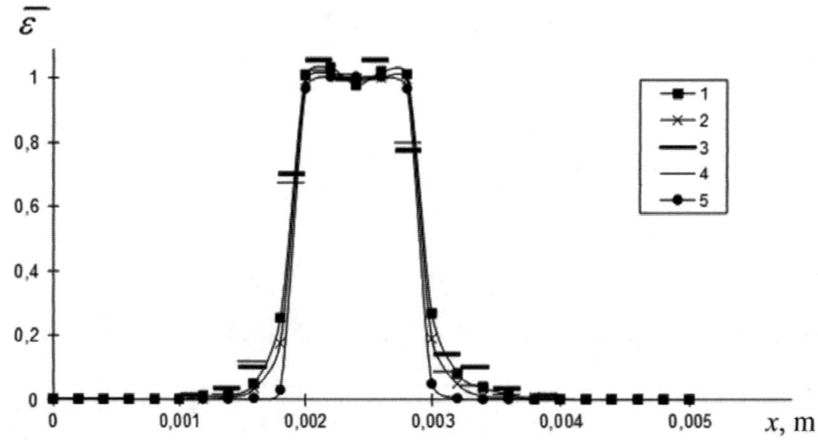

Figure 10. Estimated values of $\varepsilon(x)$: 1 – B-splines with N_ε = 13 (minimization without criterion (19)), 2 – B-splines with N_ε = 13 (stopping criterion (19)), 3 – approximation by piecewise functions N_ε = 20 (minimization without criterion (19)),4 - approximation by piecewise functions N_ε = 20 (stopping criterion (19)), 5 – "exact" value $\tilde{\varepsilon}(x)$.

The executed study confirm sufficient stability of the developed algorithm to the various disturbance, but of course it cannot be considered as completed study and just confirm the reasonability of computational experiments technique for analysis of effectiveness of inverse problems algorithm.

3. Experimental testing

The purpose of acoustics testing is measurements of:

- transient acoustics pressure $p^{\exp}(0,\tau),\ \tau \in (0,\tau_{\max})$ at the surface of specimen;
- gradient of pressure $\frac{\partial p}{\partial x}(0,\tau),\ \tau \in (0,\tau_{\max}]$.

The scheme of experimental facility is presented at figure 11. It includes:
- generators of high frequency signals of special shape 4Mhz Synthesized Function Generator SFG-2004;
- US receiver;
- US transmitter;
- digital oscillograph LECROY WaveAce 204;
- experimental module EMA-1.

Using technology of specimen's manufacturing provides installation of artificial defects at specimens during manufacturing [10-13]. To approve developed technique it is necessary to provide maximum accuracy of a-priori measurement of coordinates of defects by X-ray facilities. A few specimens were manufactured; parameters and defects of one of them are presented in table 1. The results of pressure measurements $p^{\exp}(0,\tau),\ \tau \in (0,\tau_{\max})$ are presented in figure 12.

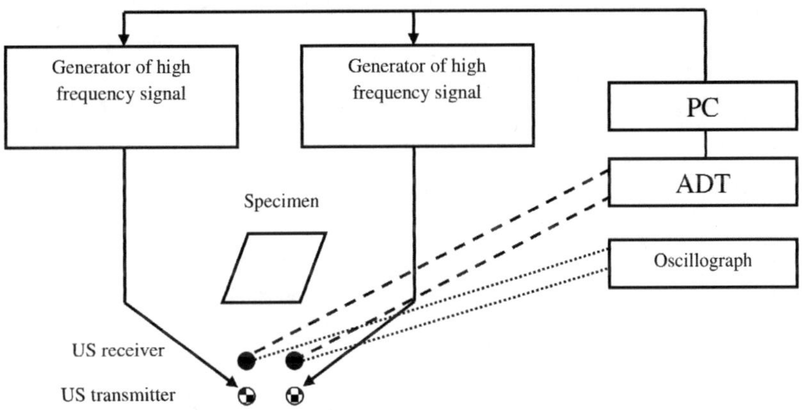

Figure 11. The scheme of experimental facility.

Table 1. Parameters and characteristics of defects of specimens

# of specimen	length × width × thickness *[mm]*	Mass *[g]*	Density *[kg/m³]*	Description of defects
2	60.0 × 60.0 × 5.65	19.72	972.87	Silicon sphere with diameter 1.8 *mm* is installed at the center of the slab and at depth 1.7 *mm* from back surface.

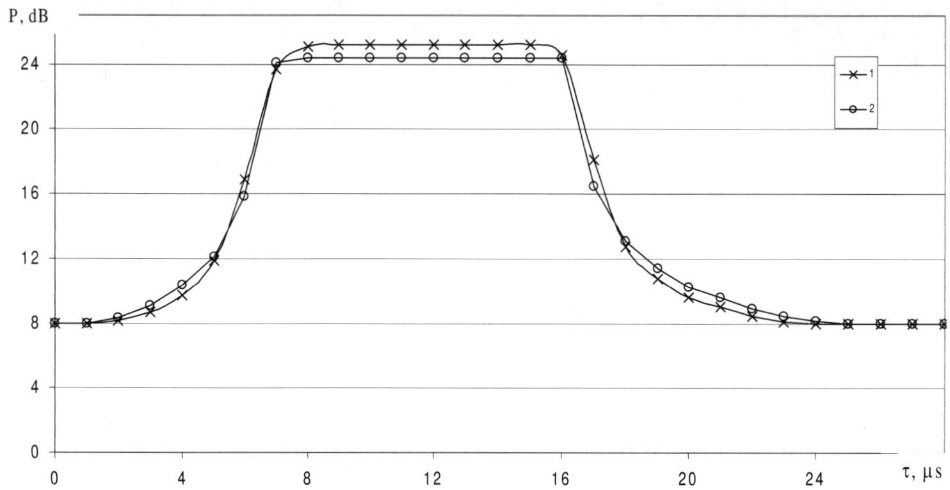

Figure 12. Experimental and calculated pressure at the front side of the specimen as function of time (test #2): 1- calculated, 2 – experimental.

The result of inverse problems solving are presented in figure 13, where $\bar{\varepsilon}(x)$ is

$$\bar{\varepsilon}(x) = \frac{\varepsilon(x) - \varepsilon_0}{\varepsilon_{max} - \varepsilon_0} \tag{20}$$

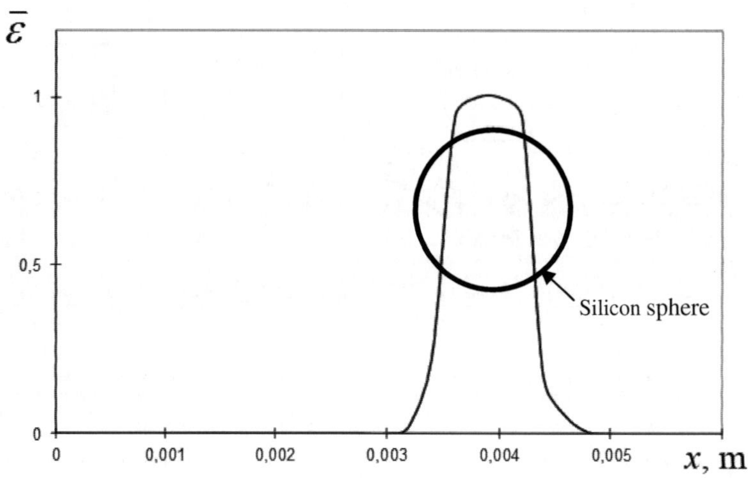

Figure 13. Value of undimensional coefficient $\bar{\varepsilon}(x)$ (test # 2)

The comparing of experimental and calculated data are presented in figures 12. Maximum discrepancy of calculated pressure from experimental measured are calculated as

$$\delta_{max} = \max_j \left| p(0, \tau_j) - p^{exp}(0, \tau_j) \right| \tag{21}$$

where $p^{\exp}(0,\tau_j)$ - acoustic pressure measured at j -th time, $p(0,j)$- acoustic pressure calculated by (1)-(5) based on inverse problem solving. Calculated values are presented in table 2. The integral discrepancy was calculated as

$$\delta^2 = \sum_{j=1}^{N_\tau} \left(p(0,\tau_j) - p^{\exp}(0,\tau_j)\right)^2 / N_j \tag{22}$$

Calculated values are presented in table 3.

Table 2. Maximum errors of inverse problems results

Test #	$\delta_{\max}$
2	1,7

Table 3. Integral errors of inverse problems results

Test #	δ^2
2	0,12
3	0,04
5	0,05

Conclusions

The executed tests have shown the corresponded increasing of pressure according defects of structure. The experimental facility and corresponded software were developed, allowing with the high accuracy to define the defects in elastic TPS materials. The executed tests have shown the corresponded increasing of pressure according defects of structure. The exception is test #4, where the power is not enough to provide resonance oscillation of high density inclusion. The deviations of the calculated pressure (using $\varepsilon(x)$ estimations) from the pressure measured in the experiments are insignificant showing sufficient accuracy in the estimations of $\varepsilon(x)$ of the analyzed materials.

Acknowledgments

This work was supported by the Russian Ministry of Science and Education in the frame of the Project Part of the financial support (grant # 9.3917.2017).

References

[1] Bogdanov V, Rodimov R, Pichkhadze K, Terterashvili A, Finchenko V, Kouznetsov V, Inflatable Ballutes to Provide Aerodynamic Shape to the Payload Bus Enabling its Atmosphere Entry. 48-th International Aeronautical Congress, 6-10 Oct., 1997, Turin, Italy, IAF-97-I.6.03

[2] Norton S J 1999 Iterative Inverse Scattering Algorithms: Methods of Computing Frechet Derivatives *Journal of Acoustical Society of America* **106**, 2653

[3] Valdivia N P, Williams E G and Herdic P C 2008 Approximations of Inverse Boundary Element Methods with Partial Measurements of the Pressure Field *Journal of Acoustical Society of America* **123** 109

[4] Astral D E and Gustafsson M G 2003 Optimal Detection of Crack Echo Families in Elastic Solides *Journal of Acoustical Society of America* **113** 2732

[5] Simonetti F and Cawley P 2003 A Guided Wave Technique for the Characterization of Highly Attenuative Viscoelastic Materials *Journal of Acoustical Society of America* **114** 158

[6] Bucaro J and Romano A J 2004 Detection and Localization of inclusions in Plates Using Inversion of Point Actuated Surface Displacement *Journal of Acoustical Society of America* **115** 201

[7] Romano A J, Bucaro J A, Vignola J F and Abraham P B 2007 Detection and Localization of Rib Detachment in Thin Metal and Composite Plates by Inversion of Laser Doppler Vibrometry Scans *Journal of Acoustical Society of America* **121** 2667

[8] Rudenko O V 1993 Non-linear methods in acoustic diagnostics De Haller K. Nonlinear acoustics applied to nondestructive testing *Defectoscopia* N8 24

[9] Alifanov O M, Artyukhin E A and Rumyantsev S V 1995 *Extreme Methods for Solving Ill-Posed Problems with Applications to Inverse Problems* (New York: Begell House)

[10] Alifanov O M, Nenarokomov A V, Nenarokomov K A , Terentieva A V, Titov D M and Finchenko V S 2014 Experimental-computational system for noncontact diagnostics of elastic materials *Proceedings of 8th International Conference on Inverse Problems in Engineering (May 12–15, 2014, Poland).- Silesian TU Publ., Gliwice-Wrocław, Poland* 13

[11] Alifanov O M, Nenarokomov A V and Nenarokomov K A 2011 Non-Contact Parameter Estimation of Elastic Materials by Inverse Methods of Nonlinear Acoustics *Proceedings of 2nd African Conference on Computational Mechanics (January 2011), UoK Publ.*

[12] Alifanov O M, Nenarokomov A V and Nenarokomov K A 2011 Non-Contact Testing of Elastic Materials by Inverse Methods of Nonlinear Acoustics *Proceedings of 7th International Conference on Inverse Problems in Engineering: Theory and Practice (May 4 – 6, 2011, Orlando, Florida, USA).- University of Central Florida Publ.*

Optimal design of ECT sensors using prior knowledge

M Neumayer[1], M Flatscher[1], T Bretterklieber[1], S Puttinger[2]

[1] Institute of Electrical Measurement and Measurement Signal Processing, Graz University of Technology, Inffeldgasse 23/II, A-8010 Graz, Austria

[2] Department of Particulate Flow Modelling, Johannes Kepler University, Altenbergerstrasse 69, A-4040 Linz, Austria

E-mail: neumayer@tugraz.at

Abstract. Electrical capacitance tomography (ECT) is a noninvasive imaging modality aiming on the reconstruction of the dielectric material properties within a region of interest. The estimation of the permittivity distribution forms a nonlinear inverse problem, which is marked by a severe ill-posed nature. As a consequence, reconstruction algorithms rely on the use of prior knowledge or regularization techniques to overcome this difficulty. In this paper we want to make use of prior knowledge to optimize the design of ECT sensors. Typical ECT sensors maintain a number between 8 to 16 electrodes, which are mounted in an equidistant scheme on the circumference of the region of interest (e.g. a pipe). This is a suitable approach for sensors with a large number of electrodes. In this work we consider sensors with a considerable smaller number of electrodes and investigate the design of such sensors using prior knowledge. We discuss the incorporation of all modeling aspects and formulate a stochastic optimization function which will lead to optimized sensor designs. We present a simulation study and discuss the design results with respect to the physical properties of the sensing scheme.

Keywords: Optimal sensor design, electrical capacitance tomography, prior knowledge, optimization

1. Introduction

Electrical capacitance tomography (ECT) is an indirect measurement system, which maintains capacitive sensing for imaging [5]. ECT is closely related to the electrical imaging systems of electrical impedance tomography (EIT) [21] and electrical resistance tomography [6], as the sensing effect of all three imaging systems is governed by the potential equation, which can be derived from the Maxwell's equations [7]. Figure 1(a) provides a sketch of an ECT sensor. The sensor assembly consists of measurement electrodes, which are attached to the outside of a nonconductive pipe [22]. The capacitances between the electrodes are measured by means of suitable electronics. To avoid interference from outside, the assembly is shielded by a screen. The measurement capacitances are influenced by the dielectric properties of the material distribution inside the pipe, which is depicted by the inclusion inside the pipe in Figure 1(a). The tomographic problem of ECT is to reconstruct the material distribution from measurements [33, 9]. Figure 1(b) shows a photography of an ECT sensor. The plastic rods and pipes inside the sensor should be reconstructed from capacitive measurements. Figure 1(c) depicts a reconstruction result for the experiment. The imaging problem of ECT is an ill-posed inverse problem [20]. Statistical inversion theory incorporates prior knowledge in the design estimators to provide suitable results [18]. Deterministic inversion methods maintain regularization techniques to obtain a stable solution [17]. In this work we want to apply methods from statistical inversion theory and maintain the Bayesian framework for the discussion. A review article about the statistical solution of the inverse problem of ECT within the

IOP Conf. Series: Journal of Physics: Conf. Series **1047** (2018) 012012 doi:10.1088/1742-6596/1047/1/012012

Bayesian framework is given by [34].

(a) Sketch of an ECT sensor. The inter-electrode capacitances are measured.

(b) Photography of different plastic rods and pipes in an ECT sensor.

(c) Reconstructed material distribution for the material distribution depicted in Figure 1(b).

Figure 1. Figure 1(a): Exemplary illustration of an ECT sensor [22]. The sensor assembly consists of measurement electrodes, which are attached to the outside of a nonconductive pipe. A shield protects measurement the process. Figure 1(b): Photography of different plastic rods and pipes within an ECT sensor. Figure 1(c): Reconstruction result for the material distribution depicted in Figure 1(b).

In this work we address the optimal design of the electrode arrangement of ECT sensors for applications in pneumatic conveying [30, 31]. Optimal design (of experiments) has seen vast fundamental and applied research effort over the years [13], [11], [12]. Publications for applied research address topics like optimal sensor positioning [10], the design of experiments on optical tomography [19] and many others. The Bayesian framework is also a favorable tool in optimal design. Within the Bayesian framework all quantities are considered to be random variables and characterized by means of probability density functions (pdf) $\pi(\cdot)$. The measurements are summarized by the vector $\tilde{\boldsymbol{d}} \in \mathbb{R}^M$, the variables to be estimated are summarized by the state vector $\boldsymbol{x} \in \mathbb{R}^N$. The posteriori distribution $\pi(\boldsymbol{x}|\tilde{\boldsymbol{d}}, \boldsymbol{\xi}_{\text{Sensor}})$ is formed, which provides a probability measure for the solution of an estimation problem, given all the existing information. A detailed description for the Bayesian modeling approach for inverse problems is presented in section 2.1. $\boldsymbol{\xi}_{\text{Sensor}}$ summarizes the variables of the optimal design problem. For our problem, $\boldsymbol{\xi}_{\text{Sensor}}$ summarizes the arrangement of the electrodes, e.g. by means of the angular position of the individual electrodes. The optimization of the electrode arrangement of a tomographic system has been addressed in several publications [4], [25] or [2], [8]. Authors have also proposed to optimize the excitation signal strength at the measurement electrodes, e.g. [15] [3] [16].

The determination of $\boldsymbol{\xi}_{\text{Sensor}}$ forms an optimization problem. For these problems, suitable optimality criteria are required [12]. Measurement uncertainty is a key parameter to access the quality of a measurement system. In the Bayesian framework the uncertainty of a measurement result can be characterized by the dispersion of the posteriori distribution $\pi(\boldsymbol{x}|\tilde{\boldsymbol{d}})$. The covariance matrix $\boldsymbol{\Sigma}_{\boldsymbol{x}}$ can be used as a summary statistic to express the uncertainty. In [12] different optimality criteria are listed, which are constructed from the covariance matrix $\boldsymbol{\Sigma}_{\boldsymbol{x}}$. For example the A criterion represents the average mean square error (MSE), which is evaluated from the trace of $\boldsymbol{\Sigma}_{\boldsymbol{x}}$ [10]. The evaluation of the covariance matrix $\boldsymbol{\Sigma}_{\boldsymbol{x}}$ can be done in different ways. For linear problems and for Gaussian distributed random variables, $\boldsymbol{\Sigma}_{\boldsymbol{x}}$ can be evaluated analytically [11]. For nonlinear problems, such as the inverse problems of ECT or EIT, a linearization can be applied to approximate $\boldsymbol{\Sigma}_{\boldsymbol{x}}$ [12, 18]. This approach has mostly been used in the cited literature. However, due to the linearization, the approach is limited by the nature of the applied Taylor series expansion. Further this approach maintains a Gaussian summary statistics only.

In [10, 35] the unscented transformation (UT) [28] has been suggested to approximate $\mathbf{\Sigma}_{\boldsymbol{x}}$ for nonlinear problems. The UT is known to provide more accurate estimates of the covariance matrix, even in case of strong nonlinear model behavior. The UT still maintains Gaussian statistics.

In this paper, we present a sampling based approach for optimal electrode design of ECT sensors with a small number of electrodes [37]. Our application is in monitoring of flow regimes [31] in pneumatic conveying [30]. Distinct flow regimes can show vastly different characteristics, making a Gaussian approximation and linearization techniques unsuitable for a proper evaluation of the resulting uncertainty. We propose the use of sample based priors [18] for uncertainty evaluation. We discuss all required modeling steps and show the construction of an optimization framework. We finally demonstrate results from numerical studies and discuss the resulting electrode topology with respect the prior information for the flow profiles.

This paper is structured as follows. In section 2 we summarize the inverse problem of ECT. This discussion includes the concept of sample based priors and a sample based reconstruction algorithm of linear-back-projection type. In section 3 we present a design approach for optimal sensor design based on prior knowledge and discuss implementation details for the algorithm. In section 4 we will present a simulation study for our approach and discuss the properties of the proposed strategy.

2. Electrical Capacitance Tomography

In this section we address the inverse problem of electrical capacitance tomography (ECT). Figure 1(a) illustrates a sketch of the cross section of an ECT sensor. A photography of an ECT sensor with different inclusions and an exemplary reconstruction result for the given distribution are presented by Figure 1(b) and Figure 1(c), respectively. Figure 2 depicts a sketch of an ECT sensor with a mathematical notation of the domains and boundaries. The interior of the pipe is referred to as the region of interest Ω_{ROI}. The boundaries of the electrodes are denoted as Γ_i.

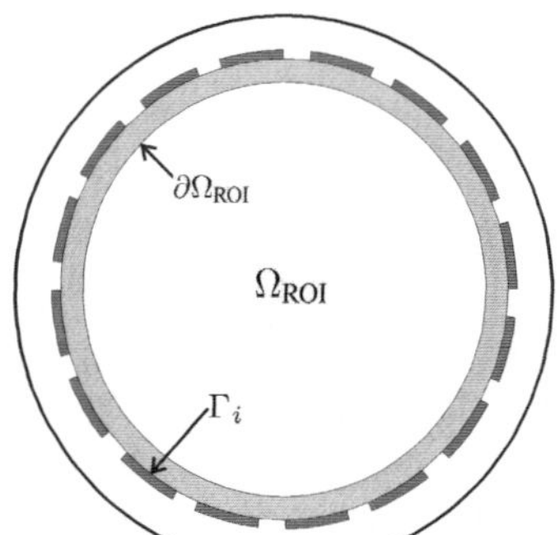

Figure 2. Scheme of an ECT sensor with notation of the domains and the boundaries.

ECT is based on the measurement of the $M = \frac{N_{\mathrm{elec}}(N_{\mathrm{elec}}-1)}{2}$ inter-electrode capacitances. This measurement process can be modeled by means of the potential equation, which can be derived from the Maxwell's equations [20]. Electric fields within the screen are governed by the potential equation

$$\nabla \cdot (\varepsilon_0 \varepsilon_{\mathrm{r}} \nabla V), \tag{1}$$

where V is the electric scalar potential and ε_0 and ε_{r} are the vacuum permittivity and the relative permittivity, respectively. To measure the inter-electrode capacitances, a voltage V_0 is applied to one electrode, while all other electrodes are connected to ground potential ($V = 0$). This can be denoted by the boundary conditions

$$V_{\Gamma_i} = V_0, \tag{2}$$

$$V_{\Gamma_j} = 0 \quad j \neq i. \tag{3}$$

Given the potential distribution V_i, the inter-electrode capacitances are given by

$$C_{j,i} = -\frac{1}{V_0} \oint_{\Gamma_j} \varepsilon_0 \varepsilon_r \vec{n} \cdot \nabla V_i d\Gamma \qquad j \neq i, \tag{4}$$

where $\vec{n}$ is the normal vector on the surface of the electrodes. The inter-electrode capacitances form the $N_{\text{elec}} \times N_{\text{elec}}$ capacitance matrix $\boldsymbol{C} = [C_{j,i}]$. The forward map for ECT is a numerical simulation of the measurement process. Hence it comprises the solution of the potential equation (1) for the N_{elec} excitation scheme and the computation of the elements of $\boldsymbol{C}$. In our work, we use a finite element method to simulate the sensor characteristics. Subsequently, we assign the dielectric permittivity values of the finite elements in the domain Ω_{ROI} to be the elements of the state vector $\boldsymbol{x} \in \mathbb{R}^N$. This type of data representation is referred to as a low level image representation [34]. For typical ECT applications the discretization of Ω_{ROI} requires several hundred finite elements [36], resulting in a rank deficient problem as typically $N > M$ holds.

2.1. *Bayesian Framework and Sample based Priors*

The estimation of $\boldsymbol{x}$ from measurements $\tilde{\boldsymbol{d}}$ requires dedicated algorithms, which are often referred to as reconstruction algorithms [26]. The forward map F is an essential part in the design of reconstruction algorithms to solve inverse problem. The Bayesian framework outlines a formidable mathematical tool to formulate suitable estimation algorithms for inverse problems [24, 18]. The fundamental nature of the Bayesian philosophy is to treat all quantities as random variables with a probability density function $\pi(\cdot)$ [18]. Bayes law is given by

$$\pi(\boldsymbol{x}|\tilde{\boldsymbol{d}}) \propto \pi(\tilde{\boldsymbol{d}}|\boldsymbol{x})\pi(\boldsymbol{x}), \tag{5}$$

where $\pi(\tilde{\boldsymbol{d}}|\boldsymbol{x})$ is termed the likelihood function and $\pi(\boldsymbol{x}|\tilde{\boldsymbol{d}})$ is the posteriori distribution. The likelihood function $\pi(\tilde{\boldsymbol{d}}|\boldsymbol{x})$ comprises the forward map F, the measurements $\tilde{\boldsymbol{d}}$ and a noise model for the measurement noise. The prior distribution $\pi(\boldsymbol{x})$ incorporates knowledge about the solution. The posteriori distribution provides a measure for $\boldsymbol{x}$ being the solution given the data $\tilde{\boldsymbol{d}}$, the forward map F and the prior knowledge.

Prior knowledge is a key ingredient for solving ill-posed inverse problems [18]. Reviews for the common prior designs used for tomographic applications like ECT and EIT are presented in [34, 38]. Gaussian priors are of form $\pi(\boldsymbol{x}) = \mathcal{N}(\boldsymbol{\mu}_{\boldsymbol{x}}, \boldsymbol{\Sigma}_{\boldsymbol{x}})$. Prior knowledge is summarized by means of Gaussian statistics using the mean vector $\boldsymbol{\mu}_{\boldsymbol{x}}$ and the covariance matrix $\boldsymbol{\Sigma}_{\boldsymbol{x}}$. A rich class for the design of prior distributions are Markov random fields (MRFs). Authors have presented smoothness priors [20] to stabilize the estimation process. Another popular prior in the field of MRFs are total variation priors [18].

In this work we maintain the concept of sample based priors [18]. Hereby a representative ensemble of images is generated from knowledge about the underlying process, which defines an empirical prior density [34]. Figure 3 depicts exemplary prior samples for expected material distributions. Figure 3(a) depicts samples from a distribution, which we refer to as rod-type distribution. Considering the ECT sensor filled with plastic rods as depicted in Figure 1(b), the samples of the rod-type distribution can be considered realistic samples for this type of material distributions. The generation of this samples is done by randomly placing circular objects in the ROI. In a similar way samples for as Gaussian-type distribution can be constructed, as depicted in 3(b). With respect to the rod-type distribution, the Gaussian-type distribution incorporates smoothness for the material distribution. Note that sample based priors can also serve to formulate Gaussian summary priors, e.g. by estimating $\boldsymbol{\mu}_{\boldsymbol{x}}$ and $\Sigma_{\boldsymbol{x}}$ from an ensemble [18, 39].

In the next section we will show the construction of a back-projection type reconstruction algorithm using sample based priors.

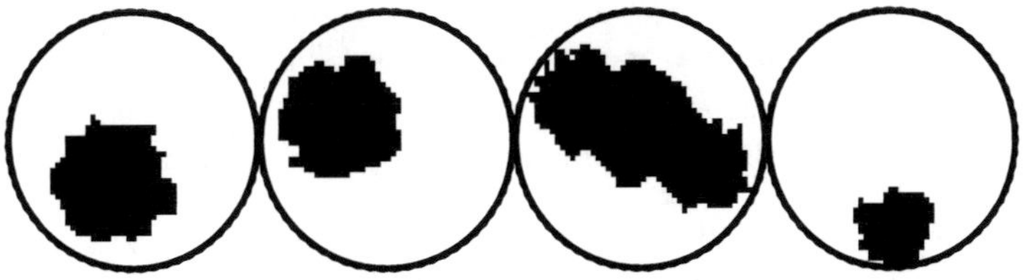

(a) Samples from a rod-type distribution.

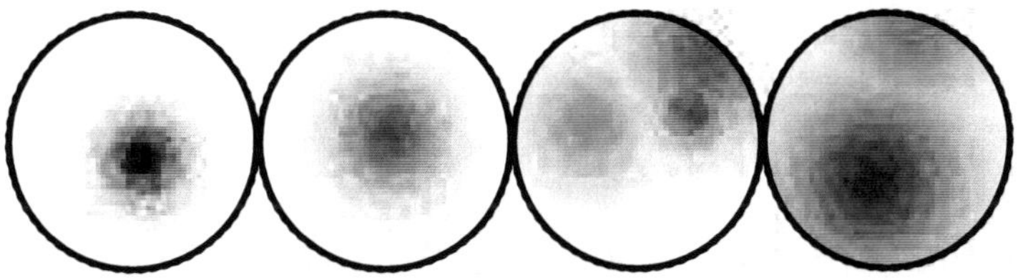

(b) Samples from a Gaussian-type distribution.

Figure 3. Exemplary samples from two different prior distributions.

2.2. Sample Based Linear Back-Projection Methods
In this subsection we summarize the construction of a linear-back-projection (LBP) reconstruction algorithm using sample based priors [27]. LBP methods are reconstruction methods, which estimate the state vector $\boldsymbol{x}$ from data (measurements) $\tilde{\boldsymbol{d}}$ by means of a matrix vector multiplication. Standard LBP methods typically maintain the Jacobian matrix to create a relation between the quantity of interest and the measurements [29]. The image quality of these methods is typically low. However, the computational effort of LBP methods is small, making them suitable for online imaging applications.

An adapted version of LBP has the form

$$\hat{\boldsymbol{x}} = \boldsymbol{P}_{\mathrm{LBP}} \tilde{\boldsymbol{d}}_{\mathrm{a}}, \tag{6}$$

where $\tilde{\boldsymbol{d}}_{\mathrm{a}}$ is an augmented data vector and $\hat{\boldsymbol{x}}$ denotes the estimation result [27]. Note that we estimate the relative permittivity of the finite elements in Ω_{ROI}. We do not estimate a deviation of the relative permittivity from a background permittivity. Also the data $\tilde{\boldsymbol{d}}$ holds the actual measurement values. We maintain reference measurements to calibrate the data [22].

The matrix $\boldsymbol{P}_{\mathrm{LBP}}$ is the reconstruction matrix and can be found by solving

$$\boldsymbol{H} \boldsymbol{P}_{\mathrm{LBP}}^{T} = \boldsymbol{X}_{\mathrm{Samples}}^{T}. \tag{7}$$

The matrix $\boldsymbol{X}_{\mathrm{Samples}}$ holds samples from the prior distribution and the rows of $\boldsymbol{H}$ are given by the augmented data vectors for the individual samples. The augmented data vector $\tilde{\boldsymbol{d}}_{\mathrm{a}}$ is formed from the data. An often used variant is given by $\tilde{\boldsymbol{d}}_{\mathrm{a}} = \begin{bmatrix} 1 & \tilde{\boldsymbol{d}}^{T} & (\tilde{\boldsymbol{d}}^{2})^{T} \end{bmatrix}^{T}$, where $\tilde{\boldsymbol{d}}^{2}$ denotes a vector with the pointwise squared elements of $\tilde{\boldsymbol{d}}$ [27].

The reconstruction result depicted in Figure 1(c) was created with a sample based LBP type algorithm for the data obtained from the measurement depicted in Figure 1(b). As the algorithm falls into the class of LBP-methods it is generally more suitable for tomographic problems with a small contrast ratio, i.e. the range of the permittivity distribution is small. This however holds for many industrial problems.

3. Optimal Sensor Design

In this section we present our approach for the optimal design of ECT sensors using prior knowledge. As stated in the introduction, we want to find an optimized electrode arrangement for the application of an sensor to monitor flow regimes in pneumatic conveying. The vector $\boldsymbol{\xi}_{\text{Sensor}}$ of our design optimization problem holds the positions of the electrodes. Figure 4 exemplifies the parameterization of the electrode geometry of an ECT sensor.

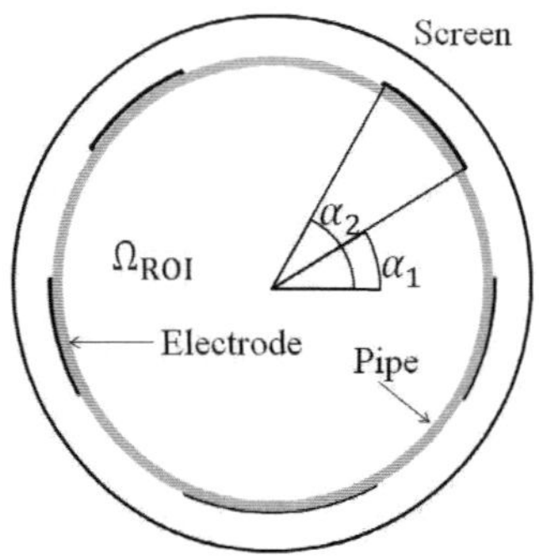

Figure 4. Exemplary parametrization of the electrode geometry.

Finding an electrode arrangement forms an optimization problem. As for any optimization problem, the design of the objective function is the most crucial decision. While classical sensor optimization often focuses on aspects like sensitivity or linearity, main objectives in measurement problems are to reduce the measurement error or the uncertainty of a measurement. The measurement error is defined as the difference between the measurement result and the true value of the measurement. For the inverse problem of ECT the image reconstruction error is given by

$$\boldsymbol{e} = \hat{\boldsymbol{x}} - \boldsymbol{x}_{\text{true}}, \tag{8}$$

where $\boldsymbol{x}_{\text{true}}$ is the true distribution and $\hat{\boldsymbol{x}}$ is the estimate. Using the image error as a criteria within the objective function implies that also the reconstruction algorithm is a component within the optimization problem, which increases the computational burden for the design optimization. The optimization of the sensor with respect to the prior distribution $\pi(\boldsymbol{x})$ requires the introduction of summary statistics into the objective function. Hence we use the mean square error (MSE) and the design objective function is given by

$$\begin{aligned} \boldsymbol{\xi}^*_{\text{Sensor}} &= \arg\min_{\boldsymbol{\xi}_{\text{Sensor}}} \mathcal{E}\left\{||(\hat{\boldsymbol{x}}(\boldsymbol{\xi}_{\text{Sensor}}) - \boldsymbol{x}||_2^2\right\} \\ & \quad \text{for } \boldsymbol{x} \propto \pi(\boldsymbol{x}), \end{aligned} \tag{9}$$

where $\mathcal{E}(\cdot)$ denotes the expectation operator. The evaluation of the MSE is performed based on samples from the prior distribution $\pi(\boldsymbol{x})$.

The suggested objective function present in this subsection is a possible approach to derive an optimal sensor design. It is similar to the A criterion used in optimal design. The evaluation is based on the ensembles generated by the prior distribution. Also alternative objective functions can be considered for future research by using concepts of different design objective functions [12]. With respect to parameter uncertainties also so called robust optimization approaches [35] should be considered for further research.

3.1. Algorithmic Setup

The solution of the suggested optimization problem to find an optimal sensor design requires some dedicated computational steps due to the stochastic nature of the problem. The computational structure is depicted in Figure 5. As discussed, the evaluation of the objective function requires the solution of the inverse problem. While the objective function is open to any inversion algorithm, we will use the sample base LBP algorithm discussed in subsection 2.2. Hence, a set of samples from the prior distribution is required to create the back-projection matrix $\boldsymbol{P}_{\text{LBP}}$ at first. As the evaluation of the reconstruction error should not be done with the same data, a second set of samples from $\pi(\boldsymbol{x})$ is required. This issue has to be considered for the evaluation of the MSE criteria. As a second point, the data for testing the algorithm should not be created by means of the same forward map as used for the reconstruction, as this would lead to an inverse crime [18]. In order to avoid this, a different simulation model has to be generated. We refer to the different simulation models as coarse and fine model. The coarse model is used for the reconstruction, the fine model is used for the computation of the data. Figure 5 summarizes the computational steps to evaluate the MSE. Based on the MSE a suitable optimization algorithm updates ξ_{Sensor} in order to solve the design problem.

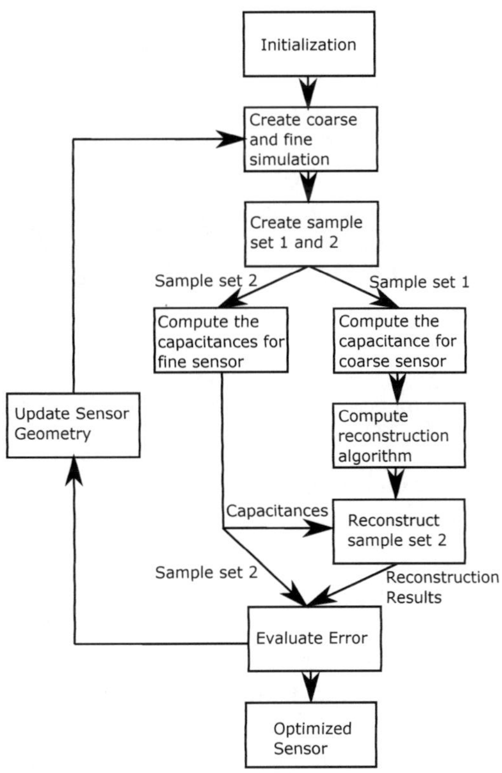

Figure 5. Flow diagram for the implementation of the optimization strategy. Two sensor simulations and different data sets are used for the evaluation of the objective function in order to avoid inverse crimes or using the same data for creating and testing of the reconstruction algorithm.

IOP Conf. Series: Journal of Physics: Conf. Series **1047** (2018) 012012 doi:10.1088/1742-6596/1047/1/012012

4. Simulation Example

In this section we present a numerical design study to investigate the behavior of our approach for finding optimal sensor designs. We consider monitoring particulate flows in pneumatic conveying using an ECT sensor. In pneutmatic conveying, granular materials are transported by means of a gas stream through a pipe [30]. Depending on the material properties and the gas stream, different distinct flow regimes appear [31]. Figure 6 depicts distinct flow regimes and the relation between gas velocity and pressure drop. Figure 7 exemplary depicts samples for different flow regimes.

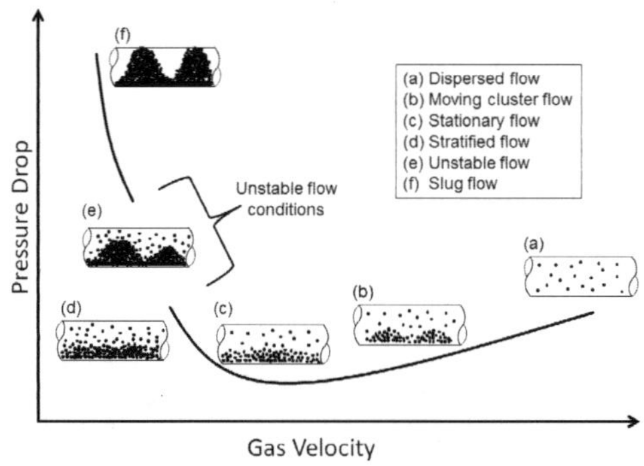

Figure 6. Different flow profiles in pneumatic conveying.

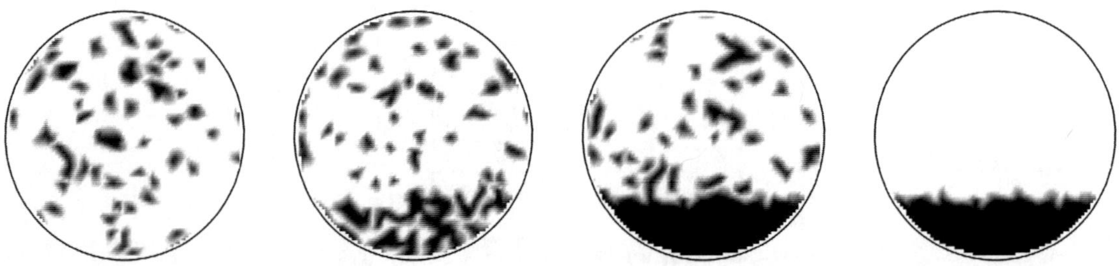

(a) Exemplary sample for a dispersed flow.

(b) Exemplary sample for a stationary flow.

(c) Exemplary sample for a stratified flow.

(d) Exemplary sample for a stratified slug flow.

Figure 7. Exemplary samples for different flow regimes in pneumatic conveying.

For our design study we consider a sensor design with only $N_{\text{elec}} = 5$ electrodes. All electrodes have the same width. We optimized the angular position on the pipe. We further applied a constraint of setting the minimum required angular distance between neighboring electrodes to $5°$. In case of no constraints the algorithm considers overlapping electrodes as infeasible solutions.

The ensemble size for the creation of the reconstruction algorithm and the validation of the MSE was set to 5000 samples each. This value was found to be a suitable large set for our investigation. For numerical optimization we used a differential evolution [32] method. We did not set a target value for the objective function, but let the algorithm run for a certain number of iterations.

We performed a multitude of test runs using the proposed design strategy. Figure 8 depicts two results. The fact that different results can be obtained leads to the conclusion, that the design problem has several local minima. This becomes clear when considering the symmetry of the prior distribution. For example all prior samples depicted in Figure 7 are symmetric with respect to a vertical center line for the cross

IOP Conf. Series: Journal of Physics: Conf. Series **1047** (2018) 012012 doi:10.1088/1742-6596/1047/1/012012

section. Considering a sensor with electrodes to the left side of the symmetry plane is an optimal sensor, also a sensor with a mirrored electrode assembly will be optimal. In fact we obtained a result with a mirrored electrode arrangement as depicted in Figure 8(a). As common feature of all results we found, that the computed results hold 3 of the 5 electrodes in the lower position of the sensor pipe. This can again be reasoned by the used prior distributions, as most flow regimes provide a dense material flow at the bottom of the transport pipe. Considering the field patterns for these electrodes, this electrode configuration is sensitive with respect to this spatial region of the sensor. The remaining two electrodes are typically placed in the upper half of the ECT sensor. These electrodes provide information about the height of the material flow. The bulk flow in the bottom of the pipe requires more electrodes, as bulk flows appear more often in the flow regimes. The result depicted in Figure 8(b) is yet an interesting outcome, as one of the two upper electrodes is on the left side and one electrode is on the right side of the sensor. In most cases the two upper electrodes are on the same side. Also this result coincides with the consideration about the sensing behavior of the electrodes.

As a test we also ran the optimization strategy for samples such as the Gaussian-type samples and the rod-type samples depicted in Figure 3. As these priors do not incorporate any symmetry, we can not expect the result to show distinct electrode arrangements. As a result we found that the optimization strategy provides the equidistant electrode configuration.

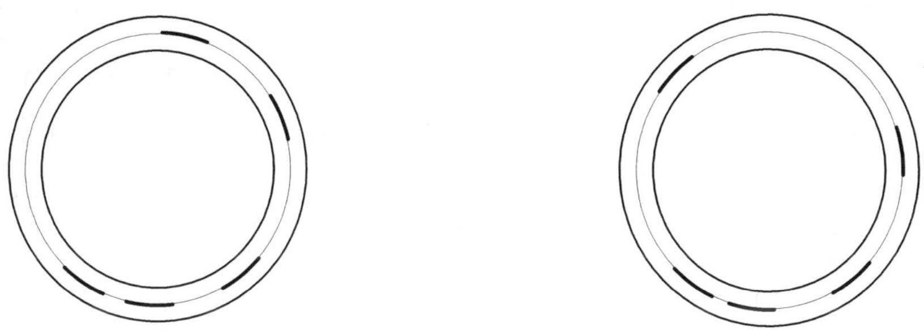

(a) Result 1. (b) Result 2.

Figure 8. Exemplary results for the sensor design problem.

4.1. Application of the Framework for Industrial Sensors

The design of ECT sensors using this approach might be of reduced importance for sensors with a larger number of electrodes. Yet we consider the design strategy as an important tool for the design of capacitive sensors in resource limited applications. Capacitive sensors are widely used to monitor environmental conditions, e.g. capacitive ice sensors are used to measure ice accretion on structure [40] like antennas or overhead supply lines. Capacitive sensors for these applications maintain a limited number electrodes to sense natural ice accretion. The presented approach can be used to design applications specific electrode structures.

5. Conclusion

In this paper we investigate an approach for optimal sensor design of ECT sensors using prior knowledge. We presented a scheme for the computational steps to be carried out for simulation and demonstrated results for an ECT sensor. In our design study we considered material distributions of pneumatic conveying. Due to the symmetry of the prior distribution, sensor designs with non-symmetric electrode arrangements can be obtained. For material distributions with no unique feature (e.g. no symmetry, or distinct center of mass), sensor with equally placed electrodes were obtained. The results coincide with the physical understanding of the measurement process. With respect to the present applications in flow imaging the authors consider the design of detectors for different flow regimes to be of significance for operators of pneumatic conveying systems.

Acknowledgement
This work is funded by the FFG Project (Bridge 1) TomoFlow under the FFG Project Number 6833795 in cooperation with voestalpine Stahl GmbH.

References

[1] Koksal A and Eyuboglu BM 1995 *Determination of optimum injected current patterns in electrical impedance tomography*, Physiol. Meas. 16, pp. 99-109.
[2] Yan W, Hong S and Chaoshi R 2006 *Optimum design of electrode structure and parameters in electrical impedance tomography*, Physiol. Meas. 27, pp. 291-306
[3] Lionheart WBR, Kaipio J and McLeod C N 2001 *Generalized optimal current patterns and electrical safety in EIT*, Physiol. Meas. 22 , pp. 85-90
[4] al Hagrey S 2012 *2D optimized electrode arrays for borehole resistivity tomography and CO2 sequestration modelling*, Pure Appl. Geophys. 169, pp. 1283-1292
[5] Jaworski AJ and Dyakowski T 2001 *Application of electrical capacitance tomography for measurement of gas-solids flow characteristics in a pneumatic conveying system*, Measurement Science and Technology, vol. 12, no. 8, pp. 1109-1119
[6] Scott DM and McCann H 2005, *Process Imaging For Automatic Control (Electrical and Computer Enginee).* Boca Raton, FL, USA: CRC Press, Inc.
[7] Ida N 2007 *Engineering Electromagnetics (2nd ed.)*, Springer Publishing Company
[8] Graham B and Adler A 2007 *Electrode placement confgurations for 3D EIT.* Physiol. Meas. 28, pp. 29-44
[9] Cui Z, Wang Q, Xue Q, Fan Q, Zhang L, Cao L, Sun B, Wang H, and Yang W 2016 *A review on image reconstruction algorithms for electrical capacitance/resistance tomography*, Sensor Review, vol. 36, no. 4, pp. 429-445
[10] Zangl H and Steiner G 2008 *Optimal Design of Multiparameter Multisensor Systems*, IEEE Transactions on Instrumentation and Measurement, Vol. 57/7, pp. 1484-1491
[11] Chaloner K and Verdinelli I 1995 *Bayesian experimental design: a review*, Stat. Sci. 10, pp. 273-304
[12] Khodja MR, Prange M D and Djikpesse HA 2010 *Guided Bayesian optimal experimental design*, Inverse Probl. 26, pp. 055008
[13] Fedorov V and Hackl P 1997 *Model-Oriented Design of Experiments*, New York, Springer, vol. 125
[14] Hyvönen N, Seppänen A and Staboulis S 2014 *Optimizing Electrode Positions in Electrical Impedance Tomography*, SIAM Journal on Applied Mathematics 74:6, pp. 1831-1851
[15] Kaipio JP, Seppänen A , Somersalo E and Haario E 2004 *Posterior covariance related optimal current patterns in electrical impedance tomography*, Inverse Probl. 20, pp. 919-936
[16] Kaipio JP, Seppänen A, Voutilainen A and Haario H 2007 *Optimal current patterns in dynamical electrical impedance tomography imaging*, Inverse Probl. 23, pp. 1201-1214
[17] Hansen P C 1998 *Rank-deficient and discrete ill-posed problems: numerical aspects of linear inversion*, Philadelphia, PA, USA: Society for Industrial and Applied Mathematics (SIAM)
[18] Kaipio JP and Somersalo E 2004 *Statistical and computational inverse problems*, Springer New York: Applied Mathematical Sciences
[19] Grauer S, Hadwin PJ and Daun KJ 2016 *Bayesian approach to the design of chemical species tomography experiments*, Appl. Opt. 55, 5772-5782
[20] Brandstätter B, Holler G, and Watzenig D 2003 *Reconstruction of inhomogeneities in fluids by means of capacitance tomography*, COMPEL, 22(3), pp. 508-519
[21] Holder D S 2005 *Electrical Impedance Tomography. Institute of Physics Publishing*, Bristol and Philadelphia, 1st edition
[22] Neumayer M, Zangl H, Watzenig D, and Fuchs A 2011 *Current reconstruction algorithms in Electrical Capacitance Tomography.* New Developments and Applications in Sensing Technology, Springer
[23] Isaksen O 1996 *A review of reconstruction techniques for capacitance tomography*, Measurement Science and Technology, vol. 7, pp. 325-337
[24] Watzenig D and Fox C 2009 *A review of statistical modelling and inference for electrical capacitance tomography*, Measurement Science and Technology, vol. 20, no. 5
[25] Polydorides N and McCann H 2002 *Electrode configurations for improved spatial resolution in electrical impedance tomography.* Meas. Sci. Technol. 13/12, pp. 1862-1870
[26] Yang WQ and Peng L 2003 *Image reconstruction algorithms for electrical capacitance tomography*, Measurement Science and Technology, vol. 14, no. 1
[27] Zangl H, Watzenig D, Steiner G, Fuchs A and Wegleiter H 2007 *Non iterative reconstruction for electrical tomography using optimal first and second order approximations*, World Congress on Industrial Process Tomography WCIPT5

[28] Julier SJ and Uhlmann JK 2004 *Unscented filtering and nonlinear estimation*, Proc. IEEE, vol. 92, no. 3, pp. 401-422, Mar.

[29] Liu S, Fu L, Yang WQ, Wang HG, Jiang F 2004 *Prior-online iteration for image reconstruction with electrical capacitance tomography*, IEE Proceedings of Science, Measurement and Technology, vol. 151, no. 3, pp. 195-200

[30] Klinzing GE, Rizk F, Marcus R and Leung L S 2011 *Pneumatic Conveying of Solids: A theoretical and practical approach*. Springer Science

[31] Jama GA, Klinzing GE and Rizk F 2000 *An Investigation of the Prevailing Flow Patterns and Pressure Fluctuation Near the Pressure Minimum and Unstable Conveying Zone of Pneumatic Transport System*, Powder Technology 112, pp. 87-93

[32] Storn R and Price K 1997 *Differential evolution - a simple and efficient heuristic for global optimization over continuous spaces*, Journal of Global Optimization 11, pp.341-359

[33] Yang WQ and Peng L 2003 *Image reconstruction algorithms for electrical capacitance tomography*, Measurement Science and Technology, vol. 14, no. 1

[34] Watzenig D and Fox C 2009 *A review of statistical modelling and inference for electrical capacitance tomography*, Measurement Science and Technology, vol. 20, no. 5, pp. 1-22

[35] Steiner G, Watzenig D, Magele C and Baumgartner U 2005 *Statistical robust design using the unscented transformation*, COMPEL, vol. 24, no. 2, pp. 606-619

[36] Watzenig D, Steiner G, Fuchs A, Zangl H and Brandstätter B 2007 *Influence of the discretization error on the reconstruction accuracy in electrical capacitance tomography*, COMPEL Volume: 26 Issue: 3

[37] Neumayer M, Bretterklieber T and Flatscher M 2015 *Reduced electrical capacitance tomography sensor for flow profile estimation*, IEEE Sensors

[38] Kaipio JP, Kolehmainen V, Somersalo E and Vauhkonen M 2000 *Statistical inversion and Monte Carlo sampling methods in electrical impedance tomography* Inverse Problems 16, pp. 1487-522

[39] Neumayer M, Bretterklieber T, Flatscher M and Puttinger S 2017 *PCA based state reduction for inverse problems using prior information*, COMPEL , Vol. 36 Issue: 5, pp. 1430-1441

[40] Neumayer M, Bretterklieber T and Flatscher M 2018, *Model Based Signal Processing for Capacitive Ice Sensing: Algorithm and Demonstration for Field Data*, accepted for I2MTC

IOP Conf. Series: Journal of Physics: Conf. Series **1047** (2018) 012013 doi:10.1088/1742-6596/1047/1/012013

A comparison of sparse Bayesian regularization methods on computed tomography reconstruction

J Ševčík and V Šmídl

Regional Innovation Center for Electrical Engineering, University of West Bohemia, Pilsen, Czech republic

E-mail: `jsevcik@rice.zcu.cz, vsmidl@rice.zcu.cz`

Abstract. Design of regularization term is an important part of solution of an ill-posed linear inverse problem. Another important issue is selection of tuning parameters of the regularization term. We address this problem using Bayesian approach which treats tuning parameters as unknowns and estimates them from the data. Specifically, we study a regularization model known as Automatic Relevance Determination (ARD) and several methods of its solution. The first approach is the conventional Variational Bayes method using the symmetrical factorization of the posterior of the vector of unknowns and the vector of tuning parameters. The second approach is based on the idea of marginalization over the vector of unknowns or the vector of tuning parameters, while the complementary vector is estimated using maximum likelihood. The resulting algorithm is thus an optimization task with non-convex objective function, which is solved using standard gradient methods.

The proposed algorithms are tested on real tomographic X-ray data and the comparison with conventional regularization techniques (Tikhonov and Lasso) is performed. The algorithm using marginalization over the tuning parameter is found to be closest to the ground truth with acceptable computational cost.

MATLAB® implementation of the reconstruction algorithms is freely available for download.

1. Introduction

Reconstruction of a tomographic image can be posed as a linear inverse problem. The choice of regularization term for such problem is essential and many possibilities are available, ranging from the basic Tikhonov regularization to total-variation [1]. Motivated by artifacts that are present at reconstructions with low number of projections, see Figure 1, we study performance of regularizations promoting sparsity. The idea is that the white pixels appearing incorrectly in the black area have no support in the data and sparse regularization should suppress them. While Lasso [2] is a prominent example of sparsity regularization, we focus our attention to Bayesian methods based on automatic relevance determination (ARD), which is the main regularization approach used in relevance vector machines [3] or sparse Bayesian learning [4].

Although the ARD principle can be applied to various transformations of the unknown vector, such as the wavelet transformation or first-order difference matrix in image processing [5, 6], we do not impose any transformation and study the effect of the sparsity assumption directly on the vector of unknowns. This is motivated by our aim to suppress artifacts in Figure 1. The probabilistic model of the ARD is based on the assumption of uncorrelated entries of the unknown vector with zero mean and unknown precision. Full Bayesian inference of such model

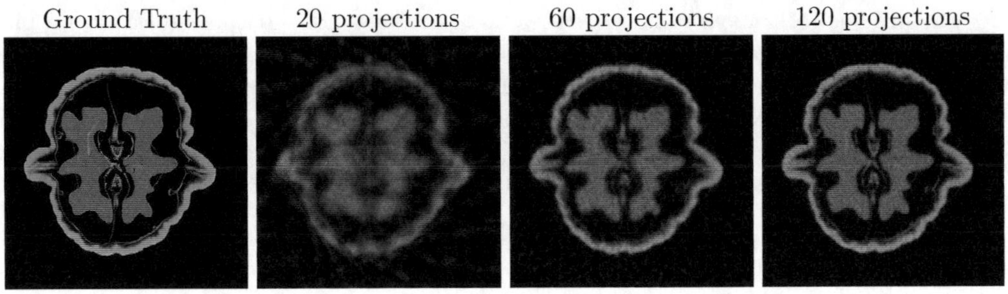

Figure 1. Illustration of the artifacts in the images reconstructed from low number of observations obtained using Tikhonov regularization (2) with $\lambda = 1$.

is analytically intractable and methods of approximate inference have to be used. We test three approximations in detail. The first approximation is the common Variational Bayes using factorization of the posterior into the product form (e.g. [7]). The second is the empirical Bayesian approach, which uses analytical marginal over the variable of interest to find best value of the tuning parameters. The third is marginalization over the tuning parameter and maximizing the marginal likelihood of the parameter of interest, which is sometimes referred to as Type II maximum likelihood [8]. In spite of many theoretical results [9], performance on a particular problem is hardly predictable. Therefore, we have implemented three methods of the model and applied them to the problem of tomographic reconstruction of a walnut [10]. Moreover, due to the need to run the methods for high dimensional problem, some standard operations such as inverse of large covariance matrices are prohibitive and are replaced by some common approximations. Therefore, good performance of the tested algorithms that is known from low dimensional problems may be violated.

As an alternative, we also test the hypothesis that the artifacts are caused by different precision of the projections at individual pixels. Relaxing the observation precision yields actually a very similar model to the relaxed penalization of the prior via the ARD model.

2. Problem formulation

In this paper, we will be concerned with a high-dimensional linear inverse problem

$$\mathbf{m} = A\mathbf{x} + \varepsilon, \tag{1}$$

where $A \in \mathbb{R}^{M\times N}$ is a known matrix, $\mathbf{x} \in \mathbb{R}^N$ is a vector of unknowns, $\mathbf{m} \in \mathbb{R}^M$ is an observation vector and $\varepsilon \in \mathbb{R}^M$ is an unknown realization of the noise or an error term.

There is a wide variety of applications that can be described by the mathematical model (1). Although the frameworks introduced in this contribution are fully general, our approaches will be illustrated on the problem of tomographic image reconstruction. From this point of view, vector $\mathbf{m}$ in (1) is composed of pixels of k measured projections (so called sinogram), matrix A stands for a measurement matrix which describes geometry of projections [11] and vector $\mathbf{x}$ is the reconstructed image in vector form.

The noise term in the computed tomography (CT) data, ε in (1), comprises two main sources – an additive (Gaussian) electrical noise and a quantum noise from the photon counts that has a Poisson statistic [12, 13]. However, we will use commonly used Gaussian approximation even for the term ε [7, 14, 15]. Since we do not model the noise correlation, we assume that the noise is not correlated. If the noise covariance matrix is known, the problem can be transformed into our formulation (1) by the change of coordinates, $\tilde{\mathbf{m}} = L^{-1}\mathbf{m}$ and $\tilde{A} = L^{-1}A$, where L^{-1} is the inverse of the Choleski decomposition of the noise covariance matrix.

IOP Conf. Series: Journal of Physics: Conf. Series **1047** (2018) 012013 doi:10.1088/1742-6596/1047/1/012013

Since the problem of tomography reconstruction is typically ill-posed, regularization plays a crucial role. Especially when the number of projections is low, the reconstructed image contains many artifacts, as illustrated in Figure 1. We investigate the use of various regularization terms to suppress the artifacts. The starting point of our research is the most basic Tikhonov regularization

$$\tilde{\mathbf{x}} = \mathrm{argmin}_{\mathbf{x}\in\mathbb{R}^N} \left\{ ||A\mathbf{x} - \mathbf{m}||_2^2 + \lambda ||\mathbf{x}||_2^2 \right\}, \tag{2}$$

where λ is a chosen constant. The method is known to provide good results for suitable choice of λ. While there is a number of recommendations and heuristics for the choice of λ, systematic estimation of this parameter can be provided by the Bayesian approach. Since the regularization methods based on Bayesian formulation are successfully used in related fields of image processing, we investigate their use in this context. The Bayesian approach treats tuning parameters as unknowns and estimates them from the data. This advantage is obtained at the cost of increased computational requirements of the resulting methods. Since exact solution of the Bayesian inversion problem including tuning parameters estimation is intractable, approximate solutions need to be used.

In this contribution, we compare several possible approximations with respect to accuracy and computational cost of the solution.

3. Bayesian approach

From the Bayesian perspective, the optimization problem (2) can be interpreted as a maximum a posteriori estimate of the following probabilistic model

$$p(\mathbf{m}|\mathbf{x}, \beta) = \mathcal{N}\left(\mathbf{m}|A\mathbf{x}; \beta^{-1} I_M\right), \tag{3}$$

$$p(\mathbf{x}|\alpha) = \mathcal{N}\left(\mathbf{x}|\mathbf{0}; \alpha^{-1} I_N\right), \tag{4}$$

where $\mathcal{N}$ denotes multivariate Gaussian distribution (defined by (A.1) in the appendix) and I_M denotes identity matrix of size $M \times M$. Parameters α, β are chosen constants, representing precision (inverse variance) of the elements of vectors $\mathbf{m}$ and $\mathbf{x}$, respectively. Logarithm of the joint distribution

$$\log p(\mathbf{m}, \mathbf{x}|\alpha, \beta) = -\frac{1}{2}\beta ||A\mathbf{x} - \mathbf{m}||_2^2 - \frac{1}{2}\alpha ||\mathbf{x}||_2^2 + c \tag{5}$$

is a sum of the same quadratic forms as in (2) and an additive constant c which aggregates all terms independent of $\mathbf{m}, \mathbf{x}$. Maximizing (5) is equivalent to minimization of (2) which is obvious when the right hand side of (5) is divided by $-\beta/2$ and assigning $\lambda = \frac{\alpha}{\beta}$. Following the Bayesian approach, the precision parameters can be treated as unknowns and thus estimated from the data. However, even the best manually tuned result of the Tikhonov regularization contains the artifacts. We conjecture that this is a consequence of the common precision parameters for all elements of the unknown vector or the residue. Therefore, we test models where precision parameters α or β are not scalars, but vectors where each element models covariance of the corresponding element of vector $\mathbf{m}$ or $\mathbf{x}$.

We investigate two possible hypotheses on the origin of the artifacts. First, pixels in the resulting image that do not have sufficient support in the inversion problem should have higher penalization towards zero value. This hypothesis corresponds to replacing scalar α by a vector. Second, the error $\boldsymbol{\varepsilon}$ is not homogeneous and some of its elements have higher variance than others. This hypothesis corresponds to replacing scalar β by a vector. These two hypotheses will be now elaborated into full probabilistic models.

The first hypotheses corresponds to replacing (4) by a different prior model

$$p(\mathbf{x}|\boldsymbol{\alpha}) = \mathcal{N}\left(\mathbf{x}|\mathbf{0}; \mathrm{diag}(\boldsymbol{\alpha}^{-1})\right), \tag{6}$$

IOP Conf. Series: Journal of Physics: Conf. Series **1047** (2018) 012013 doi:10.1088/1742-6596/1047/1/012013

where $\boldsymbol{\alpha} \in \mathbb{R}^N$ is a vector of precisions for each pixel in the image. This model is known as the Automatic Relevance Determination ARD [3]. It can be shown, that this model assigns relatively high prior probability that the elements of the vector are equal to zero. Strong information in the data is then required to increase the posterior probability of the non-zero values. In effect, estimation of the individual precisions α_i promotes sparse solutions [4], therefore this model will be called *sparse prior*.

The second hypotheses implies replacement of the likelihood (3) by a model with non-homogeneous error

$$p(\mathbf{m}|\mathbf{x}, \boldsymbol{\beta}) = \mathcal{N}\left(\mathbf{m}|A\mathbf{x}; \mathrm{diag}(\boldsymbol{\beta}^{-1})\right), \tag{7}$$

where $\boldsymbol{\beta} \in \mathbb{R}^M$ is a vector of precisions for each element of the measured vector $\mathbf{m}$. This model will be called *non-homogeneous noise model.*

Since the vectors $\boldsymbol{\alpha}$ and $\boldsymbol{\beta}$ are considered to be unknown, we define their prior distribution. The conjugate distribution for the precision parameters of a Gaussian distribution is the Gamma distribution [16]. Therefore, we choose the gamma distribution for parameter β in homogenous noise prior (3) and for all elements of vector $\boldsymbol{\alpha}$ in the sparse-prior case

$$p(\beta) = \mathcal{G}\left(\beta|\gamma_{0_\beta}; \delta_{0_\beta}\right), \tag{8}$$

$$p(\alpha_i) = \mathcal{G}\left(\alpha_i|\gamma_{0_\alpha}; \delta_{0_\alpha}\right), \quad i = 1, \ldots, N, \tag{9}$$

where $\gamma_\bullet$ and $\delta_\bullet$ are hyper-prior shaping constants. Priors for vector $\boldsymbol{\beta}$ in the non-homogenous noise prior case and scalar α in the regular prior (4) are:

$$p(\alpha) = \mathcal{G}\left(\alpha|\gamma_{0_\alpha}; \delta_{0_\alpha}\right), \tag{10}$$

$$p(\beta_j) = \mathcal{G}\left(\beta_j|\gamma_{0_\beta}; \delta_{0_\beta}\right), \quad j = 1, \ldots, M.. \tag{11}$$

We will study two specific combinations of the proposed models, namely the homogenous noise (3) with sparse prior (6)

$$p(\mathbf{m}, \mathbf{x}, \boldsymbol{\alpha}, \beta) = p(\mathbf{m}|\mathbf{x}, \beta)p(\mathbf{x}|\boldsymbol{\alpha})p(\beta)\prod_{i=1}^{N} p(\alpha_i) \tag{12}$$

and the non-homogeneous noise (7) with regular prior (4)

$$p(\mathbf{m}, \mathbf{x}, \alpha, \boldsymbol{\beta}) = p(\mathbf{m}|\mathbf{x}, \boldsymbol{\beta})p(\mathbf{x}|\alpha)p(\alpha)\prod_{i=j}^{M} p(\beta_j). \tag{13}$$

The above models can easily be extended to different versions. For example, replacement of norm $||\mathbf{x}||$ in (2) by an operator $||D\mathbf{x}||$ where D may be e.g. a differential operator. It is even possible to consider the elements of the operator to be unknown [17], however its extension to high-dimensions is challenging. Also, it is also possible to combine likelihood (7) with prior (6), we note that the resulting model exhibits extremely high flexibility due to growing number of unknown parameters. The posterior is highly multi-modal and reliable estimation is a challenge. Therefore, we focus only on the simpler models introduced above.

Graphical illustration of the proposed models is given in Figure 2 using the notation of graphical models [18].

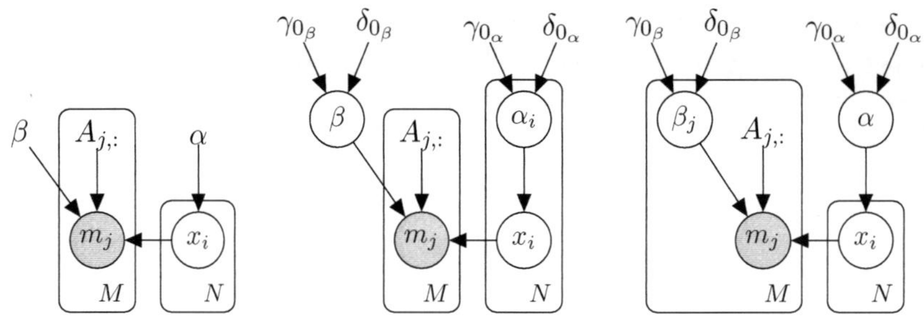

Figure 2. Directed graphical model representing the linear inverse problem with Tikhonov regularization (left), sparse prior model (middle), and non-homogeneous noise model (right).

4. Approximate Bayesian Solutions

The optimal inference for a probabilistic model is given by the Bayes rule. In our case, we seek posterior distribution of the unknown vector $\mathbf{x}$ given by measurements $\mathbf{m}$:

$$p(\mathbf{x}|\mathbf{m}) = \frac{p(\mathbf{x}, \mathbf{m})}{p(\mathbf{m})} = \frac{\int_{\boldsymbol{\theta}} p(\mathbf{x}, \mathbf{m}|\boldsymbol{\theta})p(\boldsymbol{\theta})\mathrm{d}\boldsymbol{\theta}}{p(\mathbf{m})}, \tag{14}$$

where $\boldsymbol{\theta}$ is a symbol aggregating all unknown parameters and hyperparameters. In our case, $\boldsymbol{\theta}$ is either $\boldsymbol{\theta} = [\boldsymbol{\alpha}, \beta]$ or $\boldsymbol{\theta} = [\alpha, \boldsymbol{\beta}]$. The integration of the right hand side of (14) expresses our aim to make the inference independent of the choice of the tuning parameters.

Since the exact evaluation of (14) for our choice of priors is not analytically tractable, various approximations can be used for obtaining a solution. We will briefly review the most popular principles.

Marginal likelihood also known as type II likelihood, is a technique based on evaluating only maximum of the posterior distribution (14),

$$\hat{\mathbf{x}}_{ml} = \arg\max_{x} p(\mathbf{x}|\mathbf{m}). \tag{15}$$

Since only maximum is required, it is not necessary to evaluate the marginalization constant $p(\mathbf{m})$ in (14) which makes the computation much simpler.

Empirical Bayes is a technique for choosing a fixed value of the parameters $\boldsymbol{\theta}$ via

$$p(\mathbf{x}|\mathbf{m}) \approx p(\mathbf{x}|\mathbf{m}, \boldsymbol{\theta}_{eb}), \quad \widetilde{\boldsymbol{\theta}} = \arg\max_{\boldsymbol{\theta}} \int p(\mathbf{x}, \boldsymbol{\theta}|\mathbf{m})\mathrm{d}\mathbf{x}, \tag{16}$$

which can be used if the integration over $\mathbf{x}$ is tractable.

Variational Bayes is based on approximation of the intractable joint probability (14) by a tractable approximation (denoted by symbol q to distinguish it from the exact solution) from the class of conditionally independent distributions fulfilling

$$p(\mathbf{x}, \boldsymbol{\theta}|\mathbf{m}) \approx q(\mathbf{x}, \boldsymbol{\theta}|\mathbf{m}) = q(\mathbf{x}|\mathbf{m})q(\boldsymbol{\theta}|\mathbf{m}) = q(\mathbf{x}|\mathbf{m}) \prod_{i=1}^{r} q(\boldsymbol{\theta}_i|\mathbf{m}), \tag{17}$$

where the multivariate parameter $\boldsymbol{\theta}$ is partitioned into r sub-vectors $[\boldsymbol{\theta}_1, \ldots, \boldsymbol{\theta}_r]$. The resulting $q(\mathbf{x}|\mathbf{m})$ is then used to approximate (14).

The main result of the Variational Bayes theory is that a conditionally independent approximation closest to the true posterior in the sense of the Kullback-Leibler divergence has to fulfil:

$$q^*(\mathbf{x}|\mathbf{m}) \propto \exp\left(E_{q^*(\boldsymbol{\theta}|\mathbf{m})}\left[\ln(p(\mathbf{x},\boldsymbol{\theta},\mathbf{m}))\right]\right), \tag{18}$$

$$q^*(\theta_i|\mathbf{m}) \propto \exp\left(E_{q^*(\mathbf{x}|\mathbf{m})\prod_{j=1,j\neq i}^{r} q^*(\theta_j|\mathbf{m})}\left[\ln(p(\mathbf{x},\boldsymbol{\theta},\mathbf{m}))\right]\right), \quad i = 1,\dots,r, \tag{19}$$

where $E_{q^*(\boldsymbol{\theta}|\mathbf{m})}[\cdot]$ denotes the expected value of its argument with respect to $q^*(\boldsymbol{\theta}|\mathbf{m})$ distribution. Since the approximations (18) – (19) are not given in closed form (for a computation of (18) we need to know (19), and vice-versa), an iterative algorithm for finding the solution is usually employed in the VB method.

In the following Sections, we will derive solutions of the considered models and compare their results.

5. Variational Bayes Solutions and Algorithms

In this section, we apply the Variational Bayes method [19] to probabilistic prior models discussed in previous sections and we propose two iterative algorithms suitable for high-dimension problem such as the tomography reconstruction.

First, we show a procedure for the model presuming sparse prior and homogenous noise, i.e. model described by (12).

5.1. Sparse prior

As mentioned earlier, estimation of unknown reconstruction and all unknown parameters (including hyperparametrs) can be obtained by the Bayes rule which is in this case computationally intractable. Following variational Bayesian framework we seek a posterior distribution satisfying posterior conditional independence

$$q(\mathbf{x},\boldsymbol{\alpha},\beta|\mathbf{m}) = q(\mathbf{x}|\mathbf{m})q(\beta|\mathbf{m})\prod_{i=1}^{N} q(\alpha_i|\mathbf{m}). \tag{20}$$

Application of the main VB result (18)–(19) to model (12) with partitioning (20) yields the approximate marginals in the following form:

$$q^*(\mathbf{x}|\mathbf{m}) = \mathcal{N}(\mathbf{x}|\boldsymbol{\mu};\Sigma), \tag{21}$$

$$q^*(\alpha_i|\mathbf{m}) = \mathcal{G}(\alpha_i|\gamma_{\alpha_i};\delta_{\alpha_i}), \quad i = 1,\dots,N, \tag{22}$$

$$q^*(\beta|\mathbf{m}) = \mathcal{G}(\beta|\gamma_\beta;\delta_\beta) \tag{23}$$

with shaping parameters given by

$$\Sigma = \left(A^T\widehat{\beta}A + \widehat{\mathrm{diag}(\boldsymbol{\alpha})}\right)^{-1}, \qquad \boldsymbol{\mu} = \Sigma A^T\widehat{\beta}\mathbf{m}, \tag{24}$$

$$\gamma_{\alpha_i} = \gamma_{0_i} + \frac{1}{2}, \qquad \delta_{\alpha_i} = \delta_{0_i} + \frac{1}{2}\widehat{x_i^2}, \quad i = 1,\dots,N, \tag{25}$$

$$\gamma_\beta = \gamma_{0_\beta} + \frac{M}{2}, \qquad \delta_\beta = \delta_{0_\beta} + \frac{1}{2}\overline{(\mathbf{m}-A\mathbf{x})^T(\mathbf{m}-A\mathbf{x})}, \tag{26}$$

IOP Conf. Series: Journal of Physics: Conf. Series **1047** (2018) 012013 doi:10.1088/1742-6596/1047/1/012013

where the symbol $\widehat{\cdot}$ denotes expected value of the argument with respect to its approximate distribution $q(\cdot)$. For completeness and clarity, we present the form of used moments in (24)–(26):

$$\widehat{\mathrm{diag}(\boldsymbol{\alpha})} = \mathrm{diag}\left(\left[\frac{\gamma_{\alpha_1}}{\delta_{\alpha_1}}, \dots, \frac{\gamma_{\alpha_N}}{\delta_{\alpha_N}}\right]\right) \tag{27}$$

$$\widehat{\beta} = \frac{\gamma_\beta}{\delta_\beta}, \tag{28}$$

$$\widehat{(\mathbf{m} - A\mathbf{x})^T(\mathbf{m} - A\mathbf{x})} = (\mathbf{m} - A\boldsymbol{\mu})^T(\mathbf{m} - A\boldsymbol{\mu}) + \mathrm{trace}[A^T A\Sigma], \tag{29}$$

$$\widehat{x_i^2} = \mu_i^2 + \Sigma_{i,i}, \quad i = 1, \dots, N. \tag{30}$$

A candidate solution of the set of implicit equations (24)–(30) can be found iteratively starting from an initial guess. Convergence of such algorithm to a local optimal has been proven e.g. in [20].

However, in high dimensional problems, a naive application of the iterative algorithm is prohibitive due to the necessity of covariance matrix calculation (24). The computation of inverse matrix $N \times N$, where N is large, implies high memory and time requirements. For that reason, we employ the conjugate gradient method for computation of $\boldsymbol{\mu}$ in each iteration. In addition to that, we use only diagonal elements of matrix Σ^{-1} to calculate the variances in (30). This is of course an approximation and it may have negative impact on the method performance. The resulting high-dimensional VB algorithm for sparse prior and homogenous noise (ARD–VB Algorithm) is summarized in Algorithm 1.

Algorithm 1 ARD–VB

1) Initialization:
   (i) Initialize constants $\gamma_{0_\alpha}, \delta_{0_\alpha}, \gamma_{0_\beta}, \delta_{0_\beta}$.
   (ii) Initialize hyperparameters $\boldsymbol{\alpha} = (\alpha_1, \dots, \alpha_N)^T$ and β.
2) Iterate equations (24)–(30) modified for high-dimension problem until convergence criterion is met
   (i) Solve set of linear equations (e.g. using conjugate gradient method) for $\mathbf{x}$

$$\left(A^T\beta A + \mathrm{diag}(\boldsymbol{\alpha})\right)\mathbf{x} = A^T\beta\mathbf{m} \tag{31}$$

   (ii) Invert diagonal elements of matrix Σ^{-1}, Σ_d is diagonal matrix,

$$\mathrm{diag}(\Sigma_d) \leftarrow 1/\mathrm{diag}\left(A^T\beta A + \mathrm{diag}(\boldsymbol{\alpha})\right) \tag{32}$$

   (iii) Compute hyperparameter estimates

$$\alpha_i \leftarrow \frac{2\gamma_{0_\alpha} + 1}{2\delta_{0_\alpha} + x_i^2 + \mathrm{diag}(\Sigma_d)_i} \quad i = 1, \dots, N \tag{33}$$

$$\beta \leftarrow \frac{2\gamma_{0_\beta} + M}{2\delta_{0_\beta} + (\mathbf{m} - A\mathbf{x})^T(\mathbf{m} - A\mathbf{x}) + \mathrm{tr}[A^T A\Sigma_d]} \tag{34}$$

Note that expression $\mathrm{tr}[A^T A\Sigma_d]$ in (34) is equivalent to the inner product $\langle\mathrm{diag}(A^T A), \mathrm{diag}(\Sigma_d)\rangle$. The choice of initial values of the shape and scale parameters $\gamma_\bullet$ and $\delta_\bullet$ can incorporate prior knowledge about the variances of reconstruction $\mathbf{x}$ and noise $\boldsymbol{\varepsilon}$. In case we don't know these

IOP Conf. Series: Journal of Physics: Conf. Series **1047** (2018) 012013 doi:10.1088/1742-6596/1047/1/012013

variances, it is appropriate to use values describing non-informative distribution of hyperparameters [19]. Non-informative prior is obtained for prior shape and scale parameters of the Gamma distribution equal to zero, however, for numerical reasons we set them to very low values e.g. 10^{-10}.

5.2. Non-homogeneous noise

Similar derivation can be applied to the model of non-homogeneous noise with regular prior, i.e. model described by (13). In this case, the posterior distribution is restricted to the following posterior conditional independence:

$$q(\mathbf{x}, \alpha, \boldsymbol{\beta}|\mathbf{m}) = q(\mathbf{x}|\mathbf{m})q(\alpha|\mathbf{m}) \prod_{j=1}^{M} q(\beta_j|\mathbf{m}). \tag{35}$$

The optimal approximative posterior densities found by the VB method for this model are

$$q^*(\mathbf{x}|\mathbf{m}) = \mathcal{N}(\mathbf{x}|\boldsymbol{\mu}; \Sigma), \tag{36}$$

$$q^*(\alpha|\mathbf{m}) = \mathcal{G}(\alpha|\gamma_\alpha; \delta_\alpha), \tag{37}$$

$$q^*(\beta_j|\mathbf{m}) = \mathcal{G}(\beta_j|\gamma_{\beta_j}; \delta_{\beta_j}), \quad j = 1, \ldots, M \tag{38}$$

where the equations for shaping parameters are shown in the Appendix.

In effect, the algorithm is a repetitive evaluation of the weighted least squares with Tikhonov regularization with adaptive modification of the weighting coefficients. The algorithm is summarized in Algorithm 2 and will be denoted as the WLS–VB algorithm in the remainder of this text.

Algorithm 2 WLS–VB

1) Initialization:
   (i) Initialize constants $\gamma_{0_\alpha}, \delta_{0_\alpha}, \gamma_{0_\beta}, \delta_{0_\beta}$.
   (ii) Initialize hyperparameters α and $\boldsymbol{\beta} = (\beta_1, \ldots, \beta_M)^T$.
2) Iterate equations (B.1)–(B.7) from Appendix modified for high-dimension problem until convergence criterion is met
   (i) Solve set of linear equations (e.g. using conjugate gradient method) for $\mathbf{x}$

$$\left(A^T \mathrm{diag}(\boldsymbol{\beta}) A + \alpha I_N\right) \mathbf{x} = A^T \mathrm{diag}(\boldsymbol{\beta}) \mathbf{m} \tag{39}$$

   (ii) Invert diagonal elements of matrix Σ^{-1}, Σ_d is diagonal matrix,

$$\mathrm{diag}(\Sigma_d) \leftarrow 1/\mathrm{diag}\left(A^T \mathrm{diag}(\boldsymbol{\beta}) A + \alpha I_N\right) \tag{40}$$

   (iii) Compute hyperparameter estimates

$$\alpha \leftarrow \frac{2\gamma_{0_\alpha} + N}{2\delta_{0_\alpha} + \mathbf{x}^T\mathbf{x} + \mathrm{tr}(\Sigma_d)} \tag{41}$$

$$\beta_j \leftarrow \frac{2\gamma_{0_\beta} + 1}{2\delta_{0_\beta} + (\mathbf{m} - A\mathbf{x})_j^2 + (A\Sigma_d A^T)_{j,j}} \quad j = 1, \ldots, M \tag{42}$$

The adjustments for the high dimensional problem in WLS–VB Algorithm are equivalent to those used in the ARD–VB Algorithm. Recommendations for an initialization of WLS–VB

Algorithm and its properties are the same as the recommendations and properties discussed for the ARD–VB Algorithm.

6. Methods based on analytical marginalization

In this Section, we derive algorithms for the homogenous noise models based on the marginal likelihood (15) and empirical Bayes (16). Both of these approaches are based on analytical marginalization over one variable: the marginal likelihood requires marginalization over $\boldsymbol{\theta}$, and empirical Bayes over $\mathbf{x}$.

For feasibility and clarity, we state that joint distributions used in this Section will be of the form:

$$p(\mathbf{m}, \mathbf{x}|\boldsymbol{\alpha}, \beta) = p(\mathbf{m}|\mathbf{x}, \beta)p(\mathbf{x}|\boldsymbol{\alpha}), \tag{43}$$

$$p(\mathbf{m}, \mathbf{x}, \boldsymbol{\alpha}|\beta) = p(\mathbf{m}|\mathbf{x}, \beta)p(\mathbf{x}|\boldsymbol{\alpha}) \prod_{i=1}^{N} p(\alpha_i), \tag{44}$$

where individual prior distributions on the right side are distributions (3), (6) and (9). However, precision parameter β is assumed to be fixed. Its estimation via the Variational Bayes procedure is straightforward.

6.1. Marginalization over vector of reconstruction

The idea and process introduced in this subsection was published by Wipf and Nagarajan in 2008 [4]. Following the idea of empirical Bayes (16), elements of vector $\boldsymbol{\alpha}$ can be estimated from the data using analytical marginalization of (43) over the vector of unknown reconstruction $\mathbf{x}$ and subsequent maximization of the resulting marginal likelihood function. It is equivalent to minimizing the cost function

$$\mathcal{L}(\boldsymbol{\alpha}) := -\ln \int p(\mathbf{m}|\mathbf{x}, \beta)p(\mathbf{x}|\boldsymbol{\alpha})\mathrm{d}\mathbf{x} = -\ln p(\boldsymbol{m}|\boldsymbol{\alpha}, \beta) \propto \ln|\Sigma| + \mathbf{m}^T\Sigma^{-1}\mathbf{m}, \tag{45}$$

where $\Sigma := \beta^{-1}I_M + A\mathrm{diag}(\boldsymbol{\alpha}^{-1})A^T$.

Once some $\widetilde{\boldsymbol{\alpha}} = \arg\min_{\boldsymbol{\alpha}} \mathcal{L}(\boldsymbol{\alpha})$ is computed, the estimation of vector of reconstruction $\widetilde{\mathbf{x}}$ can be obtained as the posteriori mean

$$\widetilde{\mathbf{x}} = \mathrm{diag}(\widetilde{\boldsymbol{\alpha}}^{-1})A^T\widetilde{\Sigma}^{-1}\mathbf{m}. \tag{46}$$

Several issues related to this model were observed by Wipf and Nagarajan who also proposed a solution [4]. They define a new variable $\mathbf{z} \in \mathbb{R}^N$ and an upper bounding auxiliary cost function $\mathcal{L}(\boldsymbol{\alpha}, \mathbf{z})$, for which it holds that $\mathcal{L}(\boldsymbol{\alpha}, \mathbf{z}) \geq \mathcal{L}(\boldsymbol{\alpha})$. Minimizing this cost function over $\boldsymbol{\alpha}$ for a fixed $\mathbf{z}$ can be achieved by solving

$$\mathbf{x} \leftarrow \mathrm{argmin}_{\mathbf{x}\in\mathbb{R}^N} \left\{ ||\mathbf{m} - A\mathbf{x}||_2^2 + \frac{2}{\beta}\sum_{i=1}^{N} z_i^{1/2}|x_i| \right\} \tag{47}$$

and then setting $\alpha_i^{-1} \leftarrow z_i^{-1/2}|x_i|$ for $i = 1, \ldots, N$ [4]. The method is summarized in Algorithm 3 and will be denoted as the ARD–marg**x** Algorithm.

For global convergence analysis see [4]. The initial values of all elements of vector $\mathbf{z}$ can be set to 1. Note that the optimization problem (47) can be solved by any convex optimization solver, e.g. using non-linear conjugate gradient method as the Fletcher-Reeves method in our case (see [21]), or by a special purpose Lasso solver. Computation of the optimal $\mathbf{z}$ in (48) needs to

Algorithm 3 ARD–margx

1) Initialize vector $\boldsymbol{z} = (z_i, \dots, z_N)^T$ and fixed β.
2) Iterate until convergence criterion for $\boldsymbol{\alpha}$ is met:
   (i) Solve the minimization problem (47).
   (ii) Set $\alpha_i^{-1} \leftarrow z_i^{-1/2}|x_i| \quad i = 1, \dots, N$.
   (iii) Compute the optimal $\mathbf{z}$ using

$$\mathbf{z} \leftarrow \text{diag}\left[A^T \widetilde{\Sigma}_*^{-1} A\right] \tag{48}$$

3) Compute estimation of $\mathbf{x}$ using (46).

compute matrix $\widetilde{\Sigma}_*^{-1}$. In the same way as in the previous section, we propose a simplification for high-dimensional problems. Specifically, we use only diagonal elements of matrix $\widetilde{\Sigma}$ to express the inverse matrix approximation (that is diagonal):

$$\widetilde{\Sigma}_*^{-1} = 1/\text{diag}\left(\beta^{-1} I_M + A\text{diag}(\boldsymbol{\alpha}^{-1}) A^T\right). \tag{49}$$

In the third step of the ARD–marg**x** Algorithm, we use the conjugate gradient method to solve the set of linear equations (46). First, we must solve (for unknown $\mathbf{w}$)

$$\mathbf{w} = \widetilde{\Sigma}^{-1}\mathbf{m} = \left(\beta^{-1} I_M + A\text{diag}(\boldsymbol{\alpha}^{-1}) A^T\right)^{-1} \mathbf{m} \tag{50}$$

and then set $\mathbf{x} \leftarrow \text{diag}(\boldsymbol{\alpha}^{-1}) A^T \mathbf{w}$.

6.2. Marginalization over vector of hyperparameters of precision

In previous subsection, we discuss a method based on analytical marginalization over vector of unknown reconstruction $\mathbf{x}$. In this Section, we start with analytical marginalization of (44) over the vector of hyperparameters $\boldsymbol{\alpha}$ [3], which models precision in prior distribution (6). Thus, we maximize the marginal likelihood function (15), what is in this case equivalent to minimizing the cost function

$$\begin{aligned} \mathcal{L}(\mathbf{x}) &:= -\ln \int p(\mathbf{m}|\mathbf{x}, \beta) p(\mathbf{x}|\boldsymbol{\alpha}) p(\boldsymbol{\alpha}) \text{d}\boldsymbol{\alpha} = -\ln \left(\mathcal{N}(\mathbf{m}|A\mathbf{x}, \beta^{-1} I_M) \prod_{i=1}^N \mathcal{T}(x_i|\nu_i; 0; \sigma_i^2) \right) \\ &\propto \frac{1}{2}\beta||\mathbf{m} - A\mathbf{x}||_2^2 + \sum_{i=1}^N \frac{\nu_i + 1}{2} \ln\left(1 + \frac{x_i^2}{\nu_i \sigma_i^2}\right), \end{aligned} \tag{51}$$

where $\mathcal{T}(x_i|\nu_i; 0; \sigma_i^2)$ is non-standardized Student's t-distribution (A.3) of element x_i of vector $\mathbf{x}$ with mean value $\mu_i = 0$ and with parameters $\nu_i := 2\gamma_{0_i}$ and $\sigma_i^2 := \frac{\delta_{0_i}}{\gamma_{0_i}}$, where γ_{0_i} and δ_{0_i} are parameters of prior distribution (9).

After substituting original parameters for ν_i a σ_i^2 in (51) we can express the cost function and its gradient as

$$\mathcal{L}(\mathbf{x}) \propto \frac{1}{2}\beta||\mathbf{m} - A\mathbf{x}||_2^2 + \sum_{i=1}^N \left(\gamma_{0_i} + \frac{1}{2}\right) \ln\left(1 + \frac{x_i^2}{2\delta_{0_i}}\right), \tag{52}$$

$$(\nabla \mathcal{L}(\mathbf{x}))_i \propto \left(\beta A^T(\mathbf{m} - A\mathbf{x})\right)_i + \frac{(2\gamma_{0_i} + 1)x_i}{2\delta_{0_i} + x_i^2}, \quad i = 1, \dots, N. \tag{53}$$

This model is often studied for values of $\gamma_0 = 0$ and $\delta_0 = 0$ which are appropriate for Jeffrey's prior. However, this choice implies many local extremes [9], therefore, we will study this method for a range of tuning parameters γ_0, δ_0.

Then, the unknown reconstruction $\mathbf{x}$ can be found as argument of the minimum of the cost function, $\tilde{\mathbf{x}} = \arg\min_{\mathbf{x}} \mathcal{L}(\mathbf{x})$. We can find this optimum by using e.g. non-linear conjugate gradient method as the Fletcher-Reeves method [21]. We will denote this approach as the **ARD–marg*α* Algorithm** in this paper. Initial values γ_{0_i} and δ_{0_i} can be obtained from the implied moments E_i and Var_i the meaning of which may be easier to comprehend:

$$\begin{aligned} E_i &:= \frac{\gamma_i}{\delta_i} & & \delta_i = \frac{E_i}{\mathrm{Var}_i} \\ \mathrm{Var}_i &:= \frac{\gamma_i}{\delta_i^2} & \iff & \gamma_i = \frac{E_i^2}{\mathrm{Var}_i}. \end{aligned} \tag{54}$$

7. Experimental results

In this Section, the described algorithms are applied to a real CT measurement data. For testing purpose we will use open X-ray tomographic datasets of a walnut [10] available at `www.fips.fi/dataset.php`. Our algorithms are available for download from `http://www.utia.cas.cz/linear_inversion_methods`

We compare the resulting reconstructions of all four proposed approaches and those obtained by the Tikhonov (ℓ_2) regularized reconstruction (2) and with the Lasso (ℓ_1) regularized reconstruction, $\tilde{\mathbf{x}} = \mathrm{argmin}_{\mathbf{x}\in\mathbb{R}^N} \left\{ ||A\mathbf{x} - \mathbf{m}||_2^2 + \lambda ||\mathbf{x}||_1 \right\}$, where $\lambda > 0$ is a fixed regularized parameter. Note, that the Tikhonov reconstruction can be computed simply by using conjugate gradient method to solve the set of equations, $\tilde{\mathbf{x}} = (A^T A + \lambda I_N)^{-1} A^T \mathbf{m}$. We use the optimized `pcg` function implemented in MATLAB® for solving the conjugate gradient method. The Lasso regularized reconstruction is computed by "`l1_ls`: A Matlab Solver for Large-Scale ℓ_1-Regularized Least Squares Problems" implemented by K. Koh, S.-J. Kim and S. Boyd [22].

The mentioned approaches and reconstructions will be discussed especially with respect to their ability to suppress artifacts and with respect to execution time requirements. We also quantify the error of reconstruction via the mean squared error, $\mathrm{MSE} = \frac{1}{N}\sum_{i=1}^{N}\left(x_i - x_i^{GT}\right)^2$, and the mean absolute error, $\mathrm{MAE} = \frac{1}{N}\sum_{i=1}^{N}|x_i - x_i^{GT}|$. The error is defined as a difference between the estimated image and the ground truth $\mathbf{x}^{GT}$. The ground truth image is a part of the data set and was obtained by filtered back projection computed from high-number (1200) of projections (available in the data set as well).

Exact quantification of the results is problematic especially with respect to different scales of the result provided by different method. For example, the Lasso method explains the projections using as few non-zero elements of $\mathbf{x}$ as possible. Thus, the estimated pixels have much higher value that those from other methods where contributions from many small pixels are summed up. To visualize this effect in the grayscale image (i.e. with pixels between 0..256), we present the results in two forms. First, the resulting reconstructions are shown in the obtained scale. The scale is set from the ground truth image which minimum is set to zero and maximum to 220. Methods providing estimates of $\mathbf{x}$ with higher values are truncated at the maximum allowed value of 256. The overestimated pixels should thus appear brighter than those in the ground truth image. The results are displayed in Figure 3. Second, the contrast and the brightness of the calculated images are adjusted to better fit the ground truth, i.e. images of the estimated $\mathbf{x}$ are scaled and shifted to minimize the term $||a\tilde{\mathbf{x}} + b - \mathbf{x}^{GT}||_2$, where $a, b \in \mathbb{R}$ are chosen contrast and brightness parameters. The results are displayed in Figure 4.

These results in Figures 3 and 4 visualize the effect of various methods. For example, majority of the pixels in $\mathbf{x}$ from the Lasso method is estimated in the correct scale, Figures 3, however, few pixels are much higher than those in the ground truth. Therefore, the adjustment of the

IOP Conf. Series: Journal of Physics: Conf. Series **1047** (2018) 012013 doi:10.1088/1742-6596/1047/1/012013

contrast and the brightness yields almost empty image in Figure 4. The same problem is visible for the results of the ARD-VB method. On the other hand, all pixels in the estimates of the ARD-marg**x** and the WLS-VB methods are closer to zero than those in the ground truth. After adjustment of the contrast and the brightness, the results are closer to the ground truth.

MSE and MAE values of the results of reconstruction of all discussed methods without any scaling or conversion are shown in Table 1 for selected regularization coefficient and in Figure 5 for a range or regularization coefficients. Note that due to the same pixels that cause the problem with visualization scale, the results of the Lasso are deteriorating also in terms of MSE and MAE. The values of MSE and MAE of the results adjusted for contrast and brightness are shown in Figure 6 for a range of regularization coefficients and in Table 2 for selected values.

Table 1. MSE and MAE of the estimated reconstructions from the ground truth depending on the number of projections (size of the known vector **m**), *without any scaling or shift* of reconstructed images.

approaches	20 projections		60 projections		120 projections	
	MSE ($\times 10^{-5}$)	MAE ($\times 10^{-3}$)	MSE ($\times 10^{-5}$)	MAE ($\times 10^{-3}$)	MSE ($\times 10^{-5}$)	MAE ($\times 10^{-3}$)
Tikhonov reg. ($\lambda = 1$)	7.025	**6.182**	3.416	4.124	2.741	3.706
Lasso reg. ($\lambda = 10^{-2}$)	175.125	13.216	62.338	10.955	36.555	10.048
ARD–VB Alg.	98.411	7.769	148.595	14.373	64.241	11.960
WLS–VB Alg.	9.820	6.940	5.819	5.061	3.553	3.830
ARD–marg**x**	15.620	7.243	7.778	5.061	6.139	4.605
ARD–marg$\boldsymbol{\alpha}$ $\left(\substack{E_i = 10^{-2} \\ \mathrm{Var}_i = 10^{-4}}\right)$	**6.944**	6.221	**3.210**	**4.021**	**2.168**	**3.098**

Table 2. MSE and MAE of estimated reconstructions from the ground truth *after adjustment of the contrast and the brightness* depending on the number of projections (size of the known vector **m**).

approaches	20 projections		60 projections		120 projections	
	MSE ($\times 10^{-5}$)	MAE ($\times 10^{-3}$)	MSE ($\times 10^{-5}$)	MAE ($\times 10^{-3}$)	MSE ($\times 10^{-5}$)	MAE ($\times 10^{-3}$)
Tikhonov reg. ($\lambda = 1$)	6.955	**6.206**	3.395	4.130	2.739	3.710
Lasso reg. ($\lambda = 10^{-2}$)	18.610	10.639	15.951	9.378	13.810	8.399
ARD–VB Alg.	17.968	10.384	18.122	10.396	15.920	9.327
WLS–VB Alg.	8.744	6.846	5.146	4.951	3.236	3.772
ARD–marg**x**	10.695	6.763	5.783	4.655	4.664	4.319
ARD–marg$\boldsymbol{\alpha}$ $\left(\substack{E_i = 10^{-2} \\ \mathrm{Var}_i = 10^{-4}}\right)$	**6.938**	**6.206**	**3.205**	**4.011**	**2.166**	**3.104**

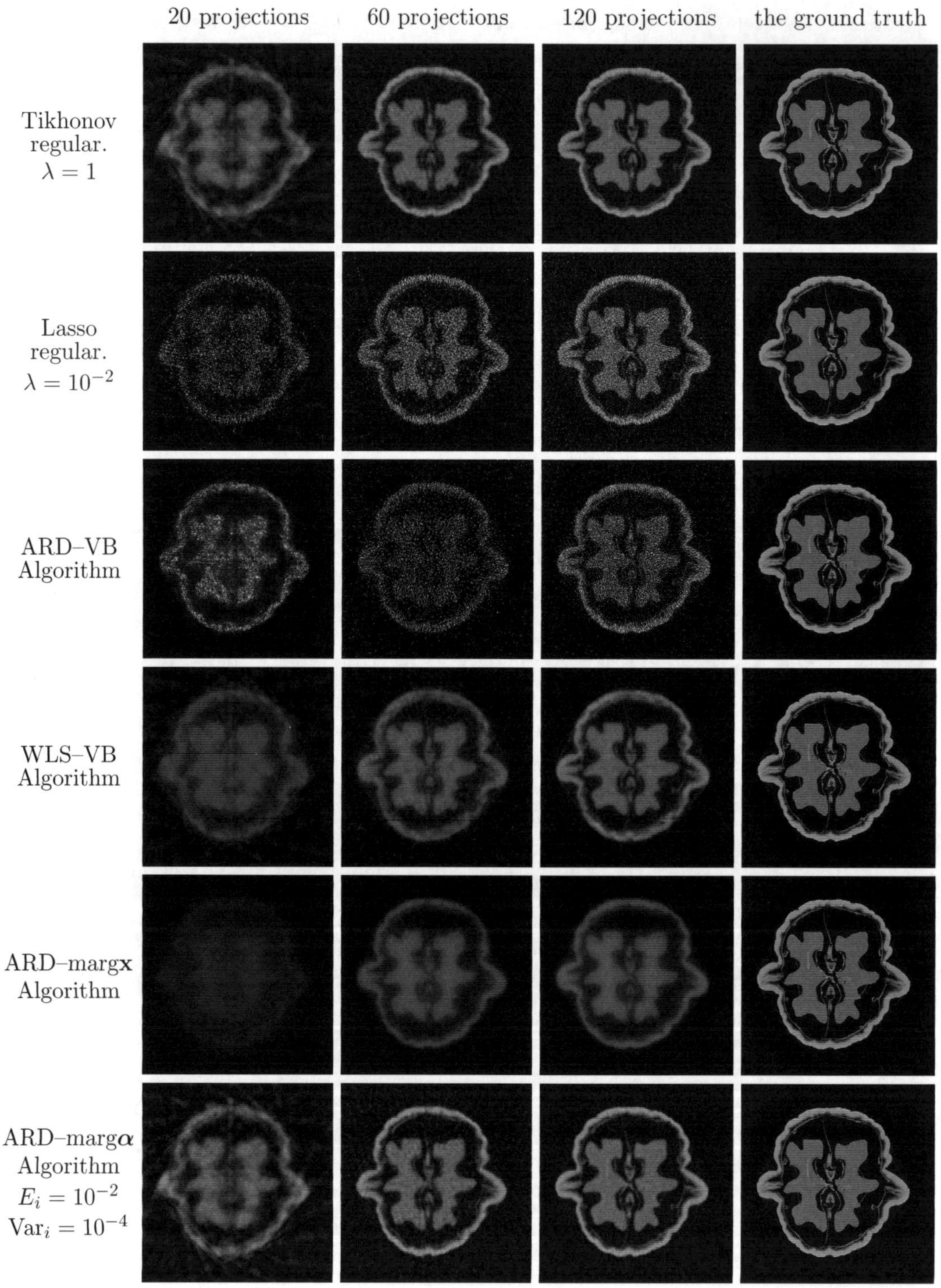

Figure 3. Resulting reconstructions of the walnut truncated to the interval $[0; 256]$ where the ground truth image scale is $[0; 220]$ obtained by: the Tikhonov regularization ($\lambda = 1$), the Lasso regularization ($\lambda = 10^{-2}$), ARD–VB Algorithm, WLS–VB Algorithm, ARD–marg$\mathbf{x}$ Algorithm and ARD–marg$\boldsymbol{\alpha}$ Algorithm ($E_i = 10^{-2}$, $\mathrm{Var}_i = 10^{-4}$) from 20, 60 and 120 projections.

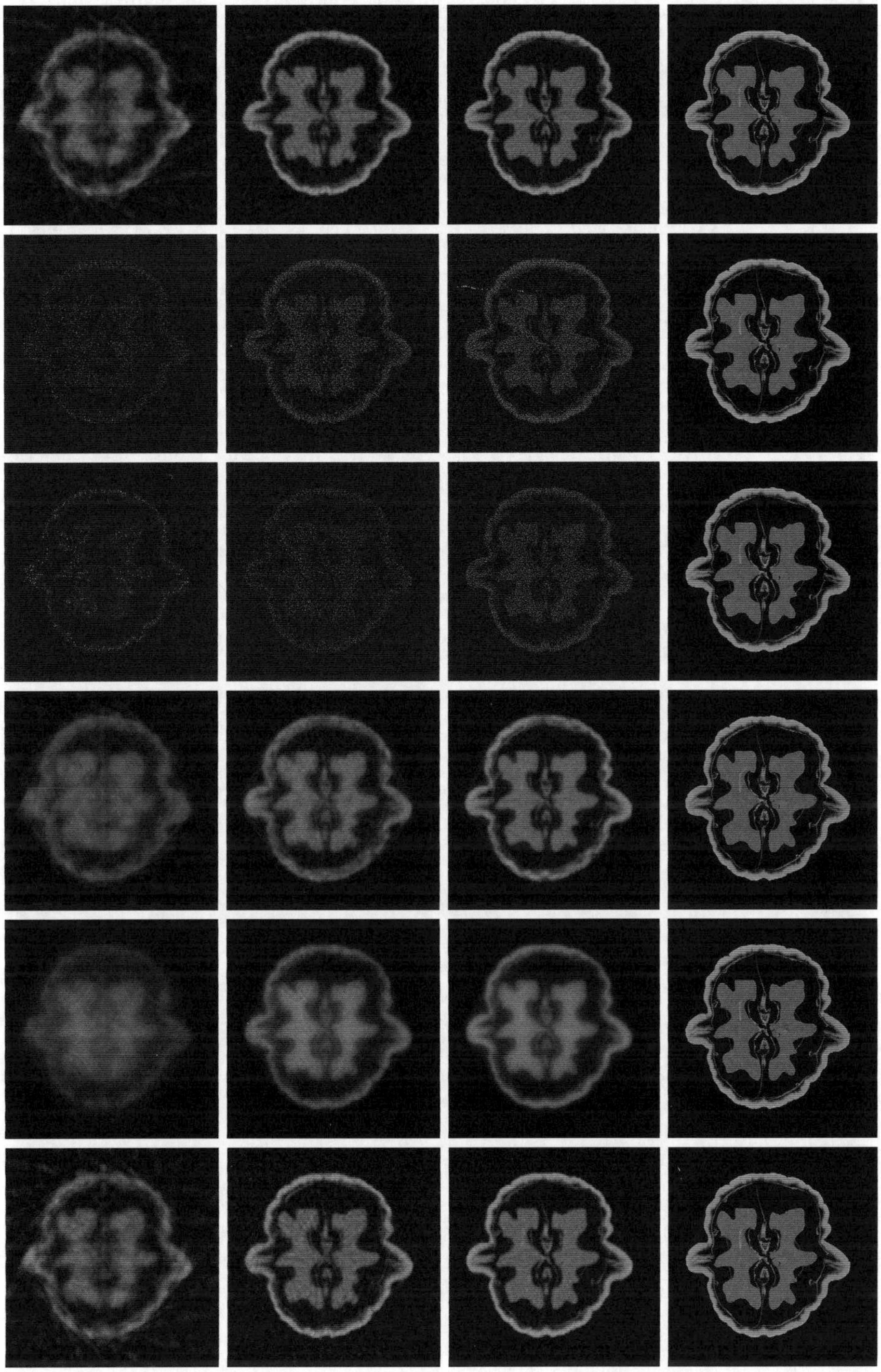

Figure 4. Resulting reconstructions of the walnut *with the adjusted contrast and brightness* and truncated to the interval$[0; 256]$ where the ground truth image scale is $[0; 220]$ obtained by: the Tikhonov regularization ($\lambda = 1$), the Lasso regularization ($\lambda = 10^{-2}$), ARD–VB Algorithm, WLS–VB Algorithm, ARD–marg$\mathbf{x}$ Algorithm and ARD–marg$\boldsymbol{\alpha}$ Algorithm ($E_i = 10^{-2}$, $\mathrm{Var}_i = 10^{-4}$) from 20, 60 and 120 projections.

Table 3. Time requirements [s] of the tested algorithms. The algorithms have been implemented in MATLAB® (v. R2015b) and computed on a PC (Windows® 7, Intel® Core™ i5-2410M CPU 2.30 GHz, RAM 4 GB) with the initialization used in the provided scripts.

approaches	20 projections	60 projections	120 projections
Tikhonov reg. ($\lambda = 1$)	0.43	0.98	1.71
Lasso reg. ($\lambda = 10^{-2}$)	467.50	451.78	465.73
ARD–VB Alg.	68.82	211.34	393.90
WLS–VB Alg.	53.72	169.95	366.72
ARD–margx	184.13	445.57	879.55
ARD–marg$\boldsymbol{\alpha}$ ($^{E_i = 10^{-2}}_{\mathrm{Var}_i = 10^{-4}}$)	11.26	30.75	72.97

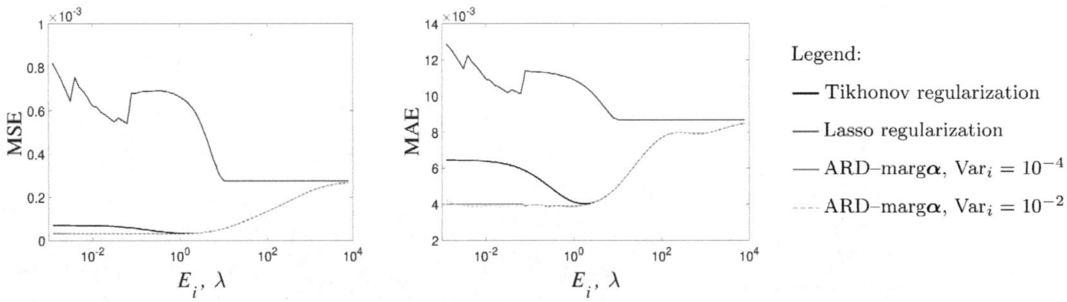

Figure 5. Mean squared error (MSE) and mean absolute errors (MAE) of reconstructions of the slice of the walnut *without any scaling or shift* obtained from 60 projections by the Tikhonov regularization, the Lasso regularization and ARD–marg$\boldsymbol{\alpha}$ Algorithm depending on the regularization parameter λ or prior moments E_i and Var_i.

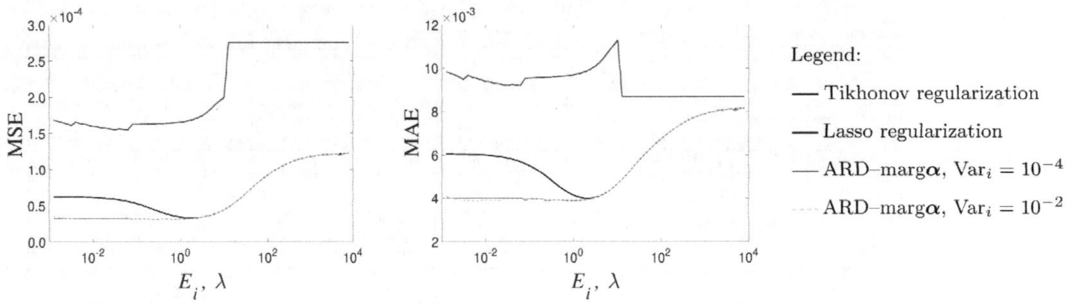

Figure 6. Mean squared error (MSE) and mean absolute errors (MAE) of reconstructions of the slice of the walnut *with the adjusted contrast and brightness* obtained from 60 projections by the Tikhonov regularization, the Lasso regularization and ARD–marg$\boldsymbol{\alpha}$ Algorithm depending on the regularization parameter λ or prior moments E_i and Var_i.

The best methods with respect to artifact suppression are the ARD–marg**x** algorithm and the WLS–ARD algorithm. However, these algorithms also seems to suppress the signal since the difference from the ground truth is greater than that of the Tikhonov regularization. The only algorithm achieving better MSE and MAE is the ARD–marg$\boldsymbol{\alpha}$ algorithm for the reported tuning parameters.

On the other hand, methods using unscaled ℓ_1 norm such as Lasso do not provide good estimated on this data. Similar results are obtained by the ARD-VB algorithm which seems to suffer from simplifications of the covariance matrix that were introduced to achieve computational feasibility in these dimensions.

The best method in our experiments turns out to be the ARD–marg$\boldsymbol{\alpha}$ method. It achieves marginally better MSE and MAE than the best tuned Tikhonov method with much lower effort in manual tuning. However, at the cost of increased computational cost (Table 3)). We conjecture that this good performance is due to the fact, that it is the only method from the Bayesian approaches that does not require any additional approximations (such as approximate matrix inversions).

8. Conclusion

The focus of this paper is on comparison of various solutions of the hierarchical Bayesian model known as ARD for a linear inverse problem. The studied solutions are based on different approximations and provide remarkably different results on the testing data. Due to the high dimensionality of the testing problem, additional simplifications had to be made in original algorithms to achieve computational feasibility. The only exception was the algorithm based on analytical marginalization over the vector of prior precisions, which also provides results closest to the ground truth. Moreover, it had also the lowest execution time of all algorithms using ARD. The algorithm using non-homogeneous noise model has some promising features, such as artifact supression, but its error is worse than that of the basic Tikhonov regularization. Perhaps a combination of the non-homogeneous noise with the ARD prior on the reconstructed vector may yield an improved solution. However, a more restricted prior parametrization of such model will be needed to cope with multi-modality of the resulting posterior.

Acknowledgments

This research has been supported by the Ministry of Education, Youth and Sports of the Czech Republic under the RICE New Technologies and Concepts for Smart Industrial Systems, project No. LO1607.

Appendix A. Probability density function

Probability density function (PDF) of normal (Gaussian) distribution, denoted as $\mathcal{N}$, of N-dimensional random vector $\mathbf{x}$ is defined as

$$\mathcal{N}(\mathbf{x}|\boldsymbol{\mu};\Sigma) = \frac{1}{(2\pi)^{N/2}|\Sigma|^{1/2}} \exp\left(-\frac{1}{2}(\mathbf{x}-\boldsymbol{\mu})^T \Sigma^{-1} (\mathbf{x}-\boldsymbol{\mu})\right). \tag{A.1}$$

PDF of gamma distribution, denoted as $\mathcal{G}$, of random variable x is defined as

$$\mathcal{G}(x|\gamma;\delta) = \frac{\delta^\gamma}{\Gamma(\gamma)} x^{\gamma-1} \exp(-x\delta), \tag{A.2}$$

where $\Gamma(\gamma)$ is the gamma function evaluated at γ. PDF of (non-standardized) Student's t-distribution, denoted as $\mathcal{T}$, of random variable x is defined as

$$\mathcal{T}(x|\nu;\mu;\sigma^2) = \frac{\Gamma\left(\frac{\nu+1}{2}\right)}{\Gamma\left(\frac{\nu}{2}\right)\sqrt{\pi\nu\sigma^2}} \left(1+\frac{(x-\mu)^2}{\nu\sigma^2}\right)^{-\frac{\nu+1}{2}}, \tag{A.3}$$

where $\mu \in \mathbb{R}$ is mean value, $\nu > 0$ expresses number of degrees of freedom and $\sigma^2 > 0$ is a scale parameter.

Appendix B. Shaping parameters of the WLS-VB algorithm

Shaping parameters of posterior distributions (36)–(38) were recognized as

$$\Sigma = \left(A^T \widehat{\mathrm{diag}(\boldsymbol{\beta})} A + \widehat{\alpha} I_N\right)^{-1}, \qquad \boldsymbol{\mu} = \Sigma A^T \widehat{\mathrm{diag}(\boldsymbol{\beta})}\mathbf{m}, \tag{B.1}$$

$$\gamma_\alpha = \gamma_{0_\alpha} + \frac{N}{2}, \qquad \delta_\alpha = \delta_{0_\alpha} + \frac{1}{2}\widehat{\mathbf{x}^T\mathbf{x}}, \tag{B.2}$$

$$\gamma_\beta = \gamma_{0_i} + \frac{1}{2}, \qquad \delta_\beta = \delta_{0_i} + \frac{1}{2}\widehat{(\mathbf{m} - A\mathbf{x})_i^2}, \quad i = 1, \ldots, M. \tag{B.3}$$

where relevant moments are

$$\widehat{\mathrm{diag}(\boldsymbol{\beta})} = \mathrm{diag}\left(\left[\frac{\gamma_{\beta_1}}{\delta_{\beta_1}}, \ldots, \frac{\gamma_{\beta_M}}{\delta_{\beta_M}}\right]\right), \tag{B.4}$$

$$\widehat{\alpha} = \frac{\gamma_\alpha}{\delta_\alpha}, \tag{B.5}$$

$$\widehat{\mathbf{x}^T\mathbf{x}} = \boldsymbol{\mu}^T\boldsymbol{\mu} + \mathrm{trace}[\Sigma], \tag{B.6}$$

$$\widehat{(\mathbf{m} - A\mathbf{x})_i^2} = (\mathbf{m} - A\boldsymbol{\mu})_i^2 + (A\Sigma A^T)_{i,i}, \quad i = 1, \ldots, M. \tag{B.7}$$

References

[1] Hämäläinen K, Harhanen L, Hauptmann A, Kallonen A, Niemi E and Siltanen S 2014 *International Journal of Tomography & Simulation* **25** 1–25
[2] Tibshirani R 1996 *Journal of the Royal Statistical Society. Series B (Methodological)* 267–288
[3] Tipping M E 2001 *Journal of Machine Learning Research* **1** 211–244
[4] Wipf D P and Nagarajan S S 2008 *Advances in Neural Information Processing Systems* **20** 1625–1632
[5] Wang L, Mohammad-Djafari A and Gac N 2017 *AIP Conference Proceedings* vol 1853 (AIP Publishing) p 060003
[6] Dumitru M, Li W, Gac N and Mohammad-Djafari A 2017 *2017 IEEE International Conference on Image Processing*
[7] Wang L, Mohammad-Djafari A, Gac N and Dumitru M 2016 *Acoustics, Speech and Signal Processing (ICASSP), 2016 IEEE International Conference on* (IEEE) pp 883–887
[8] Wipf D P, Rao B D and Nagarajan S 2011 *IEEE Transactions on Information Theory* **57** 6236–6255
[9] Buchgraber T 2013 *Variational sparse Bayesian learning: Centralized and distributed processing* Ph.D. thesis Technical University of Denmark
[10] Hämäläinen K, Harhanen L, Kallonen A, Kujanpää A, Niemi E and Siltanen S 2015 (*Preprint* `arXiv:1502.04064`)
[11] Mueller J L and Siltanen S 2012 *Linear and Nonlinear Inverse Problems with Practical Applications* (Philadelphia: Society for Industrial & Applied Mathematics)
[12] Schulze R, Heil U, Groβ D, Bruellmann D, Dranischnikow E, Schwanecke U and Schoemer E 2011 *Dentomaxillofacial Radiology* **40** 265–273
[13] Tu S J, Shaw C C and Chen L 2006 *Physics in medicine and biology* **51** 1283
[14] Hamalainen K, Kallonen A, Kolehmainen V, Lassas M, Niinimaki K and Siltanen S 2013 *SIAM Journal on Scientific Computing* **35** B644–B665
[15] Charbonnier P, Blanc-Féraud L, Aubert G and Barlaud M 1997 *IEEE Transactions on image processing* **6** 298–311
[16] Robert C P 2007 *The Bayesian choice: from decision-theoretic foundations to computational implementation* 2nd ed (New York: Springer)
[17] Tichý O, Šmídl V, Hofman R and Stohl A 2016 *Geoscientific Model Development* **9** 4297
[18] Bishop C 2006 *Pattern recognition and machine learning* vol 1 (Springer New York)
[19] Šmídl V and Quinn A 2006 *The variational Bayes method in signal processing* (Berlin: Springer)
[20] Sato M 2001 *Neural Computation* **13** 1649–1681
[21] Nocedal J and Wright S (eds) 2006 *Numerical Optimization* 2nd ed (New York: Springer)
[22] Kim S J, Koh K, Lustig M, Boyd S and Gorinevsky D 2007 *IEEE journal of selected topics in signal processing* **1** 606–617

IOP Conf. Series: Journal of Physics: Conf. Series **1047** (2018) 012014 doi:10.1088/1742-6596/1047/1/012014

A Study of the discrepancy function when modeling a packed bed using Variational Bayes and Gaussian processes

Sang-Ien Sien and A. F. Emery

Department of Mechanical Engineering, University of Washington, Seattle, WA 98195

E-mail: Emery@u.washington.edu

Abstract. Estimating properties using maximum likelihood (MLE) gives point values and a rough idea of the uncertainty in the parameters if their distributions are approximately normal. More accurate results require a Monte Carlo approach, often using the Markov Chain Monte Carlo (MCMC) method, which for complex models is unrealistic because of the computational expense. The Variational Bayes approach gives results comparable to MCMC with only a few evaluation of the model. When the model is imprecise or expensive to evaluate, Gaussian processes provide one means of analyzing the noise to determine the uncertainties.

Keywords: *Parameter Estimation, Uncertainty, Bayesian Inference,MCMC, Variational Bayes, Gaussian Processes*

1. Introduction

It is rare that a physics based model accurately represents a real situation. Firstly because the parameters of the model are usually known only approximately, but more likely because the sub-models are only approximations of the true situation. Probably the most well known deficient sub-model is the turbulence model for which any one of the several different choices may be only a crude approximation to the actual turbulence in the flow.

Let $\mathbf{D}$ be the measured response of a system and $\mathbf{M}(x, t, \theta)$ be the output of our physics based model where x, t, θ represent spatial position, time and parameters respectively, then we write

$$\mathbf{D} = \mathbf{M} + \delta + \epsilon \tag{1}$$

where δ represents the shortcomings of the model and ϵ is the measurement noise. δ is termed the 'discrepancy'. Clearly as a model is improved, δ should diminish. The improvement can be due to more accurate parameters or properties or by improved sub-models. On the other hand, it is unlikely that the model will improve if correlations are used to represent aspects of the model since most correlations, particularly in heat transfer or fluid flow, are inherently imprecise because they are developed from experimental conditions that often do not accurately relate to the model under question. Using simulations based upon discrepant models to estimate parameters of the model is common, but clearly leads to imprecise estimates. Many papers that treat parameter estimation are often based upon simulated experiments in which the 'experimental data' are either the output of the model corrupted by noise or the output of a model clearly known to be inadequate. For example in heat transfer ignoring known losses, e.g., treating a case with prescribed surface heat fluxes by a model that considers only Dirichlet boundary conditions.

Clearly if we know δ, one can always improve the model. The problem is that all that we learn from the data is $\delta+\epsilon$ and it will be impossible to separate them. Of course when using simulated experiments with known ϵ, δ is known. The problem becomes more severe when our model is used to estimate parameters. Typically, one scrutinizes the residuals, *res* defined by

$$res = \mathbf{D} - \mathbf{M}(\Theta) \tag{2}$$

where $\mathbf{D}$ are the data and $\Theta = (\theta_1, \cdots, \theta_n)$ are the parameters to be estimated. If the residuals exhibit desirable traits, e.g., zero mean, constant standard deviation, and near zero autocorrelation we can assign statistical information to our parameter estimates [17]. If the residuals do not have these traits, then much less can be said about our parameter estimates.

A good fitting model $M(x_i, t_i, \theta)$ is one for which the discrepancy is of the same order or less than the measurement error. It is important to recognize that there is no assurance that the estimated parameters, $\hat{\theta}$, of our model, $M(x_i, t_i, \theta)$ are the true values.

Consider an experiment in which $\mathbf{D}$ are time series data, for example temperatures measured as a function of time at a specific spatial point. One approach is to assume the form of $\delta(t)$, for example a polynomial, and to increase the order until the residuals exhibit the desired qualities. When the data are time series at several different spatial points, X, then one is faced with deriving $\delta(t, X)$. This is rarely effective, particularly when estimating parameters. On the other hand, if we can determine that the residuals, $res(t, X)$ exhibit similar characteristics for all X and all t, the problem may be simpler.

In this paper we discuss the estimation of the convective heat transfer coefficient and specific heat of a packed bed using a model that is clearly an approximation. Our interest is in determining if the discrepancies of the model when predicting the behavior of the temperature at different points in the bed share common traits as a function of the Reynold's number by using a Bayesian approach.

2. Packed Beds

Our packed bed, Figure 1, is a collection of small spheres of solid material contained in a circular tube. Such beds are used extensively in the chemical industry to promote chemical reactions such as absorption, stripping and distillation or as catalysts to promote reactions. Packed beds are also used extensively in thermal regenerators and for energy storage. In this latter case, the beds are often operated in a cyclical manner and a knowledge of the heat transfer coefficient for the interphase heat transfer is important to optimize their design. Because of the complexity of the system due to local variations in the fluid velocity, the transient nature of the solid material temperatures, contact resistance between the particles, and often the existence of strong, but highly irregular radial conduction, and the variation of the void fraction both axially and laterally, it is difficult to analyze the thermal behavior and particularly to optimize their design [32]. We are particularly interested in being able to predict the effects of cyclical operation which requires a knowledge of the heat transfer coefficient and the effective specific heat of the particles [1,34]. Designs are often based upon one dimensional flow and a two temperature model (fluid and packing). In reality, the flow is seldom one-dimensional and a number of experimental studies have been conducted to determine the radial velocity profile and the effective radial thermal conductivity. Recently, several CFD and multi-physics simulations based upon the Navier Stokes equations have been used to more explicitly study these effects [10, 5, 16, 4, 15]. These studies have used as many as 30 million cells and required up to 11 hours of cpu time.

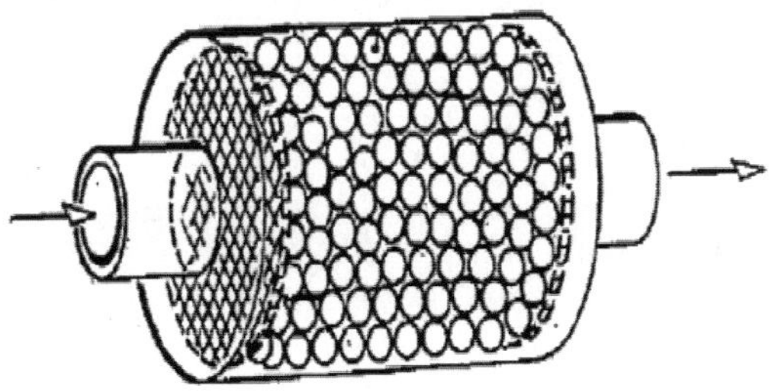

Figure 1. Schematic of Packed Bed

2.1. One Dimensional Analysis

Most designs are based upon the one dimensional two temperature models, usually referred to as the 'plug flow' method [1]. This approach is similar to the analysis of porous beds and ignores the effect of the radial conductivity and the radial variation of velocity. Reference [27] used this approach to consider rock beds and tubes with phase change and [7] treated a similar problem, but with particles that were very small relative to the bed diameter. The model is clearly deficient and entails a significant discrepancy, $\delta(r, z)$. Adebiyi [2, 23] compared the two equation model with a model that considered two dimensional flow, radial conduction, and losses to the walls for a bed for which the times of heating (the charging time) and cooling (the recovery time) were approximately of the same duration and showed that the two equation model sufficed.

The packed bed used in this experiment consists of a hollow glass cylinder, 0.614 m long, 114.8 mm OD and 102 mm ID packed with 1359 glass marbles each 15 mm in diameter. Air is heated by flowing over an electrical resistance heater to simulate a step change in supply temperature. Data are taken during the transient heating since the sensitivity of the parameters sought is highest then. Typical tests last 15 minutes for the high flow rates to several hours for the low flow rates. Reynolds numbers (see Eq. 4a) ranged from 240 to 7300. The temperatures were measured as a function of time as the bed was heated and then cooled by the flowing air. Along the bed vertical centerline, instrumented marbles were placed approximately every 7 cm from the inlet to the outlet. Each instrumented marble had a small diametral hole with a thermocouple located at its center and another at its surface to measure the air temperature. Figure 2 shows the time history of the marble center temperatures (every 10th increment in time) as a function of the marble position (z in cm) in the bed with the z=0 denoting the supply air temperature.

Because the efficiency of the bed in storing and recovering energy is strongly affected by the convective heat transfer between the flowing air and the marbles, we were interested in examining the convective heat transfer coefficient (h) during the heating and cooling periods. and the thermal capacitance (c_s) of the marbles (of diameter d) for a bed whose void fraction was f=0.50. The model follows that of Riaz [28] in which the air and bed temperatures are described by

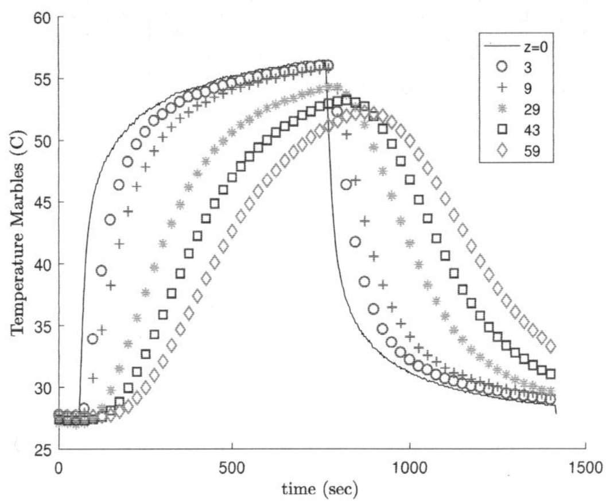

Figure 2. Packed Bed Temperatures as function of Axial Position (z, cm)

$$\text{Solids} \qquad \rho_s c_s (1-f)\frac{\partial T_s}{\partial t} = h_v (T_s - T_a) \tag{3a}$$

$$\text{Air} \qquad f\rho_a c_a (\frac{\partial T_a}{\partial t} + U_{ave}\frac{\partial T_a}{\partial z}) = h_v (T_s - T_a) \tag{3b}$$

where z represents the axial distance along the bed measured from the entrance of the bed, $h_v = h(1-f)\frac{6}{d}$ is the volumetric heat transfer coefficient, U_{ave} is the average air velocity, and the subscripts a and s refer to air and solids respectively.

2.2. Experimental Results

We analyzed 21 experiments in which the Reynolds number ranged from 240 to 7300 and compared the resulting estimates of the heat transfer coefficient obtained using Least Squares (Sec 3) to the correlations given by Mills [22], Incropera [6] and Wakao [32] assuming that the noise in the measured temperatures was normally distributed. The results presented are based upon an improved model of the bed heat losses and differ from those in [11]. Figure 3 compares the values of h estimated as described in Section 3.4 with those predicted using Mills' correlation showing that the agreement is within 20% but with a slightly stronger effect of the Reynolds number. The cause of the stronger effect of Re is not clear, but we suspect that it is likely due to the transient heating and cooling combined with the effect of heat transfer to the enclosing glass cylinder.

The packed bed correlations, [22, Eq. 4.131], [6, Eq. 7.72], [32] are respectively

$$Nu = (0.5\ Re_M^{0.5} + 0.2\ Re_M^{2/3})Pr^{1/3}(\frac{\mu_s}{\mu_b})^{-0.14} \text{with } Re_M = \frac{\rho_a U_{ave} d}{(1-f)\mu_a} \tag{4a}$$

$$f\ jh = 2.06\ Re^{-0.573} \qquad \text{with } Re = \rho_a U_{ave} d/\mu_a \tag{4b}$$

$$Nu = 2.0 + 1.1 Pr^{1/3} Re^{0.6} \qquad \text{with } Re = \rho_a U_{ave} d/\mu_a \tag{4c}$$

$$Nu = 1.2 Re_M^{0.728} \qquad \text{fitted values} \tag{4d}$$

in which the correlation, Eq. 4a, is reported to be accurate to ±25% [34]. Over the range of Reynolds numbers used, Wakao's correlation, Eq.4c is uniformly about 5% higher than that of Mills, Eq.4a. That of Incropera, Eq.4b, ranges from approximately twice at low Reynolds numbers to just slightly less at high values with an average value being approximately 20% higher. The coefficients of the fit, Eq.4d, have standard deviations of 0.225 and 0.020 respectively which are comparable with those reported by [30]. The correlations are for steady state. Kuwahara [19] suggested that Wakao's correlation needed some adjustment for the transient cases.

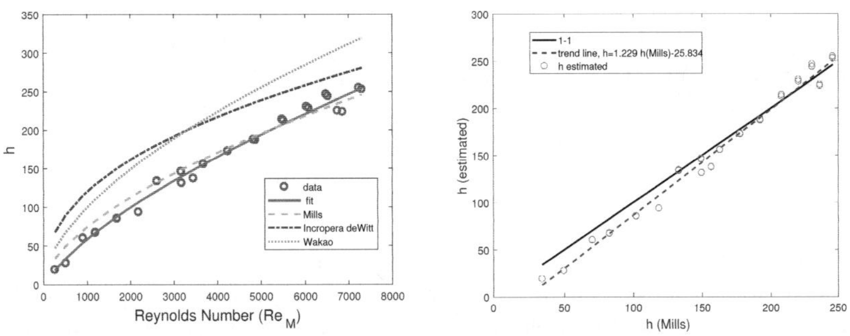

Figure 3. Estimated values of h (left) and compared to Mills' correlation (right)

3. Estimation of the Parameters

The most common method of parameter estimation is that of least squares which gives point estimates of the model parameters. If the measurement noise ϵ is normally distributed and uncorrelated and there is no discrepancy, the least squares method is identical to the maximum likelihood method (MLE) (which is equivalent to the Bayesian approach assuming a non-informative prior). Since **D** is the sum of a deterministic quantity, Eq.1, its statistical properties are those of ϵ. Different choices can be made both for the form of the error and its distribution. The most common is the normal distribution but other distribution can be used, such as assuming that the errors follow the Laplace distribution which is used to describe situations where large errors are likely. When used in the usual way, [17], least squares suffers from a few problems: 1) except for linear problems the method is iterative, 2) unless reasonably accurate initial estimates of the parameters are available, the iterative method may not converge or may converge to a local but not global minimum, 3) complex models with inter-related parameters may give rise to numerical difficulties. The MLE approach does not consider θ as unobserved random variables, but as parameters for which point estimates are desired. Using a one term Taylor series representation of the model, Eq.1, with respect to the parameters, the estimates are

$$\hat{\Theta} = (S^T S)^{-1} S^T D \tag{5a}$$

$$\sigma^2 = \frac{1}{n - N_\theta}(D - M(\Theta))^T (D - M(\Theta) \tag{5b}$$

$$cov(\Theta) = \sigma(\epsilon)(S^T S)^{-1} \tag{5c}$$

where S is the sensitivity of the model to the parameters, $dM(\Theta)/d\Theta$ and $\sigma(\epsilon)$ is the measurement noise and N_θ is the number of parameters.

What we really want to know is the probability distribution of the parameters that we are searching for, not just point values, i.e., the most probable value. To do this we make use of Bayes' relation [24]

$$p(\theta|D) = \frac{p(D|\theta)\pi(\theta)}{p(D)} = \frac{p(D|\theta)\pi(\theta)}{\int p(D|\theta)\pi(\theta)d\theta} \tag{6}$$

where D represents data (experimental results), $p(D|\theta)$ is the likelihood of obtaining the data given values of the parameters θ and $\pi(\theta)$ represents our prior knowledge (belief) about the parameters. The difficulty with Eq. 6 is the evaluation of the denominator (often termed the 'evidence') which must be evaluated if $p(\theta|D)$ is to be a proper probability distribution, i.e, the area under the curve sums to one. When several parameters are to be evaluated, the evaluation of the denominator is numerically difficult and often very computationally expensive.

For linear problems with the assumption of no prior information, i.e., $\pi(\theta)$ is non-informative, the posterior distribution, $p(\theta|D)$ will be normal. For other priors or for non-linear models, the posterior can be non-normal and estimating the uncertainty in θ is difficult. It is not uncommon to make use of Laplace's approximation [29] that relates the degree of uncertainty to the Hessian of the model.

3.1. MCMC

One approach to resolving the problem of evaluating the denominator and the need to know good initial values when using least squares is to use the Markov Chain Monte Carlo (MCMC) method. In this approach, one starts with some values of the parameters, θ_i^0, evaluates $p(D|\theta_i^0)\pi(\theta_i)$, then chooses a new set of parameters, θ_i^1 using what is called a 'proposal' distribution and checks to see if the probability has increased. If it has, the new values are accepted, if not, they are accepted with some random probability. After continuing this process, usually for a great number of times, the probability $p(\theta|D)$ is found to converge to a stable distribution. How to choose the sequence of parameters, i.e., the proposal distribution, is not trivial. Typically it is either a random walk or a Gaussian step. The sequence of choices is called a 'chain', see [3] for a useful discussion of MCMC.

We used the MCMC program DREAM [31] which employs an adaptive parallel multi-chain method and assumed that the data are normally distributed to estimate the heat transfer coefficient and the specific heat. Typical chain lengths after the burn-in, i.e., number of posterior samples, is NL, where N is the number of parallel chains and L the length of each chain. For low Reynolds numbers the burn in involved about 5000 evaluations of the model while for high Reynolds numbers it was of the order of 2500. Acceptance rates were of the order of 15% and 20% respectively. In both cases it took about 20000 model evaluations to get well defined probability distributions of the parameters. The number of evaluations depends upon how accurately one wants to estimate the posterior distributions. It took only about 20% of these values if one were satisfied with an estimate of the parameters and their standard deviation. A reduction of about 50% can be achieved if one starts with initial values taken from the MLE solutions.

3.2. Variational Bayesian Inference (VB)

Variational Bayesian methods are a family of techniques that are popular in Machine Learning. They allow one to re-write the statistical inference problems as an optimization problem, thus solving parameter estimation problems using modern optimization methods. The goal of variational Bayes is to represent $p(D)$ and $p(\theta|D)$ in simpler forms.

The most commonly used variational algorithm is based upon the mean field approach in which the posterior distribution of the parameters, θ, is represented by

$$q(\theta) = \prod_{i}^{N_\theta} q_i(\theta_i) \tag{7}$$

Despite the computational convenience of the mean-field approximation it is not appropriate when there are strong dependencies between the different θ. The alternate choice is the weighted Gaussian distribution

$$q(\theta) = \sum_{i=1}^{N_\theta} w_i N(\theta|\mu_i, \Sigma_i) = \sum_{i=1}^{N_\theta} w_i N(\theta|\mu_i, \sigma_i \mathbf{I}) \tag{8}$$

where w_i represent weights and the latter form assumes isotropic covariances (see [12]). Such mixtures can model reasonably complex, even multi-modal, distributions.

The evidence, $p(D)$ is represented as

$$log(p(D) = F(q(\theta)) + KL(q(\theta), p(\theta|D)) \tag{9a}$$

where

$$F(q(\theta) = E_q\big[log\frac{p(D,\theta)}{q(\theta)}\big] \tag{9b}$$

and

$$KL(q(\theta), p(\theta|D)) = E_q\big[log\frac{q(\theta)}{p(\theta|D)}\big] \tag{9c}$$

where E_q is the expectation over the assumed distribution, $q(\theta)$. The KL term measures the dissimilarity, i.e, the divergence, between $q(\theta)$ and $p(\theta|D)$. The more similar they are the closer KL approaches zero. The KL-divergence is always positive and equals zero only when the two distributions are equal. $F(q(\theta))$ is termed the free energy and due to the non-negativity of the KL-divergence, the free energy is always less than or equal to $p(D)$ and thus is a lower bound to it. In other words, the VB method aims to find a variational distribution that is as similar as possible to the posterior distribution, which is equivalent to maximizing the free energy. Thus the VB method has converted the estimation of parameters into a maximization problem which may be simpler than the numerical evaluation of the denominator in Eq.6.

The Variational Bayesian approach can yield results equivalent to MCMC with a computational expense that is several orders of magnitude less. Particularly good explanations of VB can be found in References [14, 26, 12].

Maximizing $F(q)$ with the constraints posed by the prior probabilities is often not a trivial numerical problem. Typically some form of Newton's method based upon the Hessian or its approximation is used, e.g., BFGS or L-BFGS [21].

3.3. Comparison of Methods

The most common method for parameter estimation is least squares. Conclusions about the uncertainty in the estimated parameters and the degree of correlation are valid only if the residuals are normally distributed. The Variational Bayes method can utilize different distributions, Eq.8. On the other hand, the MCMC approach does not require one to make assumptions about the behavior of the residuals.

To compare the different parameter estimation methods consider the temperature distribution in a slab of thickness L with generation rate Q, surface heat transfer coefficient h and conductivity k that satisfies

$$T - T_\infty = \frac{QL}{h} + \frac{QL}{2k}(L^2 - x^2) \tag{10}$$

Data were simulated by adding normally distributed noise with a standard deviation equal to 10% of $T(0) - T(L)$. We chose an error set for which the residuals were close to being normally distributed to give an idea of the best performance and most comparable performance of the methods. The results presented in Table 1 are in nondimensional form, e.g., $h^* = h_{est}/h_{actual}$. The VB results were obtained using the mean field approximation using the VBA toolbox [9] and following the algorithm suggested by Gershman [12]. Since Eq.10 is non-linear in k and h, but linear in $1/k, 1/h$, the computations were done for both representations as indicated in the table. The VB approach showed a strong effect on the estimated uncertainties of k, but the estimated values of k and h are only slightly changed while the number of model evaluations was halved because of the linearity of the model as also seen in the MLE case.

Table 1. Fitting $T - T_\infty = \frac{QL}{h} + \frac{QL}{2k}(L^2 - x^2)$

	k^*	$\sigma(k^*)$	h^*	$\sigma(h^*)$	Correlation	Modeling	# model evaluations
MLE	0.938	0.109	1.023	0.024	-0.89	k,h	4
MLE	0.938	0.108	1.023	0.023	-0.89	1/k,1/h	2
MCMC	0.964	0.110	1.020	0.023	-0.88	k,h	100
MCMC	0.935	0.114	1.024	0.022	-0.86	1/k, 1/h	100
VB	1.006	0.255	1.001	0.095		k,h	66
VB	0.932	0.550	1.023	0.099		1/k, 1/h	32

Note, that the estimates of k and h are highly correlated, clearly an undesirable result. In fact, it is not possible to estimate all three parameters, Q, k, h, simultaneously because the sensitivities are so strongly related.

3.4. Estimated Packed Bed Parameters

Table 2 lists typical results obtained using the three methods. The estimated values of h and c_s from MCMC and from least squares (MLE) differ by less than 1%. The estimates of uncertainty from MLE are about one-half of the MCMC values. See Section 4 for further discussion. Figure 4 shows characteristic pdfs for h and c_s and it is clear that they are close to Gaussian, explaining the close agreement of the two methods. The success of the MLE depends upon these distributions being close to normal. The uncertainties obtained with MCMC come directly from the computations. On the other hand, those for MLE depend upon an analysis of the residuals. If the pdf of the parameters is a reasonably multivariate normal distribution (namely that the bounds of the priors do not significantly truncate $p(\theta|D)$), then one can use the quadratic approximation [29, pg 52] in which the negative of the inverse of the Hessian of

IOP Conf. Series: Journal of Physics: Conf. Series **1047** (2018) 012014 doi:10.1088/1742-6596/1047/1/012014

the likelihood combined with an estimate of the standard deviation of the measurement noise, ϵ, is the covariance matrix of the estimated parameters (see Eq. 13). If $\sigma(\epsilon)$ is not known, then one must find a way to separate ϵ from δ. We discuss this further in Section 4.

Figures 5 and 6 compare the measured and predicted responses at each of the thermocouple locations for the lowest and highest Reynolds numbers and we see that the discrepancy gradually increases as a function of thermocouple position. The predicted temperatures agree well at early times, but display increasing uncertainty as time increases, regardless of the thermocouple position.

Table 2. Estimates of the Heat Transfer Coefficient and Heat Capacity

	Re=900				Re=7000				Typical Model evaluations
Heating	h	$\sigma(h)$	c_s	$\sigma(c_s)$	h	$\sigma(h)$	c_s	$\sigma(c_s)$	
MLE	60.1	0.09	618.1	0.18	254.7	1.09	641.5	0.73	10
MCMC	60.4	0.19	618.0	0.36	254.8	2.17	641.5	1.49	20000
VB	60.4	0.01	618.0	0.32	254.7	0.14	641.5	1.19	400
Cooling									
MLE	60.4	0.08	570.5	0.05	252.7	1.06	697.4	2.74	10
MCMC	60.1	0.19	570.5	0.85	252.7	2.10	697.4	6.19	20000
VB	60.0	0.01	570.5	0.37	252.7	0.14	697.4	3.92	400

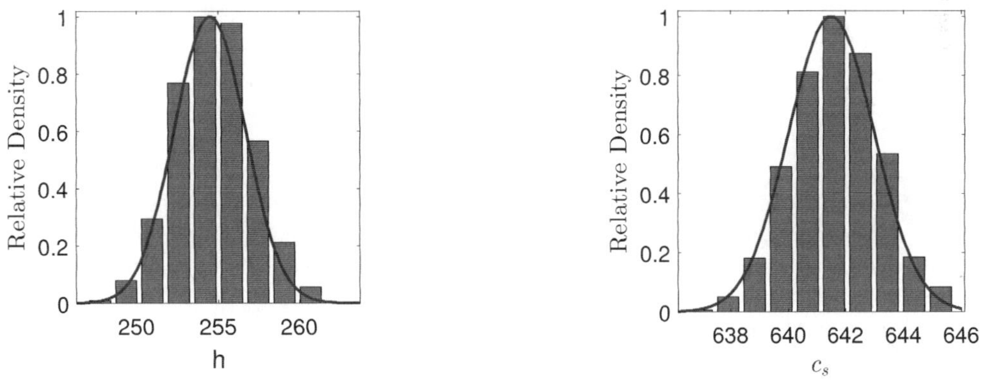

Figure 4. Probability Distributions for h and c_s compared to Equivalent Gaussian Distribution (generated by the MCMC software [31])

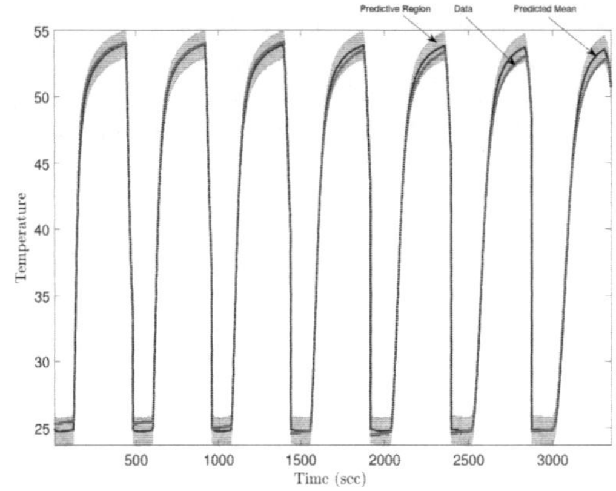

Figure 5. Showing the Regions and Means of the Predicted Temperatures and the Data for Re=900 (generated by the MCMC software [31])

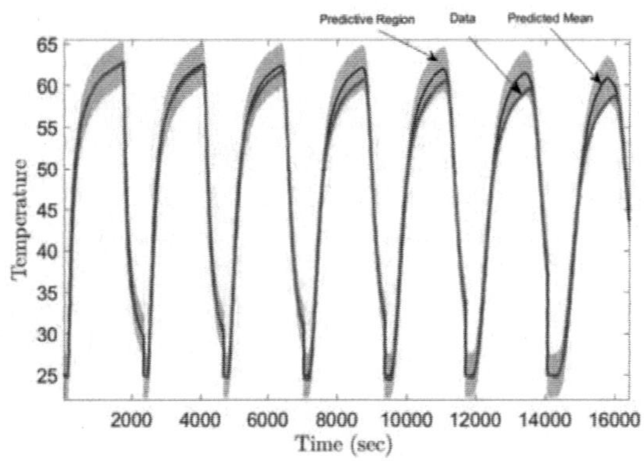

Figure 6. Showing the Regions and Means of the Predicted Temperatures and the Data for Re=7000 (generated by the MCMC software [31])

3.5. Discrepency

Figures 5 and 6 show the measured temperatures and those predicted by the MCMC approach and the region of uncertainty based on one standard deviation. The predicted temperatures agree with those obtained using parameters obtained from the MLE analysis. We see that the difference between the data and the prediction, i.e., the discrepancy, and the uncertainty, as indicated by the shaded region, gradually increases as we move up the packed bed and that it is worse for low Reynolds numbers. We are interested in determining if there is any correlation between the discrepancies. If so, it might be possible to construct an approximate form of $\delta(t, X)$ that will improve our model and yield more realistic estimates of the uncertainty in the estimated parameters.

4. Gaussian Process-Kriging

Figure 7 compares the residuals at low and high Reynolds numbers. It is clear that the residuals do not have the desired properties to enable us to separate the effect of the model errors from

IOP Conf. Series: Journal of Physics: Conf. Series **1047** (2018) 012014 doi:10.1088/1742-6596/1047/1/012014

the experimental noise to estimate the uncertainty in the estimated properties. The estimated uncertainties given in Table 2 are unrealistically small, an effect of the large amount of data. Recall that in the Least Squares approach uncertainties are inversely proportional to the square root of the number of readings if the noise is uncorrelated. For our data the residuals, Fig.7 clearly provide no information about the degree of correlation. Furthermore, we are unlikely to be able to express the residuals in a closed form that would represent the discrepancy function. Naturally the best approach would be to consider a more complex model which would include radial conduction with consideration of the wall behavior and local velocity variations. As noted in the introduction, this is not a realistic approach. In particular it would be difficult to determine which effect was the cause of the behavior shown in the figure. Instead, we are interested in seeing if there is any statistical relationship between the residuals and in particular if we could extract the measurement noise by $\epsilon = res - \delta_e$ where δ_e is a smoothed representation of the residual.

We create δ_e using a method known as Gaussian Process. For example, let a response be a function of two variables, $y(\theta_1, \theta_2)$, and consider that over a range of these variables that the responses at individual points are closely related to each other, i.e. there exists a statistical relationship between the responses. Kriging, developed by the South African engineer D. G. Krige [8, 33], assumes that the response and covariance can be expressed as

$$y(\mathbf{s}) = \mu(\mathbf{s}) + \delta(\mathbf{s}) \tag{11a}$$

$$var(\delta(\mathbf{s_1}) - \delta(\mathbf{s_2})) = \mathbf{2}\gamma(\mathbf{s_1} - \mathbf{s_2}) \tag{11b}$$

where $\mathbf{s}$ represents a specific point θ_1, θ_2, $\mu(\mathbf{s})$ represents a large scale trend and 2γ is termed the variogram. $\delta(\mathbf{s})$ is a zero mean stationary random process. The variogram would either be chosen from a number of simple forms or derived from the difference between measured data and $\mu(\mathbf{s})$. In this way, values of $y(\mathbf{s})$ could be obtained efficiently by interpolation.

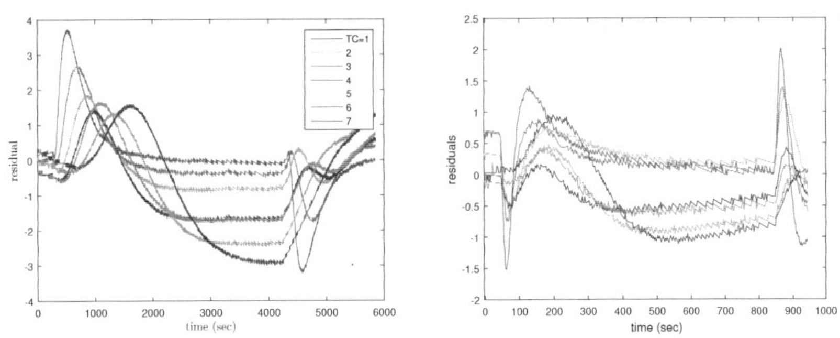

Figure 7. Residuals for Re=240 and 7300

In 2001 Kennedy and O'Hagan [18] used this idea to represent the variability of the computed response of $y(\theta_1, \theta_2)$. They proposed that an expensive-to-compute simulation model could be used to compute y at a defined set of points, $\mathbf{s} = \theta_1, \theta_2$, and $\delta(\mathbf{s})$ would represent the response at points nearby, i.e., an inexpensive interpolation model to account for the approximation errors associated with the surrogate model. In contrast to 'kriging', their model was based on a Gaussian distribution, the so called 'squared exponential' where σ represents the length scale of correlation between the data values.

$$exp(-\frac{1}{2}\frac{|\mathbf{s_1} - \mathbf{s_2}|^2}{\sigma^2}) \tag{12}$$

where $|\mathbf{s_1} - \mathbf{s_2}|$ is the distance between $\mathbf{s_1}$ and $\mathbf{s_2}$. See [25] for a useful description of the method.

In principle, the method is easy to apply, particularly for the small number of data points usually employed. For a large number of values, as exists in the residuals which are evaluated as shown in Figure 7, the derived matrices are very poorly conditioned and special techniques are needed. Fortunately, there are free versions of the algorithms available on the web that run under MATLAB [35, 36] and MATLAB itself has a kriging toolbox [21], all of which can handle a very large number of values, of the order of thousands. [However, these different programs may give slightly different results, but usually with inconsequential differences, based upon the specific optimization algorithm used.]

Figure 8 displays the result of kriging the residuals associated with one of the thermocouples. Using these smoothed functions (also known as 'trend lines' in kriging) to estimate ϵ we find that this estimated noise has a zero mean and is uncorrelated and its standard deviation is consistent with experiments conducted to evaluate the noise of the data acquisition system, $\sigma(daq)$ =0.04C.

The uncertainties listed in Table 2 for MLE were computed based on Laplace's approximation [29] and assuming an isotropic noise, $\sigma(noise)$, using

$$cov(\hat{\theta}) = \sigma(noise)^2(S^TS)^{-1} \tag{13}$$

where S is the sensitivity of the model to the parameters, Θ, and $\sigma(noise)$ is the combined effect of data acquisition noise, $\sigma(daq)$=0.04C, and the calibration noise, $\sigma(calibration) = 0.24C$.

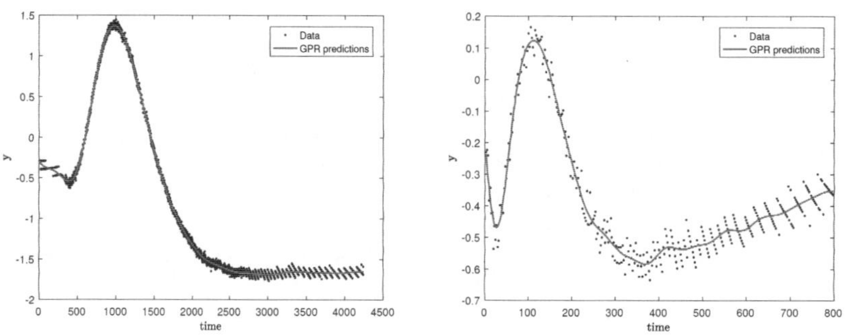

Figure 8. Residuals based on a Gaussian Process

5. Conclusions

Because the posterior distributions, Figure 4, are close to normal both the MLE and VB estimates are in good agreement with the MCMC values. Clearly both of these approaches are superior to MCMC in terms of the number of evaluations of the model needed. The MCMC required of the order of 20000 evaluations of the model, of which about half were in the burn in period and thus discarded. In contrast, the MLE needed in the order of 10 evaluations, and the VB about 400 (about 100 if only the heating or cooling data, but not both, were used). However, if the posterior were multi-modal or non-normal then, 1) while the MLE might identify the most likely value of the parameter, its estimate of the covariance will be significantly in error, 2) the VB approach will need a different form of $q(\theta)$ and a more complicated expression for $F(q)$.

In terms of the estimated parameters, all three methods give equivalent results. The MCMC method yields uncertainty estimates without any input from the analyst, in contrast to MLE. While the uncertainties vary, they refer only to the confidence in the estimate and do not reflect the inherent uncertainty associated with experiment. Table 3 gives relevant information for the set of experiments.

We had hoped that an analysis of the residuals might give us a way to estimate the discrepancy, $\delta(t, TC)$. Clearly from Figure 7 the discrepancy is a strong function of the Reynolds number and thermocouple position, but when using the Gaussian process to analyze them, we were unable to determine any features that would enable us to form any formulaic representation of $\delta(t, TC)$. As a consequence, it is not possible to separate δ and ϵ and to get reasonable estimates of the uncertainties of the estimated parameters as shown by the unrealistic values listed in Table 2.

Table 3. Uncertainties of Experimental Results based on Data from Figure 3

	$\sigma(\frac{\hat{h}-h(Mills)}{h(Mills)})$	$\sigma(\frac{\hat{h}-h(trend)}{h(trend)})$	$\sigma(\hat{c}_s)/\text{mean}(c_s)$
Heating	0.15	0.14	0.12
Cooling	0.08	0.06	0.08
Both	0.12	0.11	0.11

Regardless of the model discrepancy, the uncertainties listed in Table 3 are all of the order of 10%. The estimates $\hat{h}$ behaved the same for both the heating and cooling analyses. In contrast, most estimates of the specific heat gained from the cooling data were smaller than that from the heating data and the standard deviations are much larger, Table 2. We attribute this to the initial temperatures for each phase. When heating, all the marble temperatures are initially constant and equal to the ambient temperature. For the cooling phase, the upper section of the bed was still heating while the lower section was cooling, introducing a sizeable transient condition. In some experiments, heating was carried out long enough that the entire bed came to a constant temperature and the differences in the specific heat was less than 1%.

References

[1] Achenbach E 1995 Review: Heat and Flow Characteristics of Packed Beds *Experimental Thermal and Fluid Science* **10** pp 17-27
[2] Adebiyi G A, Nsofor E C, Steele W G and Jalalzadeh-Azar A A 1998 Parametric Study on the Operating Efficiencies of a Packed Bed for High-Temperature Sensible Energy Storage *ASME Solar Energy Engrg.* **120** pp 2-13
[3] Andrieu C, de Freitas N, Doucet A and Jordan M I 2003 An Introduction to MCMC for Machine Learning *Machine Learning* **50** pp 5-43
[4] Augier, F, Idoux F and Delenne J Y 2010 "Numerical simulations of transfer and transport properties inside packed beds of spherical particles" *Chemical Engineering Science* **65** 1055-1064
[5] Behnam M, Dixon A G, Nijemeisland M and Stitt E H 2013 A New Approach to Fixed Bed Radial Heat Transfer Modeling Using Velocity Fields from Computational Fluid Dynamics Simulations *Ind. Eng. Chem. Res.* **52** pp 15244-15261
[6] Bergman T L, Lavine A S, Incropera F P and Dewitt D P 2011 *Introduction to Heat Transfer, 6th Ed.* J. Wiley and Sons, NY, page 461
[7] Cascetta M, Cau G, Puddu P and Serra F 2015 A study of a packed-bed thermal energy storage device: test rig, experimental and numerical results *Energy Procedia* **81** pp 987-994

[8] Cressie N A C 1993 *Statistics for Spatial Date* J. Wiley and Sons, NY
[9] Daunizeau J, Adam V and Rigoux L, 2014 VBA: a probabilistic treatment of nonlinear models for neurobiological and behavioural data. PLoS Comp Biol **10** e1003441
[10] Dixon A G and Nijemeisland M 2001 CFD as a Design Tool for Fixed-Bed Reactors *Ind. Eng. Chem. Res* **40** pp 5246-5254
[11] Emery A F and Sien S-I 2017 Characterizing Uncertainties in a Thermal Storage Packed Bed Using Bayesian Inference with MCMC and Variational Approximations *Proc. 2nd Thermal and Fluids Engineering Summer Conference TFESC/IWHT-2017-17670* Las Vegas, NV
[12] Gershman S J, Hoffman M D and Blei D M 2012 Non-parametric Variational Inference *Proc. 29th International Conference on Machine Leaning* Edinburgh, Scotland, UK
[13] Gillespie B M, Crandall E D and Cargerry J J 1968 Local and Average Interphase Heat Transfer Coefficients in a Randomly Packed Bed of Spheres *AIChE J.* **14** pp 483-490
[14] Grimmer J 2011 An Introduction to Bayesian Inference via Variational Approximations *Political Analysis* **19** pp 32-47
[15] Guardo A, Coussirat M, Larrayoz M A, Recasens F and Egusquiza E 2005 Influence of the Turbulence Model in CFD Modeling of Wall-to-Fluid Heat Transfer in Packed Beds *Chem. Engrg. Science* **60** pp 1733-1742
[16] Guo X and Dai, R 2010 Numerical Simulation of flow and heat transfer in a random packed bed" *Particuology* **8** 293-299
[17] Hansen P C, Pereyra V and Scherer G 2012 *Least Squares Data Fitting with Applications, Project Muse* Johns Hopkins University Press
[18] Kennedy M C and O'Hagan A 2001 Bayesian calibration of computer models *Journal of the Royal Statistical Society: Series B (Statistical Methodology)* **63** pp 425464
[19] Kuwahara F, Shirota M and Nakayama A, 2001, A numerical study of interfacial convective heat transfer coefficient in two-energy equation model for convection in porous media, *Int. J. Heat and Mass Transfer* **44** pp 1153-1159
[20] Magnico P 2003 Hydrodynamic and Transport Properties of Packed Beds in Small Tube-to-Sphere Diameter Ratio: Pore Scale Simulation Using an Eulerian and a Lagrangian Approach *Chem. Engrg. Science* **58** pp 5005-5024
[21] MATLAB www.mathworks.com, Natick, MA
[22] Mills A T 1999 *Basic Heat and Mass Transfer* Prentice-Hall NJ page 323
[23] Nsofor E C and Adebiyi G A 2001 Measurements of the Gas-Particle Convective Heat Transfer Coefficient in a Packed Bed for High-Temperature Energy Storage *Exp. Thermal Fluid Science* **24** pp 1-9
[24] O'Hagan A and Forster J 1994 *Bayesian Inference* Oxford University Press New York
[25] O'Hagan A 2004 Bayesian Analysis of Computer Code Outputs: A Tutorial www.tonyohagan.co.uk/academic/pdf/BACCO-tutorial.pdf Last accessed 1/31/2017
[26] Ormerod J T and Wand M P 2010 Explaining Variational Approximations *American Statistician* **64** pp 140-152
[27] Pei Wen Li, Jon Van Lew, Wafaa K, Cho Lik Chan, Stephens J and OBrien J E 2011 "Transient Heat Transfer and Energy Transport in Packed Bed Thermal Storage Systems" *Developments in Heat Transfer* Dr Marco Aurelio Dos Santos Bernardes (Ed.) InTech DOI: 10.5772/20979. Available from: http://www.intechopen.com/books/developments-in-heat-transfer/transient-heat-transfer-and-energy-transport-in-packed-bed-thermal-storage-systems
[28] Riaz M 0000 "Transient Analysis of Packed Bed Thermal Storage Systems" *Solar Energy* **21** 123-128
[29] Sivia D S 1997 *Data Analysis, A Bayesian Tutorial* Oxford University Press New York
[30] Thakur, N S, Saini J S and Solanki S C, 2003, Heat Transfer and friction factor correlations for packed bed solar air heater for a low porosity system *Solar Energy* **74** pp 319-329
[31] Vrugt J 2016 Markov Chain Monte Carlo Simulation using the DREAM Software Package: Theory, Concepts and MATLAB Implementation *Environmental Modeling and Simulation* **75** pp 273-316
[32] Wakao N and Kaguei S 1982 *Heat and Mass Transfer in Packed Beds* Gordon and Breach Science Publ.
[33] Webster R and Oliver M A 2007 *Geostatistics for Environmental Scientists* J. Wiley and Sons, NY
[34] Whitaker S 1972 "Forced Convection Heat Transfer Correlations for Flow in Pipes, Past Flat Plates, Single Cylinders, Single Spheres, and for Flow in Packed Beds and Tube Bundles" *AIChE Journal* **18**pp 361-371
[35] www.gaussianprocess.org last accessed 2/6/2017
[36] http://www.sumo.intec.ugent.be/ooDACE last accessed 2/6/2017

IOP Conf. Series: Journal of Physics: Conf. Series **1047** (2018) 012015 doi:10.1088/1742-6596/1047/1/012015

Robust sparse linear regression for tokamak plasma boundary estimation using variational Bayes

Vít Škvára[1,2,3], Václav Šmídl[1,4], Jakub Urban[2]
[1]Institute of Information Theory and Automation, The Czech Academy of Sciences, Prague, Czech Republic
[2]Institute of Plasma Physics, The Czech Academy of Sciences, Prague, Czech Republic
[3]Faculty of Nuclear Sciences and Physical Engineering, Czech Technical University, Prague, Czech Republic
[4]Regional Innovation Center for Electrical Engineering, Pilsen, Czech Republic

Abstract. Precise control of the shape of plasma in a tokamak requires reliable reconstruction of the plasma boundary. The problem of boundary estimation can be reduced to a simple linear regression with a potentially infinite amount of regressors. This regression problem poses some difficulties for classical methods. The selection of regressors significantly influences the reconstructed boundary. Also, the underlying model may not be valid during certain phases of the plasma discharge. Formal model structure estimation technique based on the automatic relevance principle yields a version of sparse least squares estimator. In this contribution, we extend the previous method by relaxing the assumption of Gaussian noise and using Student's t-distribution instead. Such a model is less sensitive to potential outliers in the measurement. We show on simulations and real data that the proposed modification improves estimation of the plasma boundary in some stages of a plasma discharge. Performance of the resulting algorithm is evaluated with respect to a more detailed and computationally costly model which is considered to be the "ground truth". The results are also compared to those of Lasso and Tikhonov regularization techniques.

1. Introduction

The tokamak is one of the most promising concepts for future thermonuclear fusion reactors. In a tokamak, hot plasma is confined by a combination of strong toroidal and poloidal magnetic fields. The almost negligible weight of the plasma dictates that magnetohydrodynamic (MHD) equilibrium, i.e. a force-free state, describes its macroscopic evolution. The MHD description also includes the plasma boundary (the so-called last closed flux surface), which must be controlled during a tokamak discharge because of performance and safety reasons. Real-time feedback control is necessary since the plasma behaviour cannot be reliably predicted, particularly because the equilibrium is unstable. Such a feedback control requires the plasma boundary as an input in real-time. See e.g. [1] or [2] for more details.

One of the possibilities for real-time plasma shape reconstruction is offered by magnetic sensors near the plasma boundary. However, full reconstruction of the equilibrium magnetic field (which includes the plasma shape) is a complex task due to non-linear behavior of the plasma. In [3], it was shown that the plasma boundary can be modeled by a linear combination of a specific kind of basis functions. Since the basis functions are known, the estimation problem can be formulated as a simple linear regression with a potentially infinite number of regressors. However, many of these regressors will be uninformative or extremely sensitive to noise. Thus, our task is the selection of a relevant subset of the basis functions of the model.

This task is a well studied problem in statistics and optimization. Examples of classical methods are the Tikhonov [4] or the Lasso [5] regularizations. Crucial problem of these methods is selection of the tuning parameter. This problem is elegantly solved by the Bayesian approach which treats all tuning parameters as unknowns and estimates them from the data. Therefore, we focus on the Bayesian model known as automatic relevance determination (ARD) [6]. It is closely related to scaled $l1$ optimization [7]. In effect, the penalization forces weights of spurious regressors to zero, effectively removing them from the regression. This approach was shown to be a suitable model for estimation of the plasma boundary in [8].

In this paper, we extend the previous approach to recursive evaluation of the model in time during the full duration of a discharge. Moreover, we will assume that the model noise (or model residue) is not Gaussian distributed but that it may contain outliers or systematic errors. We will use the model of [9] to estimate the parameters. The resulting algorithm is essentially an iterated weighted least squares algorithm with adaptive estimation of the weight and regularization matrices.

The performance of the presented algorithm will be compared to traditional approaches using simulated and real measurements from a tokamak. We will attempt to provide physical interpretation of the introduced latent variables.

2. Plasma boundary model

Feedback control of plasma in a tokamak requires on-line reconstruction of the plasma boundary. The quality of the reconstruction is highly significant for tokamak operation safety and performance. The reconstruction of the boundary is done in two steps: i) the spatial distribution of scalar magnetic flux $\psi(r, z)$ is determined from the measurements, ii) the boundary is identified as the largest closed flux isoline. In this paper, we will be primarily concerned with the first step, however the quality of the reconstruction will be judged by output of the second step.

The spatial distribution of the magnetic flux $\psi(r, z)$ can be decomposed into an infinite series of basis functions f_i (see [3] for details). Using the assumption of symmetry of the three-dimensional plasma column in the toroidal direction, for each point in the vicinity of the plasma with coordinates (r, z) in the poloidal plane the flux distribution can be decomposed into an infinite series:

$$\psi(r, z) = \sum_{i=0}^{\infty} \theta_i f_i(r, z), \tag{1}$$

where f_i are functions based on Legendre functions with half integer degree [10], closely related to toroidal harmonics, and $\theta_i \in \mathbb{R}, i \in \mathbb{N}$ are multipliers associated with each basis function. For increasing i, the basis functions f_i represent higher frequencies of magnetic flux isolines. The expected shape of the boundary is relatively smooth, hence it is expected that only M lower order harmonics will be sufficient for its reconstruction. The magnetic field effects are measured by N magnetic probes around the tokamak chamber. These measurements are transformed by inverse operation into vector $y \in \mathbb{R}^N$ such that

$$y = X\theta + e, \tag{2}$$

where $X \in \mathbb{R}^{N \times M}$ is a design matrix whose rows are composed of evaluations of basis functions $f_i(r_j, z_j)$, $i = 1, \ldots, M$, where (r_j, z_j) are coordinates of the j-th probe, $j = 1, \ldots, N$, $\theta = (\theta_1, \ldots, \theta_M)$ is a vector of unknown coefficients and $e \in \mathbb{R}^N$ is the vector of noise/residue. When the θ vector is identified, we are able to evaluate $\psi(r, z)$ in an arbitrary point, enabling

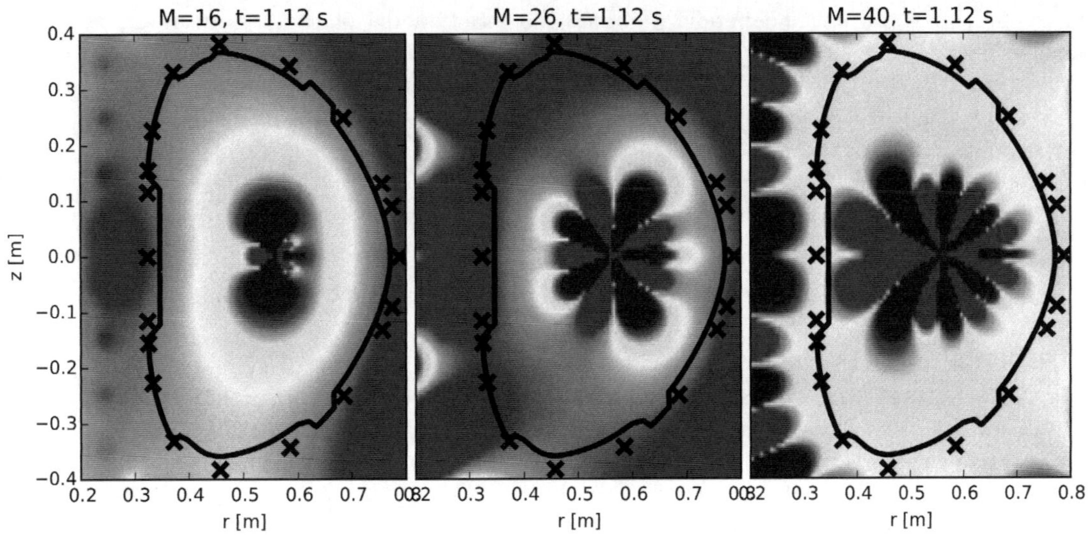

Figure 1. Illustration of the influence of the order M on the spatial distribution of magnetic flux $\psi(r, z)$ during a tokamak discharge when coefficients θ of each plot are obtained using ordinary least squares without regularization. The black line is the tokamak chamber wall, black crosses denote positions of the measurement probes.

us to identify the plasma boundary as the largest closed isoline. Illustration of the spatial distribution of the magnetic flux reconstructed from recorded data using least squares solution of (2) for different values of M are displayed in Fig. 1.

Note that the order of decomposition M significantly influences the spatial distribution of the magnetic flux making the task of plasma boundary identification sensitive to it. While the largest closed isoline in the right plot of Fig. 1 can be recognized around the yellow area inside the tokamak chamber, finding such isolines in the remaining two pictures is impossible. Therefore, we seek an optimal selection of the relevant elements of the model (2). For this purpose, we use a statistical model known as the automatic relevance determination (ARD) that identifies the parameters that are relevant and the remaining are set to zero. Using such an approach, we can then construct an ill-posed problem (2) even with $M > N$ and the algorithm should automatically selects the necessary regressors. The selected number M is thus interpreted as an upper bound on the necessary coefficients and when selected sufficiently high, the results should not be sensitive to it.

The proposed algorithm is essentially an iterative solution of the least squares with Tikhonov regularization

$$J(\theta) = ||y - X\theta||_2^2 + \sum_i \alpha_i \theta_i^2, \tag{3}$$

where the Tikhonov matrix diag(α) is adaptively updated in each iteration. This is in contrast with the classical Tikhonov regularization where the matrix diag(α) is assumed to be known. Although the proposed adaptation of α implies higher computation cost than fixed value of α, the increase is still acceptable and can be computed in real time in a closed control loop.

In this contribution, we extend the previous approach by considering unknown precision of the measurements. While the accuracy of the measuring probes is approximately similar, the error of the model may vary across the probes. Specifically, the inverse transformations required to compute vector y from the measurements is nonlinear and the variance of the model residue

IOP Conf. Series: Journal of Physics: Conf. Series **1047** (2018) 012015 doi:10.1088/1742-6596/1047/1/012015

e may be time- and space-dependent. Since the exact model of the residue is unknown, we consider the variance of each residue to be unknown and we will estimate it from the data using a variational Bayes approach. We chose a prior model that implies Student's-t distribution of the residue. In effect, the resulting variational Bayes algorithm will solve the weighted least squares with regularization

$$J(\theta) = ||T(y - X\theta)||_2^2 + \sum_i \alpha_i \theta_i^2, \tag{4}$$

with adaptively estimated weight matrix T in tandem with the Tikhonov matrix diag(α).

3. Bayesian model selection

Hierarchical priors are a commonly used tool to achieve model structure selection. The conventional least squares formulation of (2) corresponds to maximum likelihood estimation of likelihood function given by multivariate Gaussian probability distribution

$$p(y|X, \theta, \beta) = \mathcal{N}(y|X\theta, \beta^{-1}I). \tag{5}$$

where $\beta \in \mathbb{R}$ is a precision parameter of the model error. The likelihood model alone cannot decide which structure of the model is relevant. This can be achieved by Bayesian treatment with carefully chosen prior distribution on unknown parameters [11]. An extension of the model by the following prior

$$p(\theta|\alpha) = \mathcal{N}(\theta|0, A^{-1}), \tag{6}$$
$$A = \text{diag}(\alpha_1, \alpha_2, \cdots, \alpha_M),$$

leads to a maximum likelihood solution that is equivalent to Tikhonov regularization (3) where A is the Tikhonov matrix. The choice of diagonal matrix corresponds to the assumption that the elements of θ are apriori independent, each with its own prior precision α_i. In our approach, the precision parameters are treated as unknown and are estimated jointly with the parameters θ. Prior distribution of the precisions α_i and β is chosen in the form of gamma distribution:

$$p(\alpha_i) = \mathcal{G}(\alpha_i|a_0, b_0), \tag{7}$$
$$p(\beta) = \mathcal{G}(\beta|c_0, d_0). \tag{8}$$

Here, values of parameters a_0, b_0, c_0, d_0 are assumed to be known. This leads to a more complicated model but removes the burden of expert selection of the precision parameters.

Now, the task is to obtain the posterior distributions of parameters θ, A, β. This is done using the variational Bayes method. It consists of factorizing the joint posterior and then marginalizing it over individual parameters. For details see [12]. The approximate posterior densities are identified to be of forms conjugate to the priors:

$$\tilde{p}(\alpha_i|y) = \mathcal{G}(\alpha_i|a_{\alpha i}, b_{\alpha i}), \tag{9}$$
$$\tilde{p}(\beta|y) = \mathcal{G}(\beta|c, d), \tag{10}$$
$$\tilde{p}(\theta|y) = \mathcal{N}(\theta|\hat{\theta}, S), \tag{11}$$

with equations that govern the shaping parameters given in the appendix. The resulting estimation algorithm is essentially a least squares estimation with ARD adaptation of the regularization (LS-ARD) [8]. The LS-ARD algorithm is guaranteed to converge to a local minimum of the Kullback-Leibler divergence from the estimated to the true posterior. We assume that prior parameters a_0, b_0, c_0 and d_0 are chosen to be 10^{-10} to yield non-informative prior.

IOP Conf. Series: Journal of Physics: Conf. Series **1047** (2018) 012015 doi:10.1088/1742-6596/1047/1/012015

4. Robust Least Squares using Student's t noise model

To improve the results of plasma reconstruction, we alter the original algorithm from Section 3 by allowing a richer error distribution which can model outliers and systematic errors, as explained in Section 2. Specifically, we replace the Gaussian noise distribution by a Student's t-distribution with unknown degrees of freedom d. For low values of d, this distribution has heavier tails than the Gaussian. The estimation algorithm can thus identify outliers and provide more robust estimate of the parameter of interest.

While prior distributions for θ and A are the same as in the previous model, i.e. (6), (7), the likelihood model (5) is changed to

$$p(y|X,\theta,\omega,z) = \mathcal{N}(y|X\theta,(\omega \text{diag}(u))^{-1}) \tag{12}$$

$$p(\omega) = \mathcal{G}(\omega|\rho_0,\kappa_0), \tag{13}$$

$$p(u_i|d) = \mathcal{G}(u_i|d/2,d/2), \forall i \tag{14}$$

$$p(d) = \mathcal{G}(d|\phi_0,\psi_0). \tag{15}$$

The values of u_i (more exactly its square root) represent the weights in a diagonal matrix T of the weighted least square problem (4). The Student's t-distribution of the noise arises from the fact that

$$S(e|0,\omega,d) = \int_0^\infty \mathcal{N}(e|0,\omega^{-1}u^{-1})\mathcal{G}\left(u|d/2,d/2\right)du$$

Again, we use the VB framework to obtain equations for shaping parameters of the posteriors. Their forms are

$$\tilde{p}(\theta|y) = \mathcal{N}(y|\hat{\theta},S), \tag{16}$$

$$\tilde{p}(\alpha_i|y) = \mathcal{G}(\alpha_i|a_{\alpha i},b_{\alpha i}), \tag{17}$$

$$\tilde{p}(\omega|y) = \mathcal{G}(\omega|\rho,\kappa), \tag{18}$$

$$\tilde{p}(u_i|y) = \mathcal{G}(u_i|a_{ui},b_{ui}), \tag{19}$$

$$\tilde{p}(d|y) = \mathcal{G}(d|\phi,\psi), \tag{20}$$

with shaping parameters in the appendix. The final estimation algorithm – robust least squares with ARD adaptation (RLS-ARD) algorithm – is given in Algorithm 1. This more elaborate structure of the model may provide us with further information on the physical system we are describing as will be discussed in the next section.

Algorithm 1 RLS-ARD - iterative algorithm for structure selection of robust least squares problem.

Off-line:

- Select prior parameters a_{0i}, b_{0i}, ρ_0, κ_0,ϕ_0 and ψ_0.

On-line:

- Initialize $\widehat{\omega}$ and $\widehat{u}_i$ with chosen values.
- Repeat until convergence:
  - update shaping parameters of θ, using (B.1),
  - update shaping parameters of A, using (B.2),
  - update shaping parameters of ω, using (B.3),
  - update shaping parameters of u, using (B.4),
  - update shaping parameters of d, using (B.5).
- Report final estimate $\hat{\theta}$.

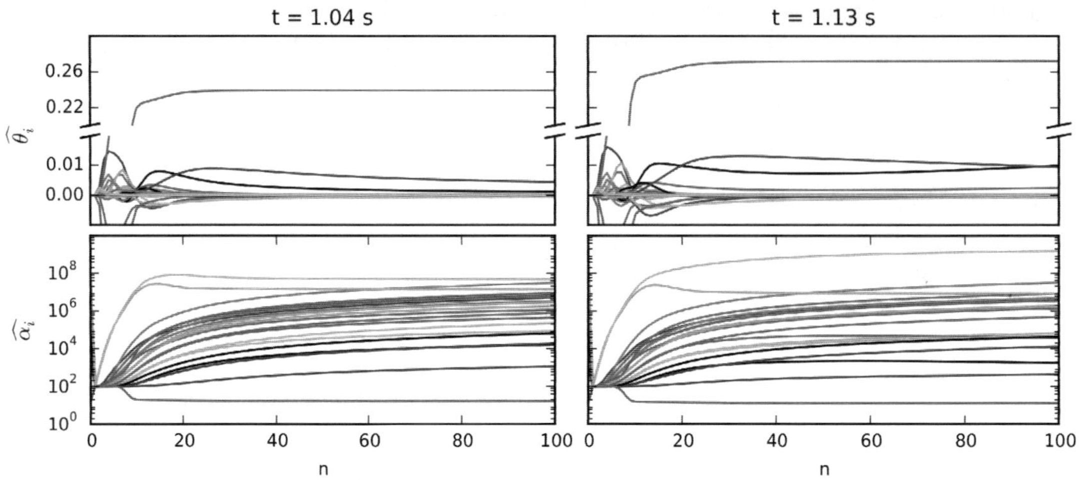

Figure 2. Convergence paths of different elements of the estimated $\widehat{\theta}_i$ during a run of the RLS-ARD algorithm for two fixed time samples during a discharge (Upper row), and convergence paths of the corresponding precisions $\widehat{\alpha_i}$ (Lower row), n denotes the iteration number.

5. Experimental Validation

5.1. Tokamak COMPASS and data from a validated numerical model

Both simulated data and measurements from operation of the tokamak COMPASS will be used in this section. Simulated data were obtained as a fit to real measurements with the EFIT++ [13] plasma reconstruction code. This code evaluates a complex numerical model of magnetohydrodynamic equilibrium which includes the full toroidal flow and anisotropy changes to the Grad-Shafranov equation. The result of the EFIT++ reconstruction is also the plasma boundary, however obtained at high computational cost. Therefore, when using the simulated measurements, we can use the boundary from the EFIT++ as "ground truth" to compare the performance of the proposed algorithm. Because we are mainly interested in the quality of the plasma boundary reconstruction, we use the Hausdorff distance [14] to compare the estimated 2D shape of the plasma boundary to the ground truth. This metric represents the maximum difference between the two boundaries evaluated on discretized angular grid.

For the purpose of this paper, all simulated data were generated with the configuration which agrees with the real geometry of the tokamak and measuring devices. The EFIT++ reconstructs the plasma field from measurements at standard sampling period 1ms, which is sufficient for real time shape control. Thus the plasma boundary ground truth is available in a time series. The proposed algorithm is designed to reconstruct the plasma boundary at each sampling time independently. Therefore, we will show results of reconstruction at selected time samples. In each time sample, an iterative algorithm (such as Algorithm 4 for RLS-ARD) will be run until convergence. The relatively slow temporal evolution of the shape can be used to warm start the iterative algorithm by a solution obtained in the previous time instant. This will be important to achieve computational cost permitting real time evaluation.

5.2. Model structure selection

First, we demonstrate how the ARD property selects the optimal regularization coefficients for reconstruction of the magnetic flux at a single time instance. We have selected data from two time samples of a discharge and run the RLS-ARD algorithm (Algorithm 4) for $n = 100$

iterations with initialization $\alpha_{0i} = 100$. Convergence of the estimated values of $\widehat{\theta_i}$ and their corresponding precision parameters (penalization coefficients) $\widehat{\alpha_i}$ are displayed in Fig. 2. With increasing precision $\widehat{\alpha_i}$ grows prior probability that $\widehat{\theta_i}$ is zero. This has a direct link to Tikhonov regularization as was hinted in section 2 where α_i is the penalization parameter that governs how much the squared value of a regression coefficient is penalized.

For these particular reconstructions, one coefficient out of 20 (i.e. $M = 20$) in the decomposition (2) seems to have a major contribution. Clearly, its precision/penalization converges to a smaller number than the common starting value of 10^2 while the other coefficients have larger precisions and thus are more penalized.

5.3. Reconstruction of a full discharge

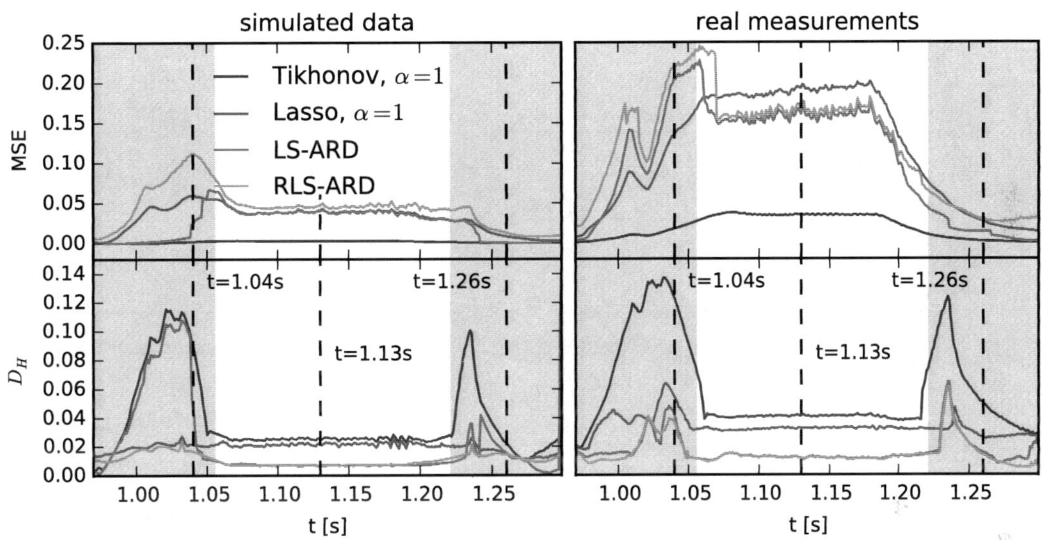

Figure 3. Comparison of model residues and Hausdorff distance for simulated data (left column) and real measurements (right column). **Top row**: mean square error of the model residue of the linear problem (2) during a plasma discharge as fitted by different methods. **Bottom row**: Hausdorff distance D_H between the reconstructed plasma boundary and ground truth (or the output of a validated numerical model in case of real measurements).
The shaded areas on the left and right distinguish the initial and terminal phase of the discharge as identified by presence of x-point computed by EFIT++. Dashed vertical lines denote time samples at which the reconstruction results are presented in Fig. 5.

Next, we compare the performance of tested algorithms on full duration of a selected plasma discharge. To understand the results better, we will describe the individual phases of a tokamak plasma discharge, which are also distinguished by shading in Fig. 3. When the discharge starts, the 2D cross section of the plasma column is elliptic. Afterwards, it undergoes a transition which, in an ideal case, ends with the plasma being elongated with a pointed tip (x-point). This part is marked by gray background color at the left side of the discharge figures. When the x-point is established, the plasma rapidly reaches the desired equilibrium state (white background). After that, the plasma collapses back to an elliptic shape before completely vanishing (gray background color on the right side of each plot).

Since the x-point is a saddle point in the $\psi(r, z)$ contours and it is usually close to the tokamak chamber wall, the identification of the plasma shape is very sensitive to its predicted position.

If a saddle point is predicted to be outside of the chamber, even just a few centimeters off its true position, the reconstructed plasma shape is elliptic and very different from the true shape.

In Fig. 3 the MSE of model residue of the linear equation (2) is plotted for the whole plasma discharge for different approaches to regularization and for both simulated and real measurements. For the Tikhonov regularization, the regularization constant α was set to be equal for all regressors. The MSE of the model residue is accompanied by values of the Hausdorff distance D_H to the ground truth for each algorithm.

Although the fit using Tikhonov has the lowest value of the residue at the measurement points during the whole discharge, the final boundary is not very close to the ground truth. The remaining methods clearly gain advantage by using stricter regularization and thus achieving a better generalization beyond the measurement points, which is crucial for identifying the boundary. Also, the plasma computed with tested methods may be predicted to have an x-point when the ground truth does not have one or vice versa, which leads to high peaks in the D_H metric. Lasso algorithm for $l1$ optimization yields results very similar to those of LS-ARD or RLS-ARD.

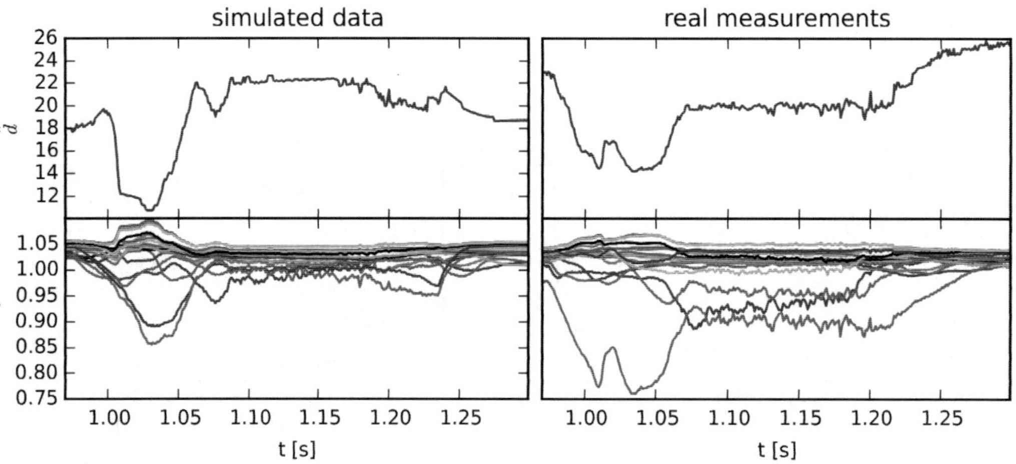

Figure 4. Measures of non-Gaussianity of the robust linear regression. **Top row**: Estimate of the degrees of freedom of the Student's noise parameter d during the discharge. **Bottom row**: estimates of the weights u_i of individual measurements.

Interestingly, the LS-ARD algorithm does not perform well on simulated data in the transition phases. The value of D_H is higher in the transition phases than for the other approaches. Higher deviation from the EFIT boundary in the transition phases means that assumptions for one of the models (EFIT or toroidal harmonics) may be violated. Also, the assumption of non-Gaussian residue of the RLS-ARD model may be more appropriate in the transition phases because of the nonlinear propagation of errors as is demonstrated by the results produced using synthetic data. Apart from the transition phases, the difference of results between LS and RLS-ARD is marginal. This similarity is also apparent from the estimated degrees of freedom of the Student's t-distribution (parameter d) from the RLS-ARD algorithm is displayed in Fig. 4. Note that it decreases during the first transition phases. Smaller values of d can be interpreted as a deviation of the measurement errors from the Gaussian distribution. This behaviour indicates the presence of outliers or a systematic error in the model. However, because we observe the deviation even on simulated data, the most likely explanation is insufficiency in the used model as discussed above. Also, we can track which measurements are suppressed from the fit since these measurements have the value of u_i smaller than 1.

The effect of the magnetic flux reconstruction on the shape of the reconstructed plasma boundary is displayed in Fig. 5, where the estimated boundary shapes are compared to the ground truth results at different moments of the discharge. These moments were selected from different stages of the discharge, see dashed lines in Fig. 3. Again, we see that better results are obtained in the stationary phase of a discharge. The value of u_i estimates at a given time is illustrated at the corresponding measurement points. Those measurement points with $u_i<1$ are highlighted with a larger circle. This can provide experiment designers with additional information about the precision of measurements.

While the RLS-ARD algorithm appeared to improve results on the simulation data, its improvement on real data is not so significant. Moreover, due to its lower computational cost, the LS-ARD was selected for implementation. The computational cost of this algorithm can be lowered by initialization of the parameters from those obtained in previous time sample (warm starts). The effect of this strategy is displayed in Fig. 6, where the number of iterations needed for convergence of the LS-ARD algorithm are shown for two variants: offline, where the initial guesses of parameters are the same for each time of the discharge, and online, where the parameters estimated in the previous time are used as initialization values for the algorithm. This results in a significant speedup, where only a few iterations are needed.

5.4. Evaluation of multiple discharges

A quantitative study of quality of reconstruction was tested on a data set of ten plasma discharges. The relative performance of the tested algorithms remains the same for all discharges. At two discharges, the RLS-ARD algorithm was worse than the LS-ARD in the initial phase of the discharge. We conjecture that this is cause by incorrect suppression of the relevant measurements (i.e. incorrectly classifying them as outliers). Statistical evaluation of the distribution of the mean and median of the Hausdorf distance for all times of the ten discharges is displayed in Fig. 7. From these results, the LS-ARD method seems to provide the most consistent and reliable results.

6. Conclusion

In this paper, we have described an inverse problem that arises in the field of tokamak plasma reconstruction and the difficulties it presents for classical methods. We have proposed a Bayesian approach to the problem resulting in an algorithm that selects a sparse and robust solution. The proposed method has been shown to provide consistently better results compared to the conventional approach. It has been shown that a model with fewer automatically selected parameters avoids the overfitting problem and thus achieves better reconstruction of the overall shape of the plasma boundary. This has been achieved on both simulated and measured data. The complexity of the computation is intentionally kept low so that the algorithm can be run in real time on existing hardware. The robust algorithm was able to improve quality of reconstruction only at the non-stationary discharge phases in simulation. However, its improvement on real data was marginal and not very reliable. Since the distribution of the error was estimated to be almost Gaussian in the most important stationary regimes, we selected the sparse LS-ARD algorithm for final implementation.

Acknowledgments

This work was partially funded by the Czech Technical University through the grant SGS15/214/OHK4/3T/14 and by the Ministry of Education, Youth and Sports of the Czech Republic under the RICE – New Technologies and Concepts for Smart Industrial Systems, project No. LO1607. COMPASS tokamak experiments are supported by MEYS projects LM2015045 and 8D15001.

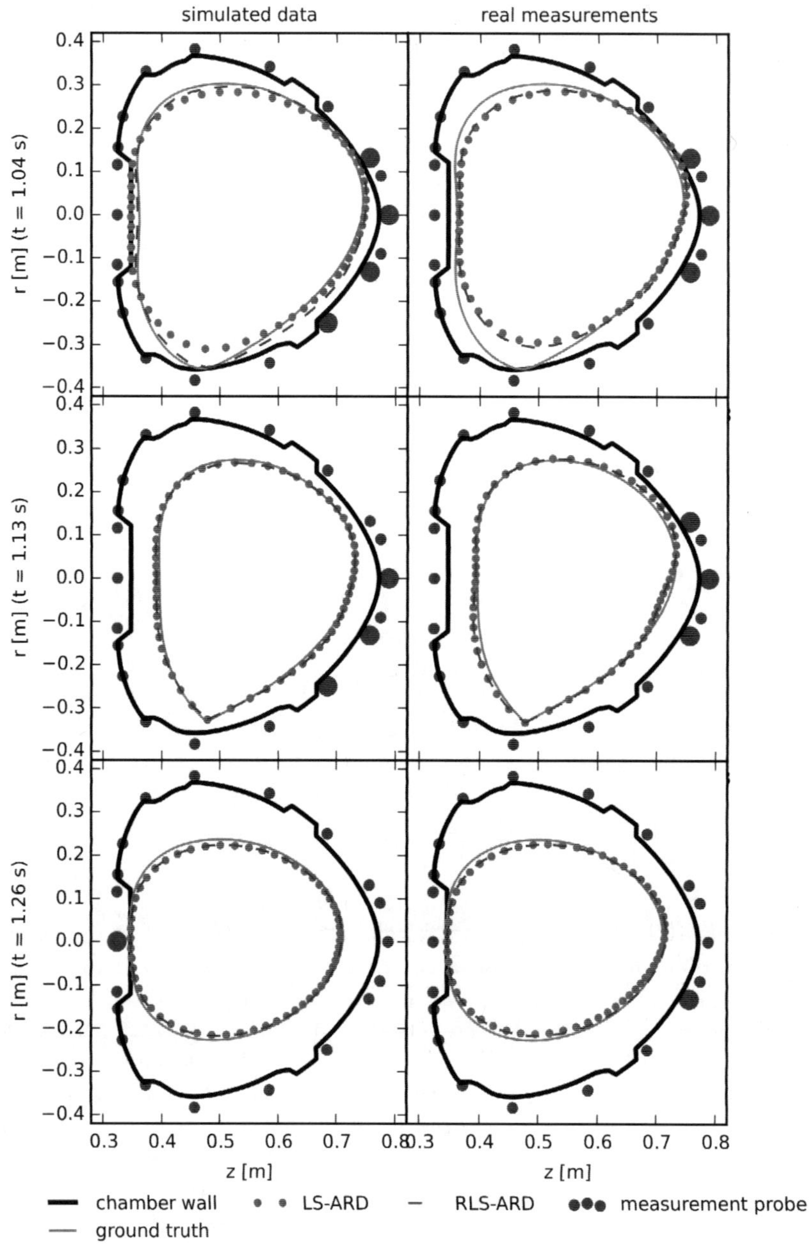

Figure 5. Estimates of the plasma boundary: ground truth from the EFIT model (red solid line), LS-ARD (green dotted line) and RLS-ARD (blue dashed line) algorithms for different data sets and different time samples of the discharge. Inner boundary of the tokamak chamber is in black. Also, the locations of the sensors are displayed (blue dots), those that have the value of estimate of u_i smaller than 1 are enlarged.

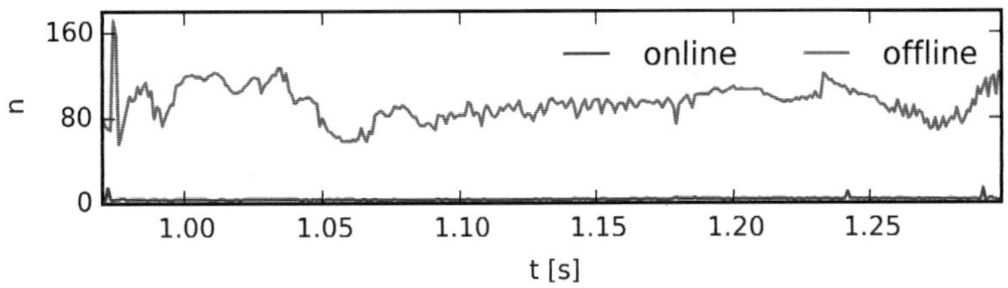

Figure 6. Number of iterations n needed for convergence of the LS-ARD algorithm during a tokamak discharge. In online mode, parameters are initialize with previous results, when in offline mode, the same initial conditions are used at every sampling time.

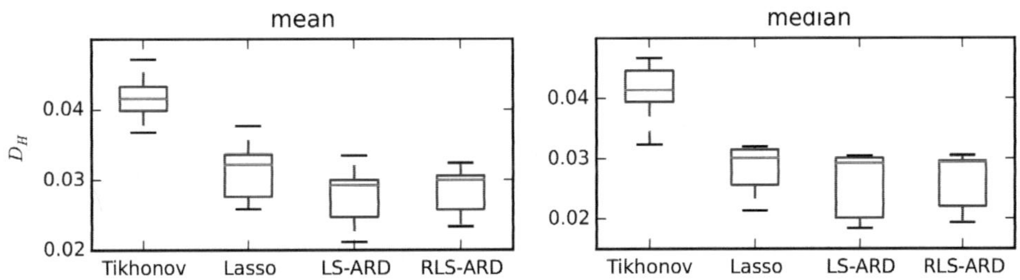

Figure 7. Mean and median Hausdorff distance D_H for 10 plasma discharges and different algorithms.

Appendix A. Parameters of least squares ARD algorithm

Hyperparameters of normal and gamma posteriors (9)-(11) are given by

$$a_{\alpha i} = a_{0i} + \frac{1}{2}, \qquad b_{\alpha i} = b_{0i} + \frac{1}{2}\widehat{\theta_i^2}. \tag{A.1}$$

$$c = c_0 + \frac{N}{2}, \qquad d = d_0 + \frac{1}{2}y^T y - y^T X\hat{\theta} + \frac{1}{2}\mathrm{tr}\left(X^T X\widehat{\theta\theta^T}\right) \tag{A.2}$$

$$\hat{\theta} = \hat{\beta} S X^T y, \qquad S = (\hat{A} + \hat{\beta} X^T X)^{-1} \tag{A.3}$$

Notation $\hat{\cdot}$ stands for expected value of the variable. Other required moments are

$$\hat{\alpha}_i = \frac{a_{\alpha i}}{b_{\alpha i}}, \qquad \hat{\beta} = \frac{c}{d},$$

$$\widehat{\theta_i^2} = S_{ii} + \hat{\theta}_i^2, \qquad \widehat{\theta\theta^T} = \hat{\theta}\hat{\theta}^T + S.$$

Appendix B. Parameters of robust least squares ARD algorithm

Hyperparameters of distributions (16)-(20) have the following form

$$\hat{\theta} = \widehat{\omega} S X^T \widehat{\mathrm{diag}(u)} y, \qquad S = (\widehat{A} + \widehat{\omega} X^T \widehat{\mathrm{diag}(u)} X)^{-1}, \tag{B.1}$$

$$a_{\alpha i} = a_{0i} + \frac{1}{2}, \qquad b_{\alpha i} = b_{0i} + \frac{1}{2}\widehat{\theta_i^2}, \tag{B.2}$$

$$\rho = \rho_0 + \frac{N}{2}, \qquad \kappa = \kappa_0 + \frac{1}{2} y^T \widehat{\mathrm{diag}(u)} y - y^T \widehat{\mathrm{diag}(u)} X \widehat{\theta} + \frac{1}{2}\mathrm{tr}(\widehat{\theta\theta^T} X^T \widehat{\mathrm{diag}(u)} X), \tag{B.3}$$

$$a_{ui} = \frac{1}{2}(\widehat{d} + 1), \qquad b_{ui} = \frac{1}{2}\widehat{d} + \frac{1}{2} y_i^2 \widehat{\omega} - y_i \widehat{\omega} (X\widehat{\theta})_i + \frac{1}{2}\widehat{\omega}(X\widehat{\theta\theta^T}X^T)_i, \tag{B.4}$$

$$\phi = \phi_0 + \frac{1}{2}N, \qquad \psi = \psi_0 - \frac{1}{2}N + \frac{1}{2}\sum_{i=1}^{N} \widehat{u_i} - \frac{1}{2}\sum_{i=1}^{N} \widehat{\ln u_i}. \tag{B.5}$$

Note that for derivation of ψ we used Stirling's formula for the term $\ln(\Gamma(d/2))$ as in [9]. The required moments of these distributions are:

$$\widehat{\theta_i^2} = S_{ii} + \widehat{\theta}_i^{\,2}, \qquad \widehat{\theta\theta^T} = \hat{\theta}\hat{\theta}^T + S, \qquad \widehat{u}_i = \frac{a_{ui}}{b_{ui}},$$

$$\hat{\alpha}_i = \frac{a_{\alpha i}}{b_{\alpha i}}, \qquad \widehat{\omega} = \frac{\rho}{\kappa}, \qquad \widehat{d} = \frac{\phi}{\psi}.$$

References

[1] Ambrosino G and Albanese R 2005 *IEEE Control Systems Magazine* **25** 76–92
[2] Ariola M and Pironti A 2008 *Magnetic control of tokamak plasmas* (Springer Science & Business Media)
[3] Faugeras B, Blum J, Boulbe C, Moreau P and Nardon E 2014 *Plasma Physics and Controlled Fusion* **56** 114010
[4] Hoerl A E and Kennard R W 1970 *Technometrics* **12** 55–67
[5] Tibshirani R 1996 *Journal of the Royal Statistical Society. Series B (Methodological)* 267–288
[6] Tipping M 2001 *The journal of machine learning research* **1** 211–244
[7] Wipf D P and Nagarajan S S 2007 A new view of automatic relevance determination *Advances in Neural Information Processing Systems* pp 1625–1632
[8] Škvára V, Šmídl V and Urban J 2015 *Journal of Physics: Conference Series* **659** 012010
[9] Christmas J and Everson R 2011 *IEEE Transactions on Signal Processing* **59** 48–57
[10] Lebedev N N and Silverman R A 1972 *Special functions and their applications* (Courier Corporation)
[11] Bishop C 2006 *Pattern recognition and machine learning* vol 1 (Springer New York)
[12] Šmídl V and Quinn A 2005 *The Variational Bayes Method in Signal Processing* (Springer)
[13] Fitzgerald M, Appel L C and Hole M J 2013 *Nuclear Fusion* **53** 113040
[14] Huttenlocher D P, Klanderman G, Rucklidge W J *et al.* 1993 *Pattern Analysis and Machine Intelligence, IEEE Transactions on* **15** 850–863

IOP Conf. Series: Journal of Physics: Conf. Series **1047** (2018) 012016 doi:10.1088/1742-6596/1047/1/012016

Inverse eigenvalue problems for checkerboard Toeplitz matrices

T H Jones[1] and N B Willms[1]

[1] Department of Mathematics, Bishop's University. Sherbrooke, Québec, Canada, J1M 1Z7.

E-mail: `tjones@ubishops.ca`, `bwillms@ubishops.ca`

Abstract.

The inverse eigenvalue problem for real symmetric Toeplitz matrices motivates this investigation. The existence of solutions is known, but the proof, due to H. Landau, is not constructive. Thus a restriction, namely the required eigenvalues are to be equally spaced, is considered here. Two types of structured matrices arise, herein termed, "checkerboard" and "outer-banded". Examples are presented. Properties of these structured matrices are explored and a full characterization of checkerboard matrices is given. The inverse eigenvalue problem is solved within the class of odd checkerboard matrices. In addition the "symmetric-spectrum" inverse eigenvalue problem is solved within a subclass of Hankel matrices. A regularity conjecture of H. Landau for the Toeplitz inverse eigenvalue problem is discussed and a similar conjecture for checkerboard Toeplitz matrices is given.

1. Introduction

1.1. Motivation

The reconstruction of a matrix to possess specified spectral data is termed an inverse eigenvalue problem. Following the classification of such problems by M. Chu [1] our attention will be directed to the class of "structured inverse eigenvalue problems" (SIEP), where the objective is to construct a matrix with a certain structure which has the prescribed set of eigenvalues. For an extensive survey of such problems, see [2], [3]. Within the class of structured inverse eigenvalue problems, the Toeplitz inverse eigenvalue problem (TIEP) has received significant attention:

(TIEP) *construct a real, symmetric Toeplitz matrix of size $n \times n$ with the prescribed eigenvalues:* $\lambda_1, \lambda_2, \ldots, \lambda_n$.

Recall that a Toeplitz matrix, $T_n \in \mathbb{R}^{n\times n}$, is one for which the entries are constant along each diagonal:

$$T_n = \begin{bmatrix} a_0 & a_1 & a_2 & \cdots & a_{n-1} \\ a_{-1} & a_0 & a_1 & \ddots & a_{n-2} \\ a_{-2} & a_{-1} & a_0 & \ddots & \vdots \\ \vdots & \ddots & \ddots & \ddots & \vdots \\ a_{1-n} & a_{2-n} & \cdots & \cdots & a_0 \end{bmatrix}.$$

When attention is restricted to symmetric matrices, the Toeplitz matrix is determined by its top row. Accordingly, (after relabeling) we denote it: $T_n = T_n(a_1, a_2, \ldots, a_n)$, $a_i \in \mathbb{R}$, $i = 1, \ldots, n$.

Given the prescribed set of distinct values,

$$\lambda_1 < \lambda_2 < \cdots < \lambda_n,$$

the existence of a real symmetric, $n \times n$ Toeplitz matrix possessing these as its eigenvalues was demonstrated by H. J. Landau [4], within a class of Toeplitz matrices having a certain additional regularity. This result is described in Subsection 5.1. Unfortunately, the proof does not present a construction, and the constructive TIEP remains, in Landau's words, "challenging because of its analytic intractability" ([4] p. 749).

While the TIEP has been solved up to size $n \leq 4$ by P. Delsarte and Y. Genin [5], forewarned by the statement of Landau, we did not expect to make headway on the analytic reconstruction of larger size general Toeplitz matrices. So we decided to impose further structure on the inverse problem. Following a strategy employed in [6] for a different SIEP, the problem we consider here is the TIEP with prescribed, *equally spaced* eigenvalues, listed in decreasing order:

$$\lambda_i = \lambda_1 - (i-1)r, \qquad i = 1\ldots, n, \quad r \in \mathbb{R}^+. \tag{1.1}$$

This particular choice of eigenvalues is motivated by the idea that the reciprocal of the minimum separation between eigenvalues can be thought of as a condition number on the sensitivity of the eigenvectors (invariant subspaces) to perturbation (see [7], Theorem 8.1.12). Therefore a solution to the TIEP with equally spaced, distinct (simple) eigenvalues will be important as a class of arbitrary size Toeplitz *test matrices* having perfect conditioning with respect to the (forward) eigenvalue-eigenvector problem. The availability of such test matrices may be helpful when comparing the efficacy of the various iterative schemes for the TIEP, such as Newton's method [8], algebraic techniques [9], [10], and isospectral flows [11], [12].

1.2. Problem Statement

Proposition 1.1. *If a symmetric Toeplitz matrix* $T\left(a_1,\ a_2,\ \ldots, a_n\right) \in \mathbb{R}^{n\times n}$ *has eigenvalues,* $\{\mu_1,\ \mu_2,\ \ldots,\ \mu_n\}$*, then the Toeplitz matrix,* $T^* = \alpha(T - a_1 I) = T^*\left(0,\ \ \alpha a_2,\ \ \ldots, \alpha a_n\right)$ *has eigenvalues:* $\{\alpha(\mu_1 - a_1),\ \alpha(\mu_2 - a_1),\ \ldots, \alpha(\mu_n - a_1)\},\ \forall\ \alpha \in \mathbb{R}$.

Proof: Adding two Toeplitz matrices produces another Toeplitz matrix, as does scaling one by a real constant. Since $a_1 I$ is Toeplitz, therefore T^* has the required Toeplitz form, and has eigenvalues equal to those of T shifted (reduced) by a_1 and then scaled by α. ■

Consequently, the prescribed set, (1.1) can be shifted for convenience to have zero mean and (equal) spacing of two. Thus we take $\lambda_1 = n - 1$ and $r = 2$, in (1.1) as the desired eigenvalue set:

$$\{(n-1),\ (n-3),\ \ldots,\ -(n-3),\ -(n-1)\}, \qquad n \in \mathbb{Z}^+. \tag{1.2}$$

So our restricted problem is:

(rTIEP) *construct a real, symmetric Toeplitz matrix,* T_n *of size* $n \times n$ *with the prescribed eigenvalues* (1.2).

A less-restricted version of the above structured inverse eigenvalue problem will be termed the *symmetric spectrum* problem:

(ssSIEP) *construct a real, symmetric matrix of size* $n \times n$*, in a given class of structured matrices, with prescribed eigenvalues symmetric about* 0*, that is, when* λ *is an eigenvalue* $-\lambda$ *is also an eigenvalue of the same multiplicity.*

IOP Conf. Series: Journal of Physics: Conf. Series **1047** (2018) 012016 doi:10.1088/1742-6596/1047/1/012016

The rest of this paper is organized as follows. In Section 2 examples of the solutions of rTIEP up to size ten are listed. In particular, two special Toeplitz forms are observed. In Section 3 we explore the properties of general symmetric matrices. In Section 4 we explore the properties of checkerboard matrices, fully characterizing them in Section 4.2. In Section 5 we return to Toeplitz matrices, describing Landau's class of "regular" Toeplitz matrices and comparing our examples to a conjecture of his ([4] p. 767). In Section 6 we solve the ssSIEP within the class of persymmetric Hankel matrices of even size. The concluding section adds another conjecture for further study.

2. Examples

2.1. Non-uniqueness

Solutions of the rTIEP are not unique. Scaling by -1 is one source of non-uniqueness, since if $T(a_1, a_2, \ldots, a_n)$ has the symmetric spectrum (1.2), then so does $T(-a_1, -a_2, \ldots, -a_n)$. Additional sign choices were observed. However for all sizes up to 10, we found, with the aid of Maple®15, at least one *non-negative* rTIEP solution (i.e. matrices with all entries non-negative). These are presented below. No claim is made that the list is exhaustive for size $n \geq 6$.

Further non-uniqueness arises from two different patterns of non-zero entries observed in rTIEP solution matrices . Up to size $n = 4$ we verified that these are the only possible patterns. The first is a *checkerboard pattern*, with zeroes occupying all positions where the sum of the row and column indices is even. So a symmetric Toeplitz checkerboard matrix, C_n is characterized as $C_{2m}(c_1,\ c_2,\ \ldots\ c_m) := T_{2m}(0,\ c_1,\ 0,\ c_2,\ 0,\ c_3,\ \ldots\ 0,\ c_m)$, when the size $n = 2m$ is even, and the same (appending a trailing 0 entry in the T matrix) when of odd size, $n = 2m+1$. The second observed pattern is termed *outer-banded.* In this pattern the non-zero elements occupy the diagonal bands furthest from the main diagonal, while zeroes occupy the main diagonal and the first $\lfloor \frac{n-1}{2} \rfloor$ off-diagonal bands. So a symmetric Toeplitz outer-banded matrix, B_n is characterized as $B_{2m}(b_1,\ b_2,\ \ldots\ b_m) := T_{2m}(0,\ 0,\ \ldots\ 0,\ b_1,\ b_2,\ \ldots\ b_m)$, when the size $n = 2m$ is even, and the same (inserting an extra leading 0 entry in the T matrix) when of odd size, $n = 2m + 1$.

2.2. Non-negative Examples by Increasing Size

For each size n, the examples listed below have the eigenvalues arranged in *decreasing* order, and the columns of the (orthogonal) eigenvector matrices, $V \in \mathbb{R}^{n \times n}$ arranged consistently. The first component of each eigenvector is chosen to be non-negative. Thus we have the eigenvalue/eigenvector decomposition: $T_n = V \Lambda V^T$ where the diagonal matrix of eigenvalues is: $\Lambda = \text{diag}(n-1,\ n-3,\ \ldots, 3-n,\ 1-n)$.

n=2, 3, 4

For size $n = 2$, the checkerboard and outer-banded patterns coincide and are necessary to solve the rTEIP.

$$B_2(1) = C_2(1) = \begin{bmatrix} 0 & 1 \\ 1 & 0 \end{bmatrix}$$

The eigenvalues are $\Lambda = \text{diag}(1,\ -1)$, and the unit eigenvectors are $V = \dfrac{1}{\sqrt{2}} \begin{bmatrix} 1 & 1 \\ 1 & -1 \end{bmatrix}$.

$$B_3(2) = \begin{bmatrix} 0 & 0 & 2 \\ 0 & 0 & 0 \\ 2 & 0 & 0 \end{bmatrix} \qquad C_3(\sqrt{2}) = \begin{bmatrix} 0 & \sqrt{2} & 0 \\ \sqrt{2} & 0 & \sqrt{2} \\ 0 & \sqrt{2} & 0 \end{bmatrix}$$

The eigenvalues are $\Lambda = \text{diag}(2,\ 0,\ -2)$, and the unit eigenvectors are:

$$V_{B_3} = \frac{1}{2}\begin{bmatrix} \sqrt{2} & 0 & \sqrt{2} \\ 0 & 2 & 0 \\ \sqrt{2} & 0 & -\sqrt{2} \end{bmatrix} \qquad V_{C_3} = \frac{1}{2}\begin{bmatrix} 1 & \sqrt{2} & 1 \\ \sqrt{2} & 0 & -\sqrt{2} \\ 1 & -\sqrt{2} & 1 \end{bmatrix}$$

Equating the characteristic polynomial of the generic $T_3(0,\ a,\ b)$ to the target polynomial with roots $\{2,\ 0,\ -2\}$, yields: $\lambda^3 - (2a^2 + b^2)\lambda - 2a^2 b = \lambda(\lambda^2 - 4)$, which implies either $a = 0$ (i.e. **outer-banded**, and $b = \pm 2$) or $b = 0$ (i.e. **checkerboard**, and $a = \pm\sqrt{2}$) .

$$B_4\left(\sqrt{3},\ 2\right) = \begin{bmatrix} 0 & 0 & \sqrt{3} & 2 \\ 0 & 0 & 0 & \sqrt{3} \\ \sqrt{3} & 0 & 0 & 0 \\ 2 & \sqrt{3} & 0 & 0 \end{bmatrix} \qquad C_4\left(\tfrac{1+\sqrt{7}}{2},\ \tfrac{3-\sqrt{7}}{2}\right) = \begin{bmatrix} 0 & \frac{1+\sqrt{7}}{2} & 0 & \frac{3-\sqrt{7}}{2} \\ \frac{1+\sqrt{7}}{2} & 0 & \frac{1+\sqrt{7}}{2} & 0 \\ 0 & \frac{1+\sqrt{7}}{2} & 0 & \frac{1+\sqrt{7}}{2} \\ \frac{3-\sqrt{7}}{2} & 0 & \frac{1+\sqrt{7}}{2} & 0 \end{bmatrix}.$$

The eigenvalues are $\Lambda = \text{diag}(3,\ 1,\ -1,\ -3)$, and the unit eigenvectors are:

$$V_{B_4} = \frac{1}{\sqrt{8}}\begin{bmatrix} \sqrt{3} & 1 & 1 & \sqrt{3} \\ 1 & -\sqrt{3} & -\sqrt{3} & 1 \\ 1 & \sqrt{3} & -\sqrt{3} & -1 \\ \sqrt{3} & -1 & 1 & -\sqrt{3} \end{bmatrix} \quad V_{C_4} = \frac{1}{\sqrt{24-8\sqrt{7}}}\begin{bmatrix} \sqrt{7}-2 & 1 & 1 & \sqrt{7}-2 \\ 1 & \sqrt{7}-2 & 2-\sqrt{7} & -1 \\ 1 & 2-\sqrt{7} & 2-\sqrt{7} & 1 \\ \sqrt{7}-2 & -1 & 1 & 2-\sqrt{7} \end{bmatrix}$$

Again equating the characteristic polynomial of the generic $T_4(0,\ a,\ b,\ c)$ to the target polynomial with roots $\{3,\ 1,\ -1,\ -3\}$, yields:

$$\lambda^4 - (3a^2 + 2b^2 + c^2)\lambda^2 - 4ab(a+c)\lambda + (a^2 + b^2 - ca - 2ba)(a^2 + b^2 - ac + 2ba) = \lambda^4 - 10\lambda^2 + 9.$$

So, either $a = 0,\ b = \pm\sqrt{3},\ c = \pm 2$, (i.e. four possibilities, all **outer-banded**) or $b = 0,\ a = \pm\left(\frac{1\pm\sqrt{7}}{2}\right),\ c = \pm\left(\frac{3\mp\sqrt{7}}{2}\right)$, where the plus/minus symbols *inside* the parentheses for a and c match, as do the plus/minus symbols *outside* the parentheses (i.e. four possibilities, all **checkerboard**), or $c = -a$ which implies that a is complex, in contradiction of assumption.

Consequently, one of the two patterns, checkerboard or outer-banded, is necessary to solve the rTIEP for all sizes $n \leq 4$, though this result has not been verified for sizes $n \geq 5$.

n=5, 6

In C_5 there are two sign choices in the non-negative solution presented: either the top symbol in every $\pm$ and $\mp$, or the bottom. The same sign choice is to be applied to the eigenvector matrix.

$$B_5\left(\sqrt{8},\ 2\right) \qquad C_5\left(\tfrac{\sqrt{22}\pm 2}{3},\ \tfrac{\sqrt{22}\mp 4}{3}\right)$$

The eigenvalues are $\Lambda = \text{diag}(4,\ 2,\ 0,\ -2,\ -4)$, and the unit eigenvectors of C_5 are:

$$V_{C_5} = \frac{1}{12}\begin{bmatrix} \pm\sqrt{22}-1 & 6 & 2\pm\sqrt{22} & 6 & \pm\sqrt{22}-1 \\ 6 & 6 & 0 & -6 & -6 \\ 2\pm\sqrt{22} & 0 & 2\mp 2\sqrt{22} & 0 & 2\pm\sqrt{22} \\ 6 & -6 & 0 & 6 & -6 \\ \pm\sqrt{22}-1 & -6 & 2\pm\sqrt{22} & -6 & \pm\sqrt{22}-1 \end{bmatrix}$$

Some of the possible size $n = 6$ matrices are: $B_{6a}\left(\sqrt[3]{15},\ \pm\sqrt{13 - 2(\sqrt[3]{15})^2 + 3\sqrt[3]{15}},\ 3 - \sqrt[3]{15}\right)$, $B_{6b}\left(\sqrt[3]{15},\ \pm\sqrt{17 - 2(\sqrt[3]{15})^2 - \sqrt[3]{15}},\ -1 - \sqrt[3]{15}\right)$, and $C_6\left(c_1,\ c_2,\ 3 - c_1 - c_2\right)$, where c_2 is a root of the 6th order polynomial $P(x) = 54x^6 - 216x^5 + 3x^4 + 612x^3 - 800x^2 + 1500x - 369$, and $c_1 = \frac{900}{2376} - \frac{17243}{3564}c_2 + \frac{2273}{792}c_2^2 - \frac{607}{594}c_2^3 - \frac{131}{132}c_2^4 + \frac{14}{33}c_2^5$. There are three positive roots of $P(x)$, but only one yields a non-negative checkerboard matrix, which is (to a floating point approximation):

$$C_6\left(2.636649799 \quad .2788346628 \quad .0845155382\right).$$

The eigenvalues are $\Lambda = \text{diag}(5,\ 3,\ 1,\ -1,\ -3,\ -5)$ (to nine significant figures).

n=7, …, 10
We have been unable, to date, to find checkerboard solutions, even numerically, for sizes $n \geq 7$. However, outer-banded examples include: $B_{7a}\left(\sqrt[3]{48},\ \pm\sqrt{28 - 2(\sqrt[3]{48})^2},\ -\sqrt[3]{48}\right)$,
$B_{7b}\left(\sqrt[3]{48},\ \pm\sqrt{20 - 2(\sqrt[3]{48})^2 + 4\sqrt[3]{48}},\ 4 - \sqrt[3]{48}\right)$,
$B_8\left(\sqrt[4]{105},\ 3.737635629,\ \sqrt{21.01070232 - 2\sqrt{105}},\ .262364371\right)$,
$B_9\left(\sqrt[4]{384},\ 3.696457542,\ .5161371538,\ .303542458\right)$, and
$B_{10}\left(\sqrt[5]{945},\ 4.605343159,\ .9009158716,\ .3392412014,\ .1628007010\right)$.

2.3. Observations for the checkerboard examples

(i) The orthogonal matrix of eigenvectors for the checkerboard solutions to rTIEP are symmetric up to size $n = 5$ and seems to remain so numerically for larger sizes.

(ii) Up to size $n = 4$, the non-negative checkerboard solution to rTIEP is unique. However, we found two such solutions at size $n = 5$, and suspect there are additional ones for size $n = 6$.

(iii) Up to size $n = 6$ we have found exactly one solution to rTIEP which is *regular* (definition by Landau, to follow in Section 5.1), all of them being non-negative, checkerboard matrices.

3. Symmetry Properties

Following is a brief review of the definitions and basic results for the various matrix symmetries.

Definition 3.1. *The following belong to $\mathbb{R}^{n\times n}$. A subscript indicates the matrix size if necessary.*

(i) $A = \text{diag}(1, -1, 1, \ldots, (-1)^{n+1})$ *is the* **alternating diagonal matrix** *with diagonal entries ± 1, starting with a positive* 1.

(ii) $J = \text{antidiag}(1, 1, 1, \ldots, 1)$ *is the* **anti-identity matrix** *with 1s on the main anti-diagonal. In other words $[J]_{i,j} = \delta_{i,n+1-j}$. J is sometimes called the* **exchange permutation** *(see Golub and Van Loan [7] p. 20), because premultiplication by J exchanges the rows of a matrix, completely reversing their order, i.e. premultiplication by J turns a matrix upside down.*

(iii) E is the **exchange** *matrix:* $\begin{bmatrix}\mathbf{e}_1 & \mathbf{e}_3 & \cdots & \mathbf{e}_{2\lceil\frac{n}{2}\rceil-1} & \mathbf{e}_2 & \mathbf{e}_4 & \cdots & \mathbf{e}_{2\lfloor\frac{n}{2}\rfloor}\end{bmatrix}^T$*, where $\mathbf{e}_i$ is the ith standard basis vector of $\mathbb{R}^n$. Note that if $n = 2k$, the matrix E is called a* **mod-2 perfect shuffle matrix**, *denoted $\mathcal{P}_{2,k}$ ([7] p. 20).*

Note that A, J, and E are all self-inverses, that is, $A^2 = J^2 = E^2 = I$.

Definition 3.2. *For $M \in \mathbb{R}^{n\times n}$ the following symmetries are defined:*

(i) M is **symmetric** *when $M = M^T$.*

(ii) M is **centrosymmetric** *when $M = JMJ$.*

(iii) M is **persymmetric** *when $M = JM^TJ$, or equivalently, $M^T = JMJ$.*

(iv) M is **super-symmetric** *when it possesses all three of the above symmetries.*

Definition 3.3. *For $M \in \mathbb{R}^{n\times n}$ the following skew-symmetries are defined:*

(i) A matrix M is **skew-symmetric** *or* **anti-symmetric** *when $M = -M^T$.*
(ii) A matrix M is **skew-centrosymmetric** *or* **anti-centrosymmetric** *when $M = -JMJ$.*
(iii) A matrix M is **skew-persymmetric** *or* **anti-persymmetric** *when $M = -JM^TJ$.*

Theorem 3.4. *Let $M \in \mathbb{R}^{n\times n}$, be a non-zero matrix.*

(i) If M possesses any two of the symmetries, (symmetric, centrosymmetric, or persymmetric), then it also has the third, and hence is super-symmetric.
(ii) If M possesses any two of the skew-symmetries, (skew-symmetric, skew-centrosymmetric, or skew-persymmetric), then it also has the third symmetry type, but not skew.
(iii) If M possesses a symmetry and a skew-symmetry, then it also has the remaining skew-symmetry.

Proof: Consider the three relations, $M \sim M^T \sim JMJ$.

When any two of these are symmetries (i.e. replace $\sim$ with equality), so is the third, proving (i). When any two of these relations are skew-symmetries, the third relation must be a symmetry, proving (ii). Finally, if any one relation is a skew-symmetry, and a second is a symmetry, then the remaining relation is a skew-symmetry, proving (iii).

■

Recall that Toeplitz matrices are persymmetric by definition. Hence the class of symmetric Toeplitz matrices is a class of super-symmetric matrices, and in particular are *centrosymmetric.*

Following [13], [14] we now recall some of the basic properties of centrosymmetric matrices.

Definition 3.5. *An n dimensional vector* $\mathbf{v}$ *is said to have* **even parity** *if $J\mathbf{v} = \mathbf{v}$, and is said to have* **odd parity** *if $J\mathbf{v} = -\mathbf{v}$. If $\mathbf{v}$ is an eigenvector with even (odd) parity, the corresponding to an eigenvalue, λ, is said to be even (odd).*

Note that some authors use the term "symmetric" to label vectors with the property $J\mathbf{v} = \mathbf{v}$, and "skew-symmetric" for vectors satisfying $J\mathbf{v} = -\mathbf{v}$.

Proposition 3.6. *The eigenspaces of a centrosymmetric matrix have bases which can be written to have parity.*

Proof: Let $M = JMJ$ and let $M\mathbf{v} = \lambda\mathbf{v}$. Then $M\mathbf{v} = JMJ\mathbf{v} = \lambda\mathbf{v}$, so $MJ\mathbf{v} = \lambda J\mathbf{v}$, so $J\mathbf{v}$ is also an eigenvector with the same eigenvalue. If $\mathbf{v} = \alpha J\mathbf{v}$ then $\alpha = \pm 1$ and so $\mathbf{v}$ has parity. Otherwise, $\mathbf{v}$ and $J\mathbf{v}$ are linearly independent and we can construct $\mathbf{u} = \mathbf{v} + J\mathbf{v}$ and $\mathbf{w} = \mathbf{v} - J\mathbf{v}$. The span of $\{\mathbf{u}, \mathbf{w}\}$ is the same as the span of $\{\mathbf{v}, J\mathbf{v}\}$, and $\mathbf{u}$ and $\mathbf{w}$ have (opposite) parity. ■

Cantoni and Butler ([13], Theorem. 2 p. 280) proved the following:

Theorem 3.7. *Every real, super-symmetric matrix (in [13] they assumed symmetric and centrosymmetric) of size n has $\lceil \frac{n}{2} \rceil$ even parity eigenvectors and $\lfloor \frac{n}{2} \rfloor$ odd parity eigenvectors.*

Such a set of eigenvectors can always be arranged alternately, with the first column even, so that the columns of the eigenvector matrix alternate in parity.

Definition 3.8. *An $n \times n$ matrix V is said to have* **alternating column parity** *when $VA = JV$, and is said to have* **alternating row parity** *when $AV = VJ$.*

Cantoni and Butler go on to assert the converse of Theorem 3.7 as well ([13] thm. 3): every real, $n \times n$ matrix with a set of n orthonormal eigenvectors, all with parity, must be super-symmetric. We present a simple proof of a slightly stronger result, and add their result as a corollary:

IOP Conf. Series: Journal of Physics: Conf. Series **1047** (2018) 012016 doi:10.1088/1742-6596/1047/1/012016

Proposition 3.9. *Let M be diagonalizable, with $M = V\Lambda V^{-1}$, where V has alternating column parity. Then M is centrosymmetric.*

Proof: Since V is assumed to have alternating column parity, we have $JV = VA$. Then

$$JMJ = JV\Lambda V^{-1}J = JV\Lambda(JV)^{-1} = VA\Lambda(VA)^{-1} = VA\Lambda AV^{-1} = V\Lambda V^{-1} = M$$

since Λ and A are diagonal and $A = A^{-1}$. ■

Corollary 3.10. *If, in addition to having alternating column parity in Proposition 3.9, V is also orthogonal, then $M = V\Lambda V^{-1}$ is also symmetric (and hence super-symmetric).*

Lemma 3.11. *Suppose V is symmetric. Then V has alternating column parity if and only if it has alternating row parity.*

4. Checkerboard Matrices

4.1. Properties of Checkerboard Matrices

Checkerboard matrices arose as observed solutions of the (rTIEP) for small sizes in Section 2, and form an interesting class. We now explore some of their basic properties.

Definition 4.1. *We say that $C \in \mathbb{R}^{m\times n}$ is an* **odd checkerboard matrix** *if its entries $(C)_{ij}$ are non-zero only where $i+j$ is odd, $i = 1,\dots,m$, $j = 1,\dots,n$. Likewise C is an* **even checkerboard matrix** *if its non-zero entries occur only where $i+j$ is even, $i = 1,\dots,m$, $j=1,\dots,n$. Denote an even checkerboard matrix by C_e, and an odd checkerboard matrix by C_o.*

Proposition 4.2.

(i) Every matrix $M \in \mathbb{R}^{m\times n}$ can be uniquely decomposed into the sum of an even checkerboard matrix and an odd checkerboard matrix.

(ii) The product of two even checkerboard matrices is an even checkerboard matrix when the product is defined.

(iii) The product of two odd checkerboard matrices is an even checkerboard matrix when the product is defined.

(iv) The product of an even and an odd checkerboard matrix is an odd checkerboard matrix when the product is defined.

Proof:

(i) Let $M \in \mathbb{R}^{m\times n}$. Define M_e to be the even checkerboard matrix with entries $(M_e)_{ij} = (M)_{ij}$ when $i+j$ is even, and 0 otherwise. Define M_o to be the odd checkerboard matrix with entries $(M_o)_{ij} = (M)_{ij}$ when $i+j$ is odd and 0 otherwise. Then $M = M_e + M_o$. Now suppose that $M = C_e + C_o$. Then $M_e - C_e = C_o - M_o$. Clearly, the left-hand side is an even checkerboard matrix, and the right-hand side is an odd checkerboard matrix. Since they are equal, they must equal the zero matrix, and so $C_e = M_e$ and $C_o = M_o$.

(ii-iv) Any even checkerboard vector in $\mathbb{R}^n$ is orthogonal to any odd checkerboard vector in $\mathbb{R}^n$. The ith row and column of an even checkerboard matrix is even when i is odd, and vis-a-versa. Similarly the ith row and column of an odd checkerboard matrix is odd when i is odd, and vis-a-versa. For checkerboard matrices, P and Q, the product $C = PQ$ has entries given by $(C)_{ij} = P(i,:)Q(:,j)$ (using the colon notation of [7]). Thus if P and Q are both even, then $(C)_{ij} = 0$ when $i+j$ is even. If P and Q are both odd, then $C_{ij} = 0$ when $i+j$ is even. Finally, if P is odd and Q is even (or vis-a-versa), then $C_{ij} = 0$ for $i+j$ odd. ■

IOP Conf. Series: Journal of Physics: Conf. Series **1047** (2018) 012016 doi:10.1088/1742-6596/1047/1/012016

Proposition 4.3. *Let M be an $m \times n$ matrix. Then:*

(i) $A_m M A_n = M$ *if and only if* M *is an even checkerboard matrix.*

(ii) $A_m M A_n = -M$ *if and only if* M *is an odd checkerboard matrix.*

(iii) $M + A_m M A_n$ *is an even checkerboard matrix, and* $M - A_m M A_n$ *is an odd checkerboard matrix.*

Proof: Note, for any matrix, $(A_m M A_n)_{ij} = (-1)^{i+j} M_{ij}$.

(i) If $A_m M A_n = M$, then $M_{ij} = -M_{ij}$ when $i + j$ is odd. Therefore $M_{ij} = 0$ when $i + j$ is odd, so M is an even checkerboard matrix. If M is an even checkerboard matrix, then since $M_{ij} \neq 0$ only when $i + j$ is even, we have $A_m M A_n = M$.

(ii) If $A_m M A_n = -M$, then $M_{ij} = -M_{ij}$ when $i + j$ is even. Therefore $M_{ij} = 0$ when $i + j$ is even, so M is an odd checkerboard matrix. If M is an odd checkerboard matrix, then since $M_{ij} \neq 0$ only when $i + j$ is odd, we have $A_m M A_n = -M$.

(iii) Consider $C = M \pm A_m M A_n$. Then $A_m C A_n = A_m M A_n \pm M = \pm C$. Thus $M + A_m M A_n$ is an even checkerboard matrix, and $M - A_m M A_n$ is an odd checkerboard matrix by properties 1 and 2 respectively.

■

Proposition 4.4. *Let C_o be an odd checkerboard matrix. If λ is an eigenvalue of C_o with associated eigenvector, $\mathbf{v}$, then $A\mathbf{v}$ is an eigenvector of C_o with eigenvalue $-\lambda$.*

Proof: Since C_o is odd checkerboard, if $C_o\mathbf{v} = -AC_oA\mathbf{v} = \lambda\mathbf{v}$, then $C_oAv = -\lambda A\mathbf{v}$. ■

Definition 4.5. *Let $\mathbf{v} \in \mathbb{R}^n$. Define $\mathbf{v}_{even} \in \mathbb{R}^{\lfloor \frac{n}{2} \rfloor}$ by $(\mathbf{v}_{even})_i = v_{2i}$, $i = 1, \ldots, \lfloor \frac{n}{2} \rfloor$, and define $\mathbf{v}_{odd} \in \mathbb{R}^{\lceil \frac{n}{2} \rceil}$ by $(\mathbf{v}_{odd})_i = v_{2i-1}$, $i = 1, \ldots, \lceil \frac{n}{2} \rceil$.*

Proposition 4.6. *If C_o is an odd checkerboard matrix of size $(2m-1) \times (2m-1)$, then C_o has an even checkerboard eigenvector corresponding to eigenvalue $\lambda = 0$.*

Proof: Let C_o be an odd checkerboard matrix of size $(2m-1) \times (2m-1)$, $m \in \mathbb{N}$, and let $\mathbf{v}$ be a vector such that $C_o\mathbf{v} = \mathbf{0}$. Also, we will choose $\mathbf{v}_{even} = \mathbf{0}$, which makes $\mathbf{v}$ an even checkerboard matrix. The odd numbered rows of C_o when multiplied by $\mathbf{v}$ clearly give 0, thus we only need to verify that the even numbered rows also give zero. There are $m - 1$ even numbered rows in C_o and m unknown entries in $\mathbf{v}$, thus we have an underdetermined homogeneous system of linear equations which is guaranteed to have non-trivial solutions. Hence we have a non-zero even checkerboard eigenvector associated with the eigenvalue $\lambda = 0$. ■

Corollary 4.7. *If C_o is an odd checkerboard matrix of size $(2m-1) \times (2m-1)$, then C_o is singular.*

Proposition 4.8. *Let C_e be a square even checkerboard matrix. If λ is an eigenvalue of C_e with associated eigenvector, $\mathbf{v}$, then $A\mathbf{v}$ is an eigenvector of C_e with eigenvalue λ.*

Proof: Since C_e is even checkerboard, if $C_e\mathbf{v} = AC_eA\mathbf{v} = \lambda\mathbf{v}$, then $C_eAv = \lambda A\mathbf{v}$. ■

Corollary 4.9. *Let C_e be a square even checkerboard matrix. Suppose that λ is an eigenvalue of C_e. Then an associated eigenvector is either an even checkerboard vector, an odd checkerboard vector, or we can construct it to be either.*

Proof: Consider eigenvectors $\mathbf{v}$ and $A\mathbf{v}$ associated with eigenvalue λ. If $\mathbf{v} = A\mathbf{v}$ then $\mathbf{v}$ is an even checkerboard vector (Prop. 4.3, part 1). If $\mathbf{v} = -A\mathbf{v}$ then $\mathbf{v}$ is an odd checkerboard vector (Prop. 4.3, part 2). If $\mathbf{v} \neq A\mathbf{v}$, then $\mathbf{u} = \mathbf{v} + A\mathbf{v}$ is also an eigenvector with the same eigenvalue, and $A\mathbf{u} = \mathbf{u}$. Also, $\mathbf{w} = \mathbf{v} - A\mathbf{v}$ has eigenvalue λ, and $A\mathbf{w} = -\mathbf{w}$. ■

Proposition 4.10. *If C_o is a square odd checkerboard matrix, then* $C_o \sim D_n := \begin{bmatrix} O_{\lceil \frac{n}{2} \rceil} & U \\ V & O_{\lfloor \frac{n}{2} \rfloor} \end{bmatrix}$.
If C_e is a square even checkerboard matrix, then $C_e \sim G_n := \begin{bmatrix} Q_{\lceil \frac{n}{2} \rceil} & O \\ O & R_{\lfloor \frac{n}{2} \rfloor} \end{bmatrix}$, *where* $U \in \mathbb{R}^{\lceil \frac{n}{2} \rceil \times \lfloor \frac{n}{2} \rfloor}$, $V \in \mathbb{R}^{\lfloor \frac{n}{2} \rfloor \times \lceil \frac{n}{2} \rceil}$, *and Q and R are square matrices of the indicated size.*

Proof: It is clear that $EC_oE^T = D_n$, and $EC_eE^T = G_n$. ■

Proposition 4.11. *The eigenvalues of a square odd checkerboard matrix C_o are symmetric about the origin.*

Proof: Since $C_o \sim AC_oA = -C_o$, the eigenvalues of C_o and $-C_o$ are identical, and hence must be symmetric about the origin. ■

Proposition 4.12. *If C_e is invertible then C_e^{-1} is even checkerboard. If C_o is invertible then C_o^{-1} is odd checkerboard.*

Proof: Suppose C_e is invertible. Now by Proposition 4.10 we know that $C_e = E^T \mathrm{diag}(Q, R)E$, so $C_e^{-1} = E^T \mathrm{diag}(Q^{-1}, R^{-1})E$. Thus C_e^{-1} is an even checkerboard matrix. Similarly, suppose C_o is invertible. C_o is similar to a block anti-diagonal matrix, the inverse of which is also block anti-diagonal, so C_o^{-1} will also be an odd checkerboard matrix. ■

4.2. Main Result

We are now able to fully characterize checkerboard matrices in Theorems 4.13 and 4.17 below.

Theorem 4.13. *Let M be an $n \times n$ diagonalizable matrix. Then M is an odd checkerboard matrix if and only if $M = V\Lambda V^{-1}$, where V can be written so that $AV = VJ$ and $J\Lambda J = -\Lambda$.*

Proof: Let M be a diagonalizable odd checkerboard matrix. From Proposition 4.11, we know that the spectrum of M is symmetric about the origin. Let $\Lambda = \mathrm{diag}((\lambda_k)_{k=1}^n)$ be skew-persymmetric, that is, $\lambda_k = -\lambda_{n+1-k}$. If n even, let V be the block matrix defined by $[U|AUJ]$ where U is the matrix whose columns are the first $\frac{n}{2}$ eigenvectors of M corresponding to $\lambda_1, \ldots, \lambda_{\frac{n}{2}}$. If n is odd, define V to be $[U|\mathbf{v}_{\frac{n+1}{2}}|AUJ]$ where $\mathbf{v}_{\frac{n+1}{2}}$ is associated with eigenvalue 0. Note that $\mathbf{v}_{\frac{n+1}{2}}$ can be chosen to be an even checkerboard vector, by Prop. 4.6. It is clear that $MV = V\Lambda$, and the rows of V have alternating parity, or in other words, $AV = VJ$.

Let M be diagonalizable, that is, $MV = V\Lambda$. Let V have rows with alternating parity, starting with even, so $AVJ = V$. And let Λ be skew-persymmetric, thus $J\Lambda J = -\Lambda$. Decompose $M = C_e + C_o$, where C_e is an even checkerboard matrix, and C_o is an odd checkerboard matrix. Then

$$MV = C_eAVJ + C_oAVJ = AVJ\Lambda.$$

Since $A^2 = I$ and $J^2 = I$, we can rearrange the above equation to

$$AC_eAV + AC_oAV = VJ\Lambda J = -V\Lambda.$$

Because $(AMA)_{ij} = (-1)^{i+j}M_{ij}$, we have that $AC_eA = C_e$ and $AC_oA = -C_o$. Therefore we have two equations

$$C_eV + C_oV = V\Lambda$$
$$C_eV - C_oV = -V\Lambda$$

Thus $C_eV = O$, and so $C_e = O$, proving that M is an odd checkerboard matrix. ■

Corollary 4.14. *If $M = V\Lambda V^{-1}$ with $J\Lambda J = -\Lambda$ and $V = AVJ = JVA$, then M is centrosymmetric odd checkerboard matrix.*

Proof: Since M is diagonalizable and $V = JVA$, M is centrosymmetric by Proposition 3.9. It is odd checkerboard by Theorem 4.13 ■

Corollary 4.15. *Under the conditions in Corollary 4.14, if in addition V is orthogonal, that is $V^{-1} = V^T$, then M is supersymmetric.*

Using the main result, we can now constructively solve the general SIEP for odd checkerboard matrices.

Corollary 4.16. *The structured inverse eigenvalue problem for an odd checkerboard matrix is solvable if and only if the spectrum is symmetric with respect to the origin.*

Proof: Since square, odd checkerboard matrices must have a symmetric spectrum of eigenvalues by the main result, Theorem 4.13, the necessity of a symmetric spectrum for a solution follows. For the converse, assume a skew-persymmetric diagonal eigenvalue matrix Λ formed with the desired eigenvalues. The Kac matrix is a diagonalizable, odd checkerboard matrix of arbitrary size. An explicit formula for the eigenvectors of the Kac matrix can be found in [15]. Use this formula to construct an eigenvector matrix of the desired size. This matrix, labeled V, can be arranged with alternating row parity by Theorem 4.13. Finally, $V\Lambda V^{-1}$ is the desired odd checkerboard matrix, by Theorem 4.13. ■

Theorem 4.17. *Let M be an $n \times n$ diagonalizable matrix. Then M is an even checkerboard matrix if and only if $M = V\Lambda V^{-1}$, where V is an even checkerboard matrix.*

Proof: Suppose that V is an invertible even checkerboard matrix. Then V^{-1} is also even checkerboard by Proposition 4.12 and we also note that a diagonal matrix is even checkerboard. Thus by Proposition 4.2 we know that $V\Lambda V^{-1}$ is also even checkerboard.

Next suppose that M is a diagonalizable even checkerboard matrix. By Proposition 4.10, M is similar to a block diagonal matrix $D = \text{diag}(Q, R)$. Both matrices Q and R are therefore diagonalizable, so that $Q = ULU^{-1}$ and $R = WMW^{-1}$, where L and M are diagonal matrices and U and W are matrices of eigenvectors of Q and R respectively. Then $\text{diag}(Q, R) = \text{diag}(U, W)\text{diag}(L, M)\text{diag}(U^{-1}, W^{-1})$. And so,

$$M = E^TDE = E^T\text{diag}(U, W)EE^T\text{diag}(L, M)EE^T\text{diag}(U^{-1}, W^{-1})E.$$

Define $V = E^T\text{diag}(U, W)E$ and $\Lambda = E^T\text{diag}(L, M)E$. Then clearly V is an even checkerboard matrix, and Λ is a diagonal matrix. ■

5. Toeplitz matrices revisited

The first non-zero entry in an outer-banded matrix satisfying rTIEP can be established.

Proposition 5.1. *Outer-banded Toeplitz matrices of even size satisfy* $|\det B_{2n}| = b_1^{2n}$.

Proof: Let B_{2m} be written with lower-triangular, Toeplitz blocks, $L \in \mathbb{R}^{m\times m}$: $B_{2m} = \begin{bmatrix} O & L^T \\ L & O \end{bmatrix}$, and let $M = \begin{bmatrix} O & I \\ I & O \end{bmatrix}$, where O, I, are $m \times m$ zero and identity matrices. Then $MB_{2m} = \begin{bmatrix} L & O \\ O & L^T \end{bmatrix}$, and $\det MB_{2m} = \det L \det L^T$. Since M is a permutation matrix, we have $\det B_{2m} = \pm(\det L)^2$. Since L is lower triangular with constant diagonal b_1, the result follows. ■

Corollary 5.2. *If* $B_{2n}\left(b_1,\ b_2,\ \ldots,\ b_n\right)$ *satisfies rTIEP then* $b_1 = \sqrt[n]{1\cdot 3\cdots(2n-1)}$. *If* $B_{2n+1}\left(b_1,\ b_2,\ \ldots,\ b_n\right)$ *satisfies rTIEP, then* $b_1 = \sqrt[n]{2\cdot 4\cdots(2n)}$.

5.1. Landau's Conjecture

In a seminal 1994 work [4], H.J. Landau demonstrated that the TIEP is solvable within a certain subclass of the symmetric Toeplitz matrices, for any prescribed set of *distinct* eigenvalues. We now describe this subclass of Toeplitz matrices ([4], p. 756), and make an observation based on our examples from Section 2.

Definition 5.3. *A real symmetric Toeplitz matrix,* $T_n\left(a_1,\ a_2,\ \ldots,\ a_n\right)$ *is* **regular**, *provided it, and every leading principal submatrix,* $T_n\left(a_1,\ a_2,\ \ldots,\ a_k\right)$, $1 \le k \le n$, *have the properties: i) distinct eigenvalues, and ii) when ordered by size, the eigenvalues alternate parity, with the largest one even. Let* $\mathcal{T}_n$ *denote the set of regular matrices* $T_n\left(0\ \ a_2,\ \ldots,\ a_n\right)$.

Landau ends his paper by conjecturing that the solution to TIEP for any set of distinct eigenvalues is unique within $\mathcal{T}_n$. Examining the examples of Section 2, we note that i) outer-banded matrices of size $n \ge 3$ are not regular, (since the principal two-by-two submatrix is the zero matrix), ii) we found exactly one regular checkerboard solution of rTIEP of each size $n \le 6$, iii) we showed that the regular solutions of rTIEP up to size 4 are unique.

Based on these observations we make the following conjecture.

Conjecture 5.4. *The rTIEP has a unique, regular, non-negative, odd checkerboard solution for all sizes.*

6. A Hankel Matrix SIEP

In [16], the authors gave a closed form construction for a real symmetric Toeplitz matrix of even size T_{2n} with arbitrary given eigenvalues, $(\lambda_i)_1^n$, each of double multiplicity. Further, each double eigenvalue was associated with one even parity eigenvector, and one odd. We observe that this matrix can be used to solve the ssSIEP within the class of persymmetric Hankel matrices.

Recall that a Hankel matrix has constant entries along each antidiagonal, so that a Hankel matrix, H_n, can be defined by exchanging the rows (or columns) in a Toeplitz matrix. If the Toeplitz matrix is symmetric, the Hankel matrix will be persymmetric, ie. $H_n = JT_n = T_nJ$, and both are supersymmetric.

Proposition 6.1. *Let* M *be centrosymmetric, and let* λ_e *and* λ_o *be even and odd parity eigenvalues of* M *respectively. Then* λ_e *and* $-\lambda_o$ *are even and odd parity eigenvalues of both of the exchanged centrosymmetric matrices,* JM *and* MJ, *for the same eigenvectors.*

Proof: By Proposition 3.9, all eigenvectors of M have parity. Suppose $(\mathbf{v}_e, \lambda_e)$ and $(\mathbf{v}_o, \lambda_o)$ are eigenvector-eigenvalue pairs for M of even and odd parity, respectively. Then $M\mathbf{v} = M(JJ)\mathbf{v}_e = \lambda_e\mathbf{v}_e$ which implies $MJ\mathbf{v}_e = \lambda_e\mathbf{v}_e$, and $M(JJ)\mathbf{v}_o = \lambda_o\mathbf{v}_o$ which implies $-MJ\mathbf{v}_o = \lambda_o\mathbf{v}_o$, so that MJ has the desired property. On the other hand, $JM\mathbf{v}_e = \lambda_e J\mathbf{v}_e = \lambda_e\mathbf{v}_e$, and $JM\mathbf{v}_o = \lambda_o J\mathbf{v}_o = -\lambda_e\mathbf{v}_e$. So JM also has the desired property. ■

Theorem 6.2. *If T_{2n} is constructed as per [16] p. 290, with given eigenvalues, $(\lambda_i)_{i=1}^n$, each of double multiplicity, then $H_{2n} = JT_{2n}$ is a supersymmetric Hankel matrix with symmetric spectrum, $((\lambda_i)_{i=1}^n \cup (-\lambda_i)_{i=1}^n)$.*

Proof: The construction ([16] p. 290) guarantees that each double eigenvalue, λ_i of the symmetric, Toeplitz matrix T_{2n} (hence supersymmetric) is associated with one even, and one odd parity eigenvector, $i = 1, \ldots, n$. Since T_{2n} is centrosymmetric, Proposition 6.1 applies, to the effect that the spectrum of the Hankel matrix, $H_{2n} = JT_{2n}$ consists of an even parity set of eigenvalues, $(\lambda_i)_{i=1}^n$ and an odd parity set, $(-\lambda_i)_{i=1}^n$. Clearly, H_{2n} is supersymmetric. ■

7. Conclusion

Small-sized examples of symmetric Toeplitz matrices with equally spaced eigenvalues were sought, with the goal of finding patterns for the entries of the general size, restricted inverse problem. Two solution forms were found: "(odd) checkerboard", and "outer banded". Diagonalizable odd checkerboard matrices were characterized by alternating row parity in their (orthogonal) eigenvector matrix, and a spectrum symmetric about zero. The solution to the inverse problem for this class of matrices was then solved by utilizing a known orthogonal matrix which can be arranged to have alternating row parity.

We conjecture that the restricted Toeplitz inverse eigenvalue problem possesses both odd checkerboard and outer banded solutions of all sizes, $n \in \mathbb{N}$, and that all solutions ($n > 2$) must be one of these two forms. We also suspect that a non-negative odd checkerboard matrix will be the unique solution belonging to Landau's class of "regular" symmetric Toeplitz matrices.

Acknowledgments

The authors wish to acknowledge financial support from the Bishop's University Senate Research Committee.

References

[1] Chu M T 1998 *SIAM Review* **40** 1–39
[2] Chu M and Golub G 2002 *Acta Numerica* **11** 171
[3] Chu M T and Golub G H 2005 *Inverse Eigenvalue Problems: Theory, Algorithms, and Applications* (Oxford University Press)
[4] Landau H J 1994 *J. Amer. Math. Soc.* **7** 749–767
[5] Delsarte P and Genin Y 1984 Spectral properties of finite Toeplitz matrices *Lecture Notes in Control and Information Sciences* vol 58 (Springer, New York) pp 194–213
[6] Gladwell G M L, Jones T H and Willms N B 2014 *J. Appl. Math.* **2015, Special Issue** 6 pages URL `http://dx.doi.org/10.1155/2014/515082`
[7] Golub G H and Loan C F V 2013 *Matrix Computations* 4th ed (The John Hopkins University Press)
[8] Chan R, Chung H and Xu S 2003 *BIT* **43** 7–20
[9] Laurie D P 1988 *SIAM J. Sci. Statist. Comput.* **9** 401–405
[10] Trench W F 1997 *SIAM J. Sci. Comput.* **18** 1722–1736
[11] Chu M T 1993 On a differential equation $dx/dt = [x, k(x)]$ where k is a Toeplitz annihilator Tech. rep. North Carolina State University
[12] Webb M 2013 Isospectral flows and the Toeplitz inverse eigenvalue problem Tech. rep. Cambridge Centre for Analysis Cambridge, UK
[13] Cantoni A and Butler P 1976 *Lin. Alg. App.* **13** 275–288
[14] Weaver J R 1985 *Amer. Math. Monthly* **92** 711–717
[15] da Fonseca C M, Mazilu D A, Mazilu I and Williams H T 2013 *Applied Mathematical Letters* **26**(12) 1206–1211
[16] Diele F, Laudiano T and Mastronardi N 2004 *SIAM J. Matrix Anal. Appl.* **26** 285294

An Error Reduction Technique in Richardson-Lucy Deconvolution Method

Hiroyuki Yamauchi[1] **and Worawit Somha**[2]

[1]Fukuoka Institute of Technology,Wajiro-Higashi, Higashi-ku, Fukuoka, Japan
[2]King Mongkut's Institute of Technology, Ladkrabang, Bangkok, Thailand

E-mail: yamauchi@fit.ac.jp

Abstract. An error reduction technique for Richardson-Lucy deconvolution (RL-deconv) is proposed. The deconvolution is indispensable technique for inversely analysing the SRAM fail-bit probability variations caused by the Random Telegraph Noise (RTN). The proposed technique reduces the phase difference between the two distributions of the deconvoluted RTN and the feedback-gain in the maximum likelihood (MLE) gradient iteration cycles. This avoids an unwanted positive feedback, resulting in a significant decrease in probability of undesired ringing occurrence. A quicker convergence benefit of the RL-deconv algorithm while avoiding the ringing is achieved. It has been demonstrated that the proposed technique reduces its relative deconvolution errors by 100 times compared with the conventional RL-deconv. This provides an increase in accuracy of the fail-bit-count prediction by over 2-orders of magnitude while accelerating its convergence speed by 33times of the conventional one.

Keywords:
Blind deconvolution, Random telegraph noise, Richardson-Lucy deconvolution

1. Introduction

Ordinary brief method for estimating the bit fail probability for the static random access memory (SRAM) is expected to be no longer applicable because the simple Gaussian based analyses method [1-6] cannot be used any more. The main reasons for this change are : (1) the tail distribution doesn't obey Gaussian any more. It does follow log-normal distribution and (2) the tail length of the distributions of the SRAM operating margin will be no longer negligible small fraction.

Figures 1(a) and 1(b) compare the tail length of the log-normal distribution **g**. Once the tail length of **g** becomes longer than that for **f**, the shape of the tail distribution of **h** is governed by **g** not **f**, as shown in Fig. 1(b). Where, **f** and **g** denote the distributions of the Random Dopant Fluctuation (RDF) and the Random Telegraph Noise (RTN), respectively [1-3]. The symbols of $\otimes^{-1}$ and $\otimes$ represent for the operators for deconvolution and convolution, respectively..

Considering that the tail length of the distribution **g** is expected to become no longer ignored, the tail distribution of **h** will have to be estimated precisely by the convolution ($\otimes$) of **f** with **g**, *i.e.,* $\mathbf{h}=\mathbf{f}\otimes\mathbf{g}$.

Figures 2(a) and 2(b) compare the relationships among **f**, **g**, and **h** between the convolution and the deconvolution, respectively. Since the shape of the distribution of **g** follows a log-normal distribution, the tail distribution of **h** cannot be approximated by the Gaussian distribution any more. Thus, the ordinary Gaussian based analyses [1-6] cannot be used for estimating the tail of the **h** distribution.

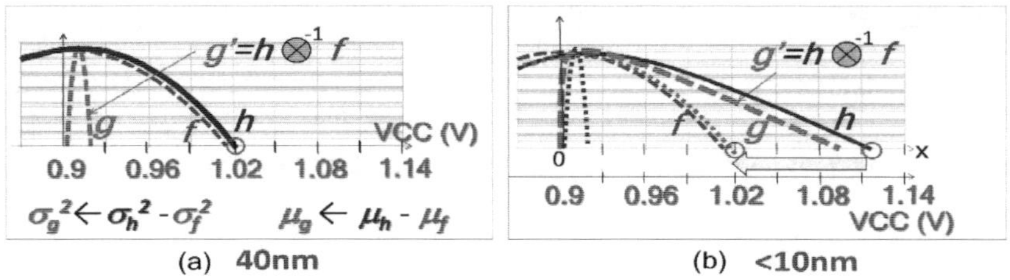

Figure 1. Comparison of deconvolution **g(x)** of **h(x)** with f(x), i.e., $\mathbf{g(x)=h(x)\otimes^{-1}f(x)}$ in case of (a) tail length of **f(x)>g(x)** and (b) **f(x)<g(x)** corresponding to the case for future beyond 10nm.

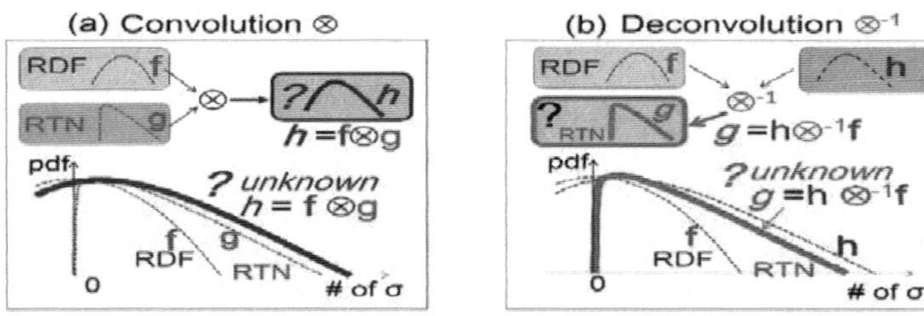

Figure 2. (a) Concept of convolution (h=f⊗g) and (b) deconvolution (g=h⊗⁻¹f).
Forward and inverse problems can be solved by convolution and deconvolution, respectively.

The most ordinary SRAM statistical analyses method have no choice but to rely on the Gaussian model for simplicity as follows:

$$\mathbf{f}=N(\mu_f, \sigma_f), \quad \mathbf{g}=N(\mu_g, \sigma_g), \quad \mathbf{h}=N(\mu_h, \sigma_h) \tag{1}$$

where N shows Normal (*i.e.,* Gaussian) distribution function and μ and σ are mean and deviation, respectively.

The convolution result $\mathbf{h}=N(\mu_h, \sigma_h)$ is simply given by

$$\mu_h = \mu_f + \mu_g, \quad \sigma_h^2 = \sigma_f^2 + \sigma_g^2 \tag{2}$$

The deconvolution $\mathbf{g}=\varphi(\mu_g, \sigma_g)=\mathbf{h}\otimes^{-1}\mathbf{f}$ is also simply given by

$$\mu_g = \mu_h - \mu_f, \quad \sigma_g^2 = \sigma_h^2 - \sigma_f^2 \tag{3}$$

Thus, it was straightforward to estimate both of convolution (**h=f⊗g**) and deconvolution ($\mathbf{g=h\otimes^{-1}f}$) under the assumption that both of the distributions **f** and **g** obey Gaussian.

On the other hand, if the tail length of the log-normal distribution **g** no longer accounts for just a fraction but a large percentage of the overall **h** (see Fig. 1(b)), the non-Gaussian inverse problem needs to be solved by a complex numerical calculation. However, it must not be easy to use in the manufacturing field.

These two main reasons mentioned above lead to a significant pressure to develop the user-friendly methods for solving the non-Gaussian inverse problem so that the user can easily figure out the unknown factors [5-7], although the SRAM designers are unfamiliar with such kind of methodology.

Thus, in this paper, we proposed and discussed the practical way to solve the non-Gaussian deconvolution process based on the Richardson-Lucy deconvolution (RL-deconv) algorithm [7-9]. This is because the RL-deconv algorithm is most widely used technique for image processing of all the MATLAB® built-in deconvolution functions.

We did the applicability evaluation for the SRAM margin analyses before [10-12] and demonstrated huge errors in the exponentially decreasing tails of the deconvoluted **g**. The papers concluded that huge errors are observed in all of the deconvolution tools provided by off-the-shelf MATLAB® [10-12] and its error level is not acceptable for the accuracy request point of view of the target application.

To the best of our knowledge, there have been very little qualified published solutions to the ringing error problems with the deconvolution for the SRAM fail-bit probability analyses [10-12].

The almost relevant ideas described in the publications rely on the human optical illusion trick for the image recognition. Thus, the required ringing suppression level is unfortunately quite low, *i.e.,* that is 10-orders of magnitude smaller dynamic range than that for the SRAM fail-bit probability.

The authors previously proposed the alternative algorithms to avoid the ringing [10-13] but it causes an intolerable time-consuming iteration cycles as a side effect of the introduced fine segmentations [6].

In that sense, we still have to rely on the maximum likelihood (MLE) gradient base RL-deconv algorithm [7-9] because it provides a much faster convergence characteristic than that for the previously introduced segmentation methods. Thus, we have to solve the ringing issues in the RL-deconv algorithm by all possible means so that the benefits from the RL algorithm can be well exploited.

In this paper, we experimentally tried to figure out the root cause for the ringing and we demonstrated the ringing elimination for the first time based on the proposed algorithm to control the key procedure and parameters for the RL deconvolution.

For specific understanding of the application examples, the following deconvolution scenarios are recounted by using Fig. 3(a): (1) A certain **h** within the product target specification ($\mathbf{SP_{prod}}$) is predefined. The **f** is truncated at a certain point (**TP**) based on the screening condition. As a result, the **f** is converted to the truncated $\mathbf{f_{TP}}$. (2) The distribution **g** is unknown. However, there is a case where **g** has to be estimated to set a target value for the RTN **g** reduction from a device characteristics improvement point of view. Figure 3(b) recounts another scenario: (1) $\mathbf{SP_{prod}}$ and **g** are predefined. (2) The truncated $\mathbf{f_{TP}}$ is unknown but should be estimated because **h** has to be set within the target specification $\mathbf{SP_{prod}}$ [6], as shown in Fig. 3(b). The distributions of **g** and $\mathbf{f_{TP}}$ can be calculated in theory by the following deconvolution: $\mathbf{g=h\otimes^{-1}f_{TP}}$ and $\mathbf{f_{TP}=h\otimes^{-1}g}$, respectively. However, the deconvolution of the **g** or $\mathbf{f_{TP}}$ is sort of ill-posed inverse problem. Thus, a troublesome operation has to be expected unlike the convolution: $\mathbf{h=f_{TP}\otimes g}$ (Fig. 2(a)) [5-7].

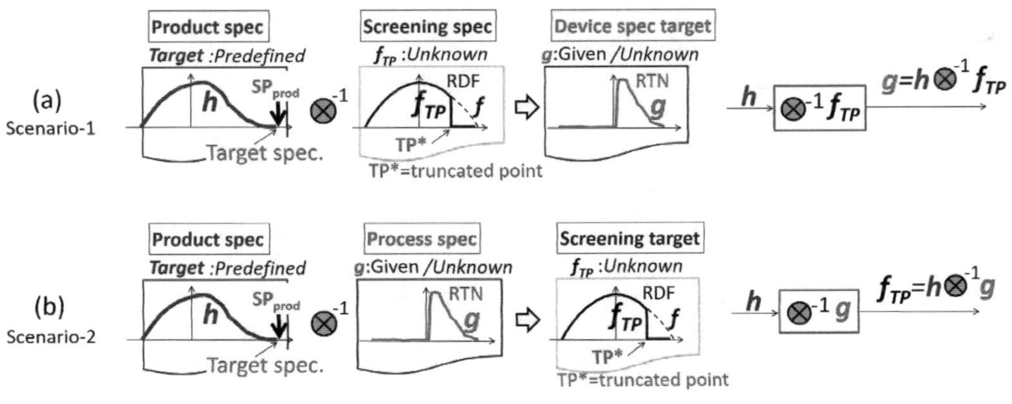

Figure 3. Potential scenarios for application of deconvolution
(a) Scenario-1: Based on product and screening specifications, process target for RTN reduction is determined.
(b) Scenario-2: Based on product and RTN specifications, screening condition TP is determined.

The rest of the paper is organized as follows. Review for the MATLAB® built-in RL-deconv algorithm in terms of what is behind the ringing in the RL iterations and what is the most sensitive procedure to the ringing noise amplification in Section 2. The proposed ringing elimination technique is proposed in Section 3. The advantages over the conventional MATLAB® built-in RL-deconv-functions are discussed based on the simulation results in section 4, followed by conclusion in section 5.

IOP Conf. Series: Journal of Physics: Conf. Series **1047** (2018) 012017 doi:10.1088/1742-6596/1047/1/012017

2. Issues of Conventional Algorithm

2.1. Richardson-Lucy deconvolution algorithm

The Richardson-Lucy (RL) deconvolution algorithm uses the maximum likelihood (MLE) gradient iterations with the multiple convolution (⊗) processes, as shown in the expressions of (4) – (8). The super script (t) for **g, h, p** and **q** denotes the number of iteration cycles.

$$\mathbf{g}^{(t+1)} = \mathbf{g}^{(t)} \times [\ (\ \mathbf{h}/(\mathbf{f} \otimes \mathbf{g}^{(t)})\ ) \otimes \mathbf{f}\hat{}\ ] \tag{4}$$

$$\mathbf{g}^{(t+1)} = \mathbf{g}^{(t)} \times \mathbf{q}^{(t)} \tag{5}$$

$$\mathbf{q}^{(t)} = \mathbf{p}^{(t)} \otimes \mathbf{f}\hat{} \tag{6}$$

$$\mathbf{p}^{(t)} = \mathbf{h}\ /\ (\ \mathbf{f} \otimes \mathbf{g}^{(t)}\ ) = \ \mathbf{h}\ /\ \mathbf{h}^{(t)} \tag{7}$$

$$\textbf{where,}\ \ \mathbf{f}\hat{} = \text{flipped}\ \mathbf{f},\ \ \mathbf{h} = \mathbf{g} \otimes \mathbf{f}\ \ \text{and}\ \ \mathbf{h}^{(t)} = \mathbf{f} \otimes \mathbf{g}^{(t)} \tag{8}$$

2.2. Issues of RL-deconv algorithm

Figure 4 shows the simulated waveforms for $\mathbf{p}^{(t)}$,$\mathbf{q}^{(t)}$,$\mathbf{g}^{(t)}$,$\mathbf{h}^{(t)}$ expressed by the above (4) – (8).
Because the convolution process plays a role of low-pass filtering, a high-frequency noises on the waveforms are eliminated. However, a low-frequency noise amplification on the $\mathbf{g}^{(t)}$ (ringing of $\mathbf{g}^{(t)}$), i.e., the ringing noise issue still remains, as shown in Fig. 4.
The probable root cause of the noise amplification is conceptually illustrated in Fig. 5.
In this paper, this noise amplification is referred to as "ringing error".

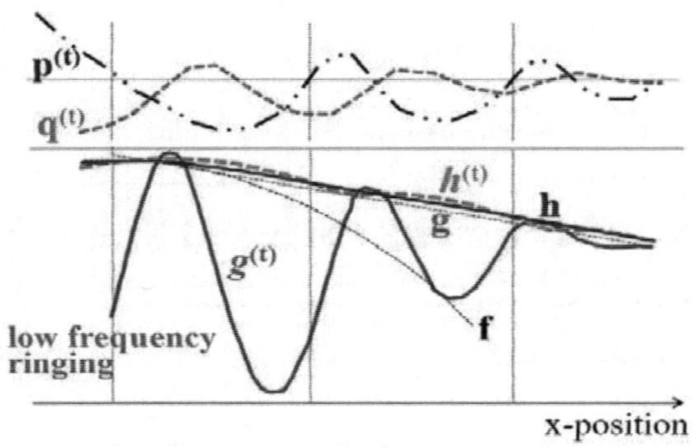

Figure 4. Example of ringing of deconvoluted $g^{(t)}$

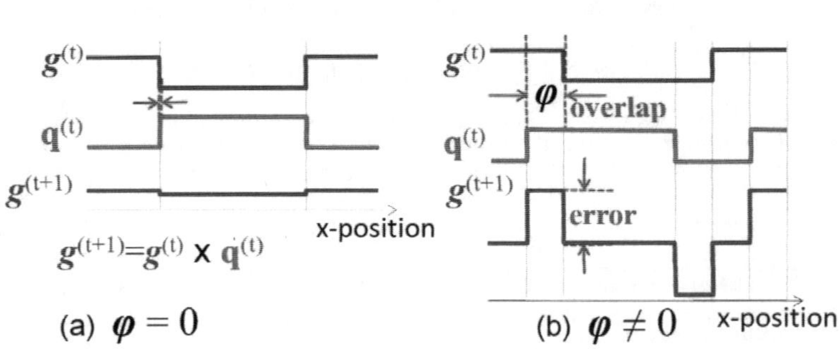

Figure 5. Comparison of the cases of (a) φ=0 (no ringing happen) and (b) $\varphi \neq 0$ (ringing can happen)

The mechanism of the ringing in the iteration cycles can be explained as follows:
(i) If the phase (x-position) difference $\boldsymbol{\varphi}$ between $\mathbf{g}^{(t)}$ and $\mathbf{q}^{(t)}$ becomes larger than a certain value (see Fig. 5(b) compared with Fig. 5(a)), the successive deconvoluted value $\mathbf{g}^{(t+1)}$ is erroneously amplified. This is caused by the unwanted overlap by $\boldsymbol{\varphi}$ between the $\mathbf{g}^{(t)}$ and $\mathbf{q}^{(t)}$, as shown in Fig. 5(b). Where, the superscript of **(t)** denotes the numbers of iteration cycles.
(ii) Mechanism of (i) is caused by a positive feedback to $\mathbf{g}^{(t+1)}$ in the next cycle. Then the noise amplification behavior can be seen on the curves of $\mathbf{g}^{(t+1)}$ in the iteration cycles.

Figure 6 shows the reason why the phase difference (x-position) between $\mathbf{g}^{(t)}$ (dash-line) and $\mathbf{q}^{(t)}$ (outer solid line) is made in the deconvolution processes expressed by equations (4)-(8).

The phase difference by $\boldsymbol{\varphi}$ is developed through the two successive convolution (⊗) processes in every iteration cycles : the 1st ⊗ process in $\mathbf{h}^{(t)}= \mathbf{f} \otimes \mathbf{g}^{(t)}$, then the 2nd ⊗ process in $\mathbf{q}^{(t)}=(\mathbf{h}/\mathbf{h}^{(t)}) \otimes \mathbf{f}^{\wedge}$ (See the expressions of (4) – (8). As a result, the x-position of $\mathbf{q}^{(t)}$ is shifted by $\boldsymbol{\varphi}$ from that for $\mathbf{g}^{(t)}$, as shown in Fig. 6. As used herein, "**x**" in Fig. 6 corresponds to the number of σ for Gaussian. The two distributions **g** and **f** differently obey the Gamma $\boldsymbol{G}(\alpha, \beta)$ and Gaussian $N(\sigma, \mu)$, respectively. To plot the two in the same x-y plot, the distribution parameters are normalized. For example, the **g** distribution is plotted normalized to the **f** distribution of $N(\sigma=1.0, \mu=0)$.

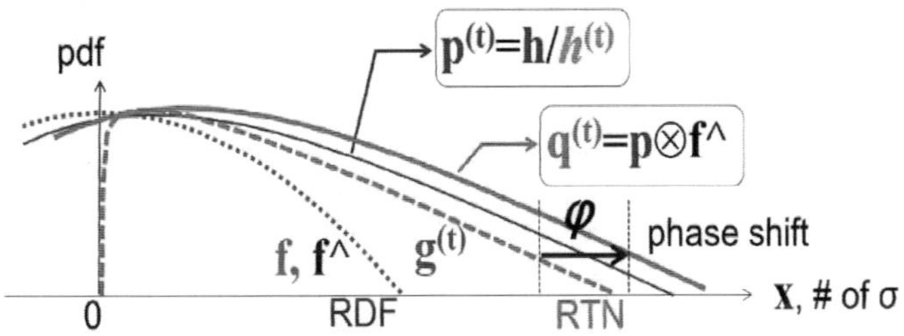

Figure 6. Concept of probable cause of phase error $\boldsymbol{\varphi}$ in iterations.

In this paper, in order to demonstrate the ringing elimination for the first time, we propose the algorithm for minimizing the phase error $\boldsymbol{\varphi}$ with the control of the key procedure and parameters in the RL deconvolution in the following sections.

3. Proposed Ringing Elimination Methods

In this section, a method for elimination of the phase shift $\boldsymbol{\varphi}$ of $\mathbf{q}^{(t)}$ is explained.
The key is to use a certain filter **ALG** that is convoluted with **p** such that the position of $\mathbf{q}^{(t)}$ $(=\mathbf{p}^{(t)} \otimes \mathrm{ALG})$ gets closer to the curve of $\mathbf{g}^{(t)}$. The $\boldsymbol{\sigma}$ value of the **ALG** distribution is the same as **f** but its mean value $\boldsymbol{\mu}$ is shifted by $\boldsymbol{\theta}$ $(\boldsymbol{\mu}=\boldsymbol{\theta})$, as shown in Fig. 7.
The operation of $\mathbf{p}^{(t)} \otimes \mathrm{ALG}$ is used for the phase shifting so that the phase of $\mathbf{q}^{(t)}$ can be aligned with $\mathbf{g}^{(t)}$ by designing the mean-shift value $\boldsymbol{\theta}$ of the **ALG** distribution, as shown in Fig. 7.

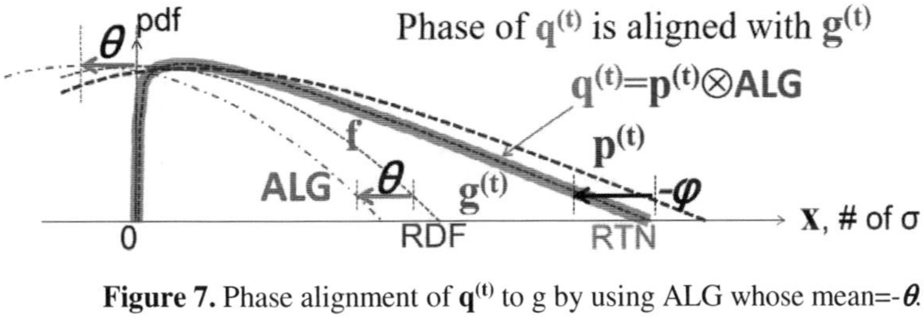

Figure 7. Phase alignment of $\mathbf{q}^{(t)}$ to g by using ALG whose mean=-$\boldsymbol{\theta}$.

The proposed deconvolution process can be expressed by the following expression (9).

$$\mathbf{g}^{(t+1)}=\mathbf{g}^{(t)}\times[(\mathbf{h}/(\mathbf{f}\otimes\mathbf{g}))\otimes\mathbf{ALG}] \tag{9}$$

The difference from the conventional expression (4) is to introduce the distribution of **ALG** for **f^** in (9).
Thus, the key to successful elimination of the phase shift φ is how to predict how much θ for **ALG** is needed for the phase alignment of $\mathbf{q}^{(t)}$ with $\mathbf{g}^{(t)}$, as shown in Fig. 7.

3.1. Design of key parameters φ and θ

In this subsection, how to determine the key parameters φ and θ are described by using Figs. 8 and 9.

The amount of phase shift φ can be predicted based on predefined distributions of **f** and **h** as follows:

(1) The average gradient values α_f and α_h for **f** and **h** are used because the value of φ depends on their values. For example, φ can be expressed by $\varphi = \mathbf{k} \times (\alpha_h - \alpha_f)$, as shown in Fig. 8.
Where, **k** is determined by the slope of the dotted line, as shown in Fig. 9(a).

(2) The range of the value of θ where the phase shift φ becomes low enough for ringing elimination is searched. Then the relationship of φ and θ is drawn, as shown in Fig. 9(b).

As a result, the required θ of ALG for ringing elimination can be designed based on the average gradient of the distribution of given **f** and **h**.

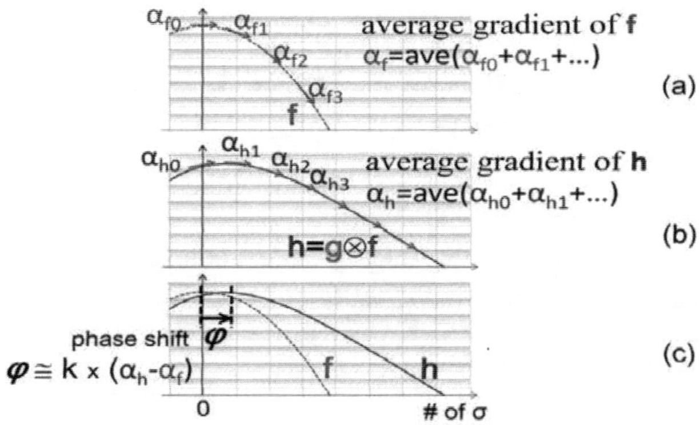

Figure 8. (a) α_h and (b) α_f are average gradient of the distributions of **f** and **h**.
(c) Relationships of φ and gradient difference (α_h-α_f) between predefined **f** and **h**.
φ depends on (α_h-α_f). **k** is determined by the slope of dotted line shown in Fig. 9(a).

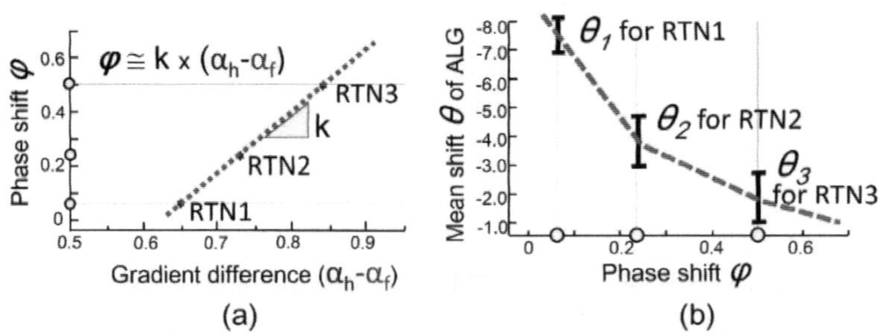

Figure 9. (a) Relationships of φ and gradient difference (α_h-α_f) between predefined **f** and **h**.
(b) Relationships of φ and required mean shift value θ by ALG.
Based on these relationships, θ can be determined for ringing elimination.

In order to discuss the tail length dependencies, three different tail lengths for RTN1, RTN2, and RTN3 distributions (see Fig. 10(a)) are assumed, which correspond to the scaling size of 40nm, 15nm, and 8nm in Fig. 10(b), respectively. Three sets of α and β for ***G***(α, β) for RTN1, RTN2, and RTN3 are assumed as shown in Fig. 10(a), respectively. As used herein, **g** and **f** obey Gamma ***G***(α, β) and Gaussian ***N***(σ, μ) distributions, respectively. The ***G***(α, β) is plotted normalized to **f** with ***N***(σ=1.0, μ=0).
"**x**" of raw score in Fig. 10(a) corresponds to the z-number (i.e., number of σ for Gaussian).
The operating voltage (VCC) is one of examples being used for "**x**", as shown in Fig. 1.

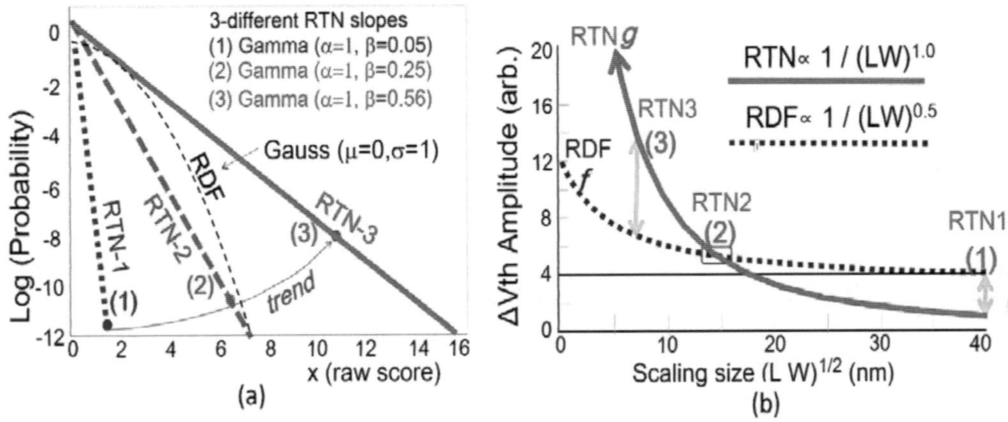

Figure 10. (a) Probability density function comparisons among RDF, RTN1, RTN2, and RTN3.
(b) Trend of variation amplitude comparison between RDF **f(x)** and RTN **g(x)**.

Figure 11 shows the RTN tail length dependencies of the required $\boldsymbol{\theta}$. It is found that the allowable range of the required $\boldsymbol{\theta}$ (*i.e.,* margin of $\boldsymbol{\theta}$) is wide enough to tolerate the error from $\boldsymbol{\Delta\theta}$=1 to 2, as shown in Fig. 11. In addition, it depends on the slope (*i.e.,* tail length) of the RTN distribution.

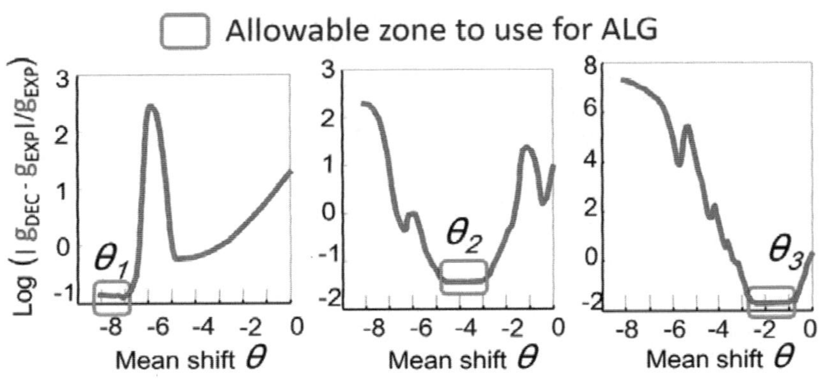

Figure 11. RTN dependencies of required mean shift value $\boldsymbol{\theta}$.
(a) RTN1, (b) RTN2, and (c) RTN3. No ringing happens in the zone for θ_1, θ_2, and θ_3.

3.2. Demonstration of deconvolution with ALG

The deconvolution results $\mathbf{g}^{(t)}$ at each iteration cycle number (#) of 1, 2, and 10 are shown in Figs. 12(a),(b), and (c) (top to bottom), while changing the mean shift ($\boldsymbol{\theta}_A, \boldsymbol{\theta}_B, \boldsymbol{\theta}_C$) for ALG (left to right), respectively.

It is exhibited that the ringing errors on $\mathbf{g}^{(t)}$ are amplified for the cases of $\boldsymbol{\theta}_A$ and $\boldsymbol{\theta}_B$ as the iteration cycles number # is increased from 1 to 10, as shown in Figs 12(a) and (b), respectively.

On the other hand, these errors are fully suppressed in the case of $\boldsymbol{\theta}_C$ set within allowable zone, as shown in Fig. 12(c). The **q** and **p** for the case of $\boldsymbol{\theta}=\boldsymbol{\theta}_C$ becomes flat at the #=10 and their value are almost 1 (*i.e.,* achieved convergence) in the overall range of x, as shown in Fig. 12(c). It is found that an appropriate design of $\boldsymbol{\theta}$ enables to eliminate the ringing behaviors.

Compared with the Fig. 12(c), the distributions for **q** and **p** in the cases of $\boldsymbol{\theta}=\boldsymbol{\theta}_A$ and $\boldsymbol{\theta}_B$ are fluctuated, as shown in Figs 12(a) and (b).

This is caused by the phase misalignment of the feedback gain $\mathbf{q}^{(t)}$ with the deconvoluted $\mathbf{g}^{(t)}$ as explained in the previous section.

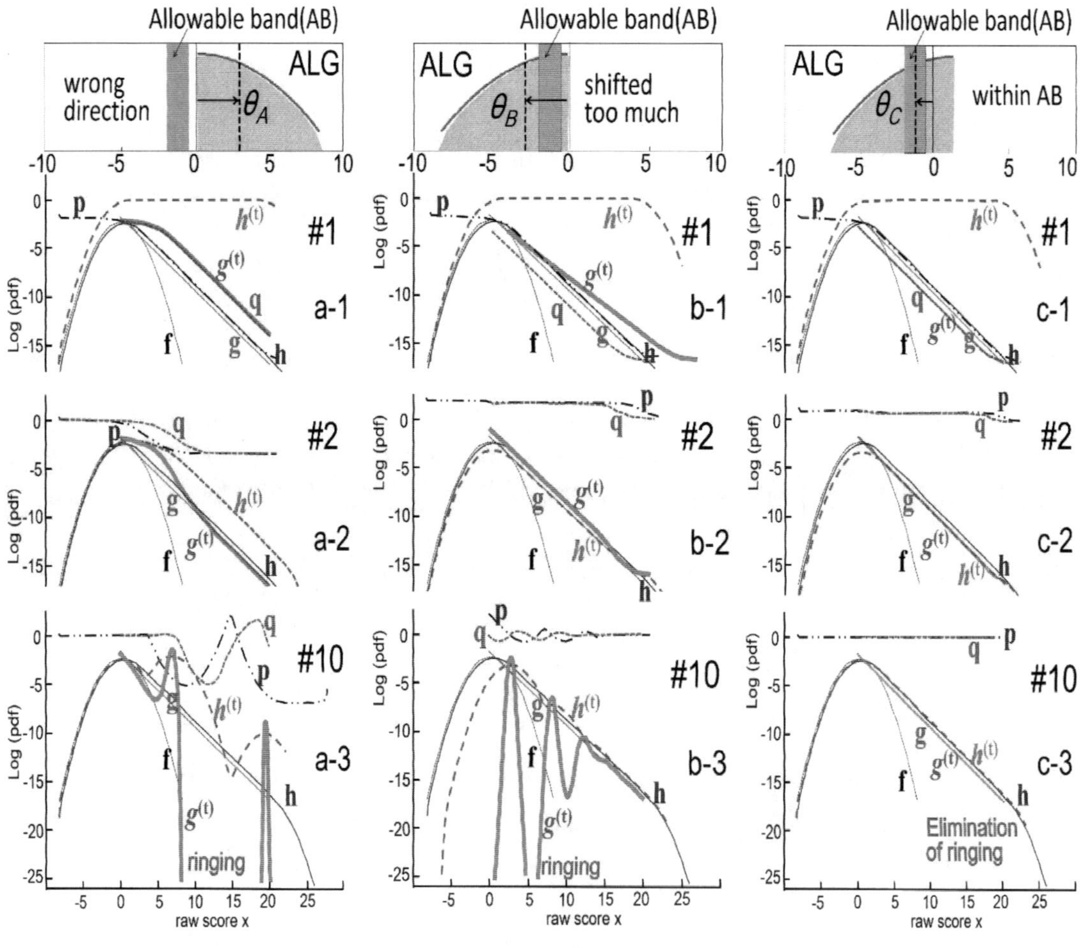

Figure 12. ALG mean shift ($\boldsymbol{\theta}$) dependencies of the ringing behaviors (left to right).
left (a): $\boldsymbol{\theta}_A = +2$, center (b): $\boldsymbol{\theta}_B = -2.8$, right (c): $\boldsymbol{\theta}_C = -1.2$
Iteration cycles dependencies of deconvolution $g^{(t)}$ distributions (up to down)
Iteration cycles # =1 (top), 2(middle), and 10 (bottom), respectively.

IOP Conf. Series: Journal of Physics: Conf. Series **1047** (2018) 012017 doi:10.1088/1742-6596/1047/1/012017

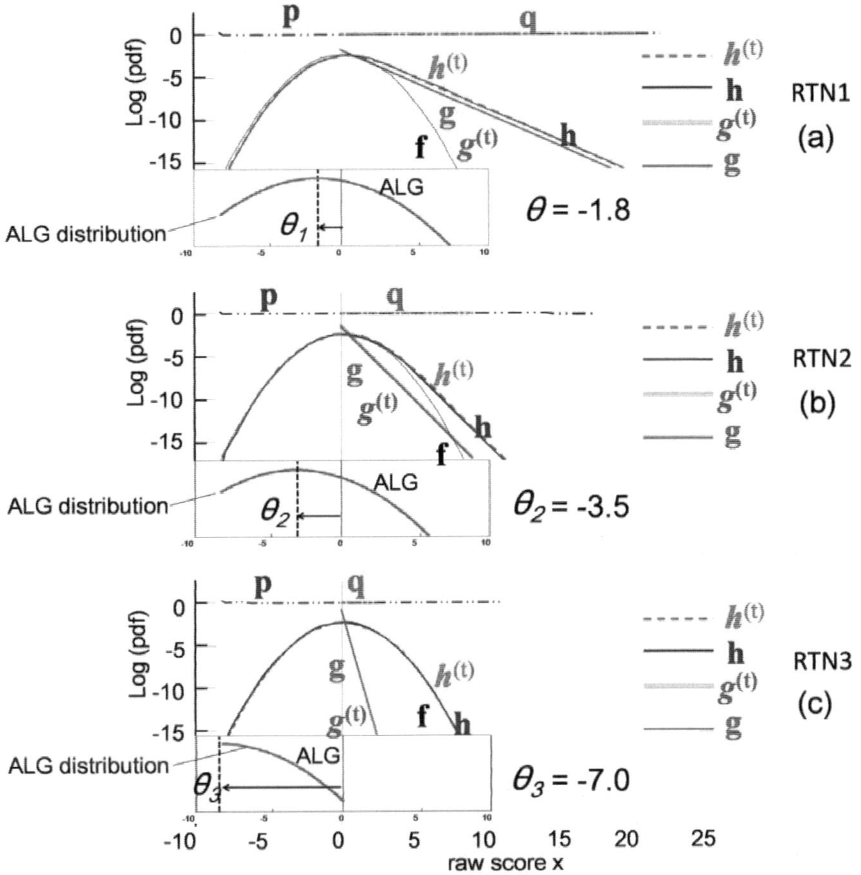

Figure 13. Demonstrations of ringing elimination for (a) RTN1, (b) RTN2, and (c) RTN3. Mean shift of ALG for RTN1, RTN2, and RTN3 are θ_1=-1.8, θ_2 =-3.5 and θ_3=-7.0, respectively

Figure 13 demonstrates the $\mathbf{g}^{(t)}$ deconvolution results for (a) RTN1, (b) RTN2, and (c) RTN3, respectively. The mean shifts of ALG for RTN1, RTN2, and RTN3 are set to θ_1= -1.8, θ_2 = -3.5 and θ_3= -7.0, respectively. Those values of θ are within allowable zones, as shown in Fig. 10.

It is found that appropriate designs of θ depending on the different RTN distributions enable to eliminate the ringing behavior for (a) RTN1, (b) RTN2, and (c) RTN3, respectively.

4. Deconvolution Errors and Its Error Impact on SRAM Fail Bit Count Prediction Accuracy

The deconvolution errors and its impact on the SRAM fail-bit count (FBC) prediction accuracy are compared between this work and the conventional one in this section.

4.1. SRAM Fail-bit Count Estimation

Figure 14 shows the concept of how does the deconvolution errors of $\mathbf{g}^{(t)}$ gives impact on the FBC prediction errors. $FBC_{EXP}(x_p)$ for the golden case (expected no $\mathbf{g}^{(t)}$ error case) represents the FBC at x_p. FBC is given by the integration of the curve of the cumulative density function (cdf) of $\mathbf{h}=\mathbf{f}\otimes\mathbf{g}$ from $+\infty$ to x_p (see Fig. 14 (a)) . This value corresponds to the fail bit probability. Thus, if the extracted $\mathbf{h}^{(t)}$ line

IOP Conf. Series: Journal of Physics: Conf. Series **1047** (2018) 012017 doi:10.1088/1742-6596/1047/1/012017

is deviated from the golden line of the **h** (see Fig. 14 (b)), the value of FBC_{EXTR} (x_p) becomes different from that for the golden value of FBC_{EXP} (x_p). This results in the FBC prediction errors.
If the value of FBC_{EXTR} (x_p) is larger or smaller than that for FBC_{EXP} (x_p), this corresponds to the case for an over-estimation or under-estimation of FBC, respectively.

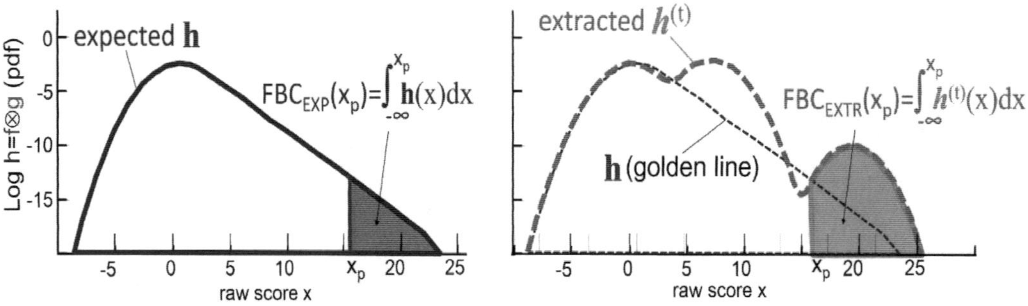

(a) no error (golden case) (b) with error (extracted)

Figure 14. Concept of fail bit count (FBC) errors caused by deconvolution errors.
(a) FBC_{EXP} at each x-point is given by integration of expected **h=f⊗g**.
(b) If extracted $\mathbf{h}^{(t)}$ is deviated from **h**, FBC_{EXTR} (right figure) has to be expected.

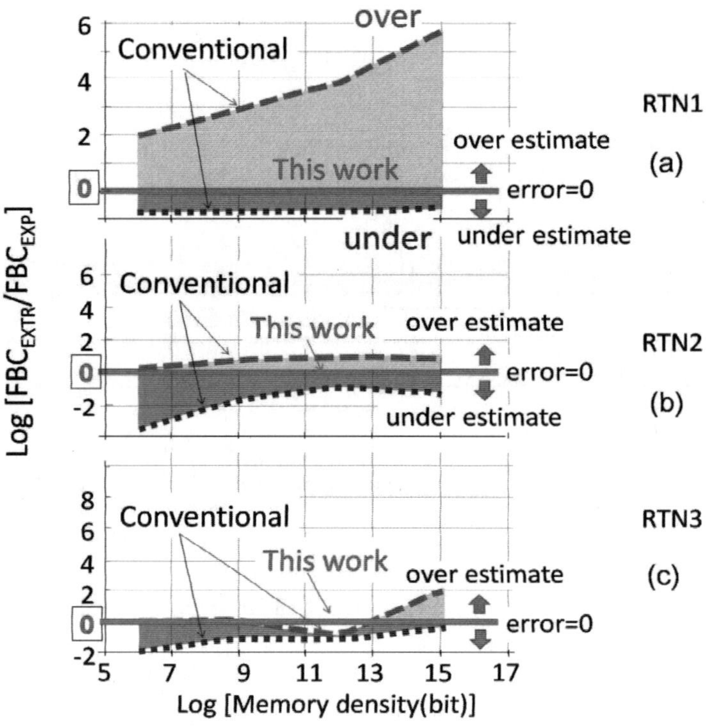

Figure 15. SRAM memory density dependencies of fail bit count (FBC) errors.

Herein, the total number of SRAM bits of 10^6 (1000chips x 1Kbit) to 10^{15} (1000 chips x 1Tbit) are assumed. Thus, the only 1bit fail probability for 1000chips of 1Kbit, 1Mbit, 1Gbit, and 1Tbit are $1/10^6$, $1/10^9$, $1/10^{12}$, and $1/10^{15}$, respectively.
Thus, the position of x_p depends on the total memory bits. For example, the position of x_p for 10^{15} is shifted to the right edge side of the tail compared with that for the case for 10^6. This means that the accuracy in the tail region becomes more important as the number of SRAM bits (density) increase. Figure 15 shows the SRAM memory density dependencies of the FBC prediction errors.
As you can see in Fig. 15, the fluctuation of $\mathbf{h}^{(t)}$ caused by $\mathbf{g}^{(t)}$ ringing increases the over-estimated or under-estimated FBC prediction errors. On the other hand, this work well eliminates the ringing, resulting in much less over/under estimation for the case of RTN1,RTN2, and RTN3 distributions.

The relative deconvolution error of $\mathbf{g}^{(t)}$ across raw score x for (a) RTN1, (b) RTN2, and (c) RTN3 are compared between conventional and this work, as shown in Fig. 16. It is found that this work can reduce the relative deconvolution error of $\mathbf{g}^{(t)}$ by 2-3 orders of magnitude compared with the conventional one. It is noticed that the polarity (plus+ and minus-) of the errors are ignored in the relative error plot.

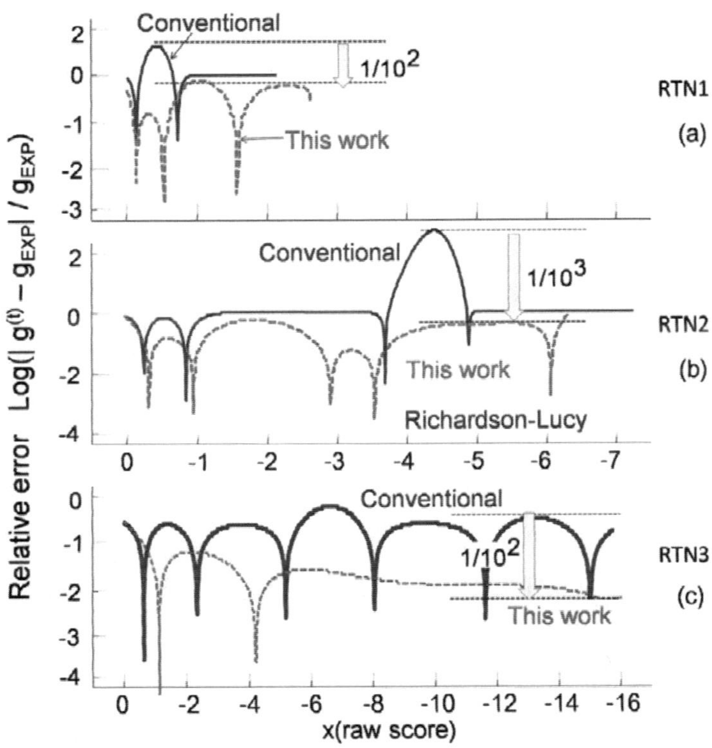

Figure 16. Comparisons of relative deconvolution error of $\mathbf{g}^{(t)}$ across raw score x for (a) RTN1, (b) RTN2, and (c) RTN3 between conventional and this work.

Figure 17 shows the comparisons of cumulative density function cdf(x_P) at x_P where the pdf of golden h(x) is 10^{-12} for RTN1, RTN2, and RTN3 between the conventional and this work. The cdf(x_P) value corresponds to the predicted fail-bit counts for 1000-chips of 1Gbit SRAM.
It is found that this work can reduce both of the errors of FBC prediction and required iteration cycles for convergence, as shown in Fig. 17.

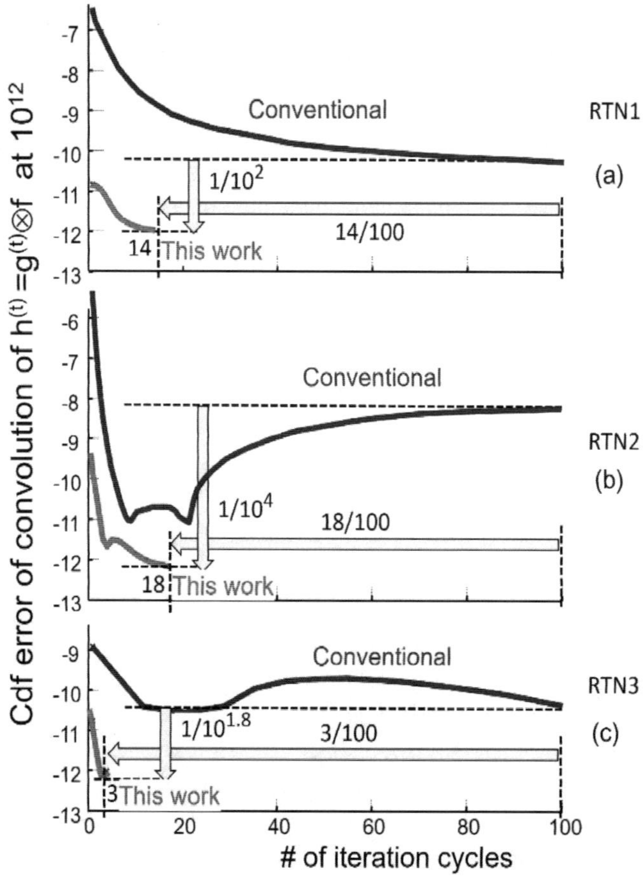

Figure 17. Comparisons of cdf error of $\mathbf{h}^{(t)} = \mathbf{g}^{(t)} \otimes f$ and its iteration cycles required for convergence for (a) RTN1, (b) RTN2, and (c) RTN3 between conventional and this work.

5. Summary and Discussions

The proposed ringing prevention technique successfully circumvents the ringing error thanks to reducing the phase difference between the feedback-gain **q** and deconvolution target distributions $\mathbf{g}^{(t)}$ in RL-deconv iteration cycles. It is found that the proposed one can avoid any unwanted positive feedbacks, resulting in no error amplification. It has been shown that the proposed technique reduces its relative errors of the RTN deconvolution by 10^2~10^3 times compared with the conventional RL-deconv. This enables to increase accuracy of the fail-bit-count prediction based on the cdf of the convolution ($\mathbf{h}^{(t)} = \mathbf{g}^{(t)} \otimes \mathbf{f}$) of the RTN $\mathbf{g}^{(t)}$ with the RDF **f** by over 2-orders of magnitude while accelerating its convergence speed by 7~30 times of the conventional one.

References

[1] K. Takeuchi, et al, "Comprehensive SRAM Design Methodology for RTN Reliability" , Digest of IEEE Symposium on VLSI Technology, (2011), pp. 130-131
[2] X. Wang, et al, "RTS amplitude distribution in 20nm SOI FinFETs subject to Statistical Variability", SISPAD 2012, (2012) pp.296-299
[3] X. Wang, et al, "Simulation Study of Dominant Statistical Variability Sources in 32-nm High-k/Metal Gate CMOS", IEEE Electron Device Letters , vol.33, no.5, pp. 643-645, 2012

[4] J. Franco, et al, "RTN and PBTI-induced Time-Dependent Variability of Replacement Metal-Gate High-k InGaAs FinFETs", IEEE Electron Device Meeting, Dec. 2014
[5] Q. Shi, et al, "Application of Iterative Deconvolution for Wire Fault Location via Reflectometry", Digest of IEEE Int. Symp. on Instrumentation & Measurement, Sensor Network and Automation (IMSNA), (2012), pp. 102-106
[6] W. Somha, H.Yamauchi, "Convolution/Deconvolution SRAM Analyses for Complex Gamma Mixtures RTN Distributions" , Digest of IEEE ICICDT-2013,(2013), pp. 33-36
[7] Fish D. A. et al, "Blind deconvolution by means of the Richardson–Lucy algorithm", Journal of the Optical Society of America A 12 (1): pp.58–65
[8] White R. L. "Image Restoration Using the Damped Richardson-Lucy Method" The Restoration of HST Images and Spectra II, Space Telescope Science Institute, 1994, pp.104-110
[9] F. Dell'Acqua1, "A Modified Damped Richardson-Lucy Algorithm to Improve the Estimation of Fiber Orientations in Spherical Deconvolution", Proc. Intl. Soc. Mag. Reson. Med. 16 (2008)
[10] H.Yamauchi, W. Somha, "Comparative Study on Deconvolution Function Dependencies of RTN/RDF Effect Estimation Errors in Analyzing Sub-nm-Scaled SRAM Margins", IEEE MWSCAS 2014, pp.230-233 (2014).
[11] H.Yamauchi, W. Somha, "Errors in Solving Inverse Problem for Reversing RTN Effects on VCCmin Shift in SRAM Reliability Screening Test Designs", IEEE SOCC 2014, pp.318-323 (2014)
[12] H.Yamauchi, W. Somha, "Deconvolution Algorithm Dependencies of Estimation Errors of RTN Effects on Subnano-Scaled SRAM Margin Variation", IFIP/IEEE VLSI-SoC 2014, pp.97-102 (2014)
[13] H.Yamauchi, W.Somha, "Ringing Error Prevention Techniques in Lucy-Richardson Deconvolution Process for SRAM Space-Time Margin Variation Effect Screening Designs" ,16th IEEE Latin-American Test Symposium (2015)